AF552752

Medicinal & Nutraceutical Plants from The Himalayas

Medicinal & Nutraceutical Plants from The Himalayas

Editors

V.L. Chopra
Padma Bhushan Awardee
Formerly with National Research Centre on Plant Biotechnology
Indian Agricultural Research Institute
New Delhi

S.K. Vats
Emeritus Scientist
CSIR-Institute of Himalayan Bioresource Technology
Palampur, Himachal Pradesh

NEW INDIA PUBLISHING AGENCY
New Delhi – 110 034

NEW INDIA PUBLISHING AGENCY
101, Vikas Surya Plaza, CU Block, LSC Market
Pitam Pura, New Delhi 110 034, India
Phone: + 91 (11) 27 34 17 17 Fax: + 91 (11) 27 34 16 16
Email: info@nipabooks.com
Web: www.nipabooks.com

Feedback at feedbacks@nipabooks.com

© 2020, Publisher

ISBN: 978-93-87973-75-6

All rights reserved, no part of this publication may be reproduced, stored in a retrieval system or transmitted in any form or by any means, electronic, mechanical, photocopying, recording or otherwise without the prior written permission of the publisher or the copyright holder.

This book contains information obtained from authentic and highly regarded sources. Reasonable efforts have been made to publish reliable data and information, but the author/s, editor/s and publisher cannot assume responsibility for the validity of all materials or the consequences of their use. The author/s, editor/s and publisher have attempted to trace and acknowledge the copyright holders of all material reproduced in this publication and apologize to copyright holders if permission and acknowledgements to publish in this form have not been taken. If any copyright material has not been acknowledged please write and let us know so we may rectify it, in subsequent reprints.

Trademark notice: Presentations, logos (the way they are written/presented), in this book are under the trademarks of the publisher and hence, if copied/resembled the copier will be prosecuted under the law.

Composed and Designed by NIPA

This work is respectfully dedicated to

Professor M.S. Swaminathan

Preface

Apart from treating ailments, derivatives of plant origin have been used for improving overall well-being by many communities around the world. In contrast to modern medicines, which in general are single compounds, plant derivatives are used as combination of many different compounds, derived at times from different species. These are believed to attack multiple targets, and through different mechanisms for generating a synergistic effect. Plants promoting the feeling of well-being may improve body resistance for tackling physical, chemical or biological stress, or help reducing disorders associated with aging, tiredness or weakness.

In recent years, such compounds have been termed as 'adaptogens', and are attributed the broad role of initiating biochemical alterations that enhance body's ability to adapt to unfavourable conditions in both ailing and healthy individuals. In brief, the adaptogens have a prophylactic (tonic or a fortifier) role. The adaptogenic plants are believed to help people handle stress. The exponentially growing information suggests that adaptogens plants have antioxidant, liver protective and antitoxin activities which improve blood sugar metabolism, reduce craving for alcohol and sugar, improve immune resistance, increase energy and stamina, improve muscle tone, induce faster recovery, impart better focus and concentration, reduce anxiety, improve sleep quality, enhance motivation & productivity and elevate mood.

The remarkable rise in trade of medicinal plant species is testimony to the growing popularity of bio-molecules. The global market in herbal medicines increased from US $ 12.4 billion in 1994 to US $ 62 billion in 2005.

Interestingly, the top ten selling herbs have such therapeutic properties as immunity enhancer, anti-stress, anti-oxidant, liver tonic, adaptogen, anti-depressant; cardiotonic, anti-inflammatory, hepatoprotective, mood elevator, etc. and qualify for similarity to adaptogenic plants.

Within Himalayas, a large number of plant species have found utility amongst different dwelling communities for enduring stress due to environmental harshness. While it is difficult to lay down rigid qualifying bench mark, some broad criteria are necessary to decide which species will merit inclusion of plants that promote wellness. A wide range of terms and expressions related to the adaptogenic action are known in literature. In this book, selected plants are targeted which, when taken by people without symptom of disease, protect the regular user from acquiring that affliction.

The following are some activities which could relate plant metabolites to improved feeling of well-being:

*Immune-modulation (Agent that balances and improves the immune response of the body in fighting bacteria, viruses, microbes, allergens etc and maintaining health)

*Enhancing strength and physical endurance

*Content of antioxidants (protection against damage caused by free radicals)

*Anti-inflammatory and vasodilator properties leading to calmness

*Improving digestion and providing important nutrients

*Promoting healthy metabolism/enhancing stamina

In recent times, herbals are increasingly being embraced as an alternative of choice for perusing a healthy life style. This trend transcends the economic or developmental status of countries of the global community. As a consequence, the global nutraceutical market size is expected to grow dramatically: from US $ 198.7 billion in 2016 to $ 285.0 by 2021 at a compound annual growth rate of 7.5% (ASSOCHAM-RNCOS, 2017). The Indian nutraceutical market, estimated at US $ 4 billion in 2017, is estimated to reach $ 10 billion in 2022 at CAGR of 21% (ASSOCHAM-MRSS, 2017). This rapidly growing market is exerting unprecedented pressure on demand of plant material on which this enterprise is based.

In trade of medicinal and aromatic plants, there has been a three-fold increase globally, from US $ 1.1 billion (1999) to US $3.0 billion (2015). India is the second largest exporter of medicinal plants in the world. Importantly, 60-90 % trade by volume of medicinal and aromatic plants is of collection from the wild. This rampant plunder is severely eroding genetic diversity and driving several important species of nutraceutical and aromatic value to brink of extinction. Quite clearly, the prevailing situation is not sustainable and calls for urgent course correction. Among the aspects requiring urgent and well planned attention are: (1) Working out designed Agro-technologies for bringing the relevant important medicinal and aromatic plant species under organised cultivation so that collecting from the wild is replaced by raising the required material by scientific cultivation and (2) genetically improving the targeted plant species for desired pharmaceutical attributes and biochemical profiles for enhancing their effectiveness. For creating a base for achieving these objectives, this book presents the current state of the art for nine plant species of the Himalayan region.

We believe that the book will be of interest to a very wide and diverse audience which includes academics and researchers, interest groups of drug industry, administrators and policy makers.

V.L. Chopra
S.K. Vats

Contents

Contributors

V.L. Chopra
Formerly, National Research Centre on Plant Biotechnology
Indian Agricultural Research Institute
New Delhi

Ram A. Vishwakarma
Director
CSIR- Indian Institute of Integrative Medicine
Jammu – 180001

Ram Krishen Ogra
Former Senior Technical Officer
CSIR-Institute of Himalayan Bioresource Technology
Palampur, Himachal Pradesh – 176061

B.D. Sharma
Former Principal, Scientist
NBPGR
Amit Lodge, Phagli, Shimla
Himachal Pradesh – 171004

J. C. Rana
National Coordinator
Diversity International
NASC Complex
Pusa, New Delhi – 110012

Sanjay Kumar
Director
CSIR-Institute of Himalayan Bioresource Technology
Palampur, Himachal Pradesh – 176061

Rakesh Rana
Principal Scientist
CSIR-Institute of Himalayan Bioresource Technology
Palampur, Himachal Pradesh – 176061

Brij Lal
Former, Senior Principal Scientist
CSIR-Institute of Himalayan Bioresource Technology
Palampur, Himachal Pradesh – 176061

Dinesh Sharma
Scientist
CSIR-Institute of Himalayan Bioresource Technology
Palampur, Himachal Pradesh – 176061

Akshay Nag
Post Doctoral Fellow
Department of Biotechnology, South Campus
Panjab University, Chandigarh – 160014

D.R. Nag
Advisor
Regional cum Facilitation Centre –North 1
NMPB, RI in ISM, Joginder Nagar
District Mandi
Himachal Pradesh

AKS Rawat
Executive Director
Ethnomedicinal Research Centre
(DST support) FEEDs
Campus, Hangbung, District Kangpokpi
Manipur

S. Srivastava
Scientist
CSIR-National Botanical Research Institute
Lucknow, Uttar Pradesh – 226001

S.S. Samant
Scientist-G & In-charge
Govind Ballabh Pant National Institute of Himalayan Environment and Sustainable Development
Himachal Regional Centre
Mohal-Kullu, Himachal Pradesh – 175126

Manohar Lal
Scientist Fellow
Govind Ballabh Pant National Institute of Himalayan Environment and Sustainable Development
Himachal Regional Centre
Mohal-Kullu, Himachal Pradesh – 175126

Dinesh Sharma
Scientist
CSIR-Institute of Himalayan Bioresource Technology
Palampur, Himachal Pradesh – 176061

S.K. Vats
Emeritus Scientist
Institute of Himalayan Bioresource Technology (CSIR)
Palampur, Himachal Pradesh – 176061

Manu Sharma
Professor
M.M. University Mullana
Dist. Ambala, Haryana – 133207

1

In Perspective : Plants of Medicinal and Nutraceutical Relevance : Approach to Quality Improvement and Productivity Enhancement for Novel Characteristics and Commerce*

V. L. Chopra and Ram A. Vishwakarma

An overwhelming proportion of plant material used for compounding herbal medicines and nutraceuticals is currently collected from the wild. With traditional health care systems gaining increasing popularity, the enormous amount of plant material extracted from the limited plant populations available in natural stands is a cause of justifiable alarm. The unscrupulous plunder of plant populations growing in the wild is pushing increasing number of important species to the endangered category and many to the verge of extinction. Also, unethical practices like adulteration and storage of medicinal plants under unacceptable conditions are rampant. All this is happening, when the consumer is becoming increasingly quality conscious and is demanding traceability of source material used for formulations and adherence to good practices for the entire chain of manufacturing process for herbal drugs and plant-based nutraceuticals. Clearly, solution to this problem lies in replacing the 'collection from the wild' paradigm to 'production by cultivation'. An essential prerequisite for meeting this objective is a well thought out research agenda which not only tackles productivity enhancement goals for the targeted plant species but also addresses the validity of the entrenched notion among healers who use the traditional medical systems, that only material collected from the wild produces the desired results. The reason advanced in support of this argument is that phytomedicines are derived from plant populations growing in the wild under organic conditions and under their natural *rhizosphere* and *rhizoplane*. The associated soil-microbial complexes, and a variety of endophytic microorganisms, may be the initiators or modifiers of the production and expression of medicinal properties.

---

*Reproduced from Chopra, V.L. and Vishwakarma, R.A. (2017). *Plants for Wellness and Vigour,* New India Publishing Agency, New Delhi

It also so happens that a majority of medicinal plants are shade loving, C4 types and will necessarily require their species-specific ambient environment for elicitation of the medicinally relevant moities.

It will be an over simplification, however, to assume that agronomic packages for cultivation that include supplements (both organic and inorganic) *per se* will interfere with metabolic processes crucial for production and accumulation of medicinally relevant compounds. While the fact remains that the comparable conditions of the wild in different locations are also not identical and uniform, for the overall effectiveness of the *ayurvedic* treatment, standardised ingredients of traceable and reproducible composition and quality are crucial. And attaining this objective is possible only when production is organized under scientifically determined and operationally controlled conditions. For standardising production agronomy, the beneficial *rhizoshere* and *rhizoplane* microbes will of course have to be appropriately factored in while formulating the package of practices. We will revert to this issue later when discussing requirements of a designed research agenda for agronomic research.

(A) PLANT BIOLOGY ASPECTS (V.L. Chopra)

Basic requirements for increasing productivity: In order to make the transition of organized cultivation instead of extraction from the wild, we will need to establish a starting seed stock. For a self-pollinated species, this will be the seed from single plants, selected from naturally growing population, which posses the required constellation of desired productivity and quality parameters. For a cross pollinated species, the starting material will be the seed resulting from inter-mating of selected plants in an isolation block.

Genetic Improvement

Genetic up-gradation is the real and sustainable route to productivity improvement. The genetic approach is based on the well established fact that plant characters are controlled by genes. And the desired genes of different parental donors can be brought together by following appropriate breeding procedures. From the resultant improved variety, we can harvest the desired products of two parental varieties. What is more, by deploying more efficient genes, we can enhance the yield of the desired compound both in quantity and quality. For genetic improvement approach to succeed and deliver the expected results will require, however, systematic work for gaining insight into a number of biological characteristics of the targeted plant species. The information collated in this book indicates that, for most plant species dealt with, critical details of key biological aspects and parameters are not available. This deficiency must be addressed on an urgent basis by executing a well planned and vigorous research agenda. The aspects that deserve detailed working out are briefly dealt with.

Phenology

Study of phenology unravels the timing of periodic developmental phenomena such as flowering, reproduction and migration. All periodic biological events of plants respond to the prevailing environment. Therefore, these events will need to be validated for the timing of their onset and duration for the environment in which the crop is grown for commerce or research. The periodicity of different biological events is sensitive to environmental variables such as changes in elevation or climate. Climate change also has implications with reference to shifts in the pollination pattern and pollinator's behaviour. For medicinal plants, phenological behaviour has special significance because they are distributed across wide altitudinal ranges. And, very little information is currently available regarding phonological shifts involving multiple scales.

Floral and Reproductive Biology

Information on several aspects of flor7al and reproductive biology of the chosen plant species is necessary for designing operational plans for improvement by breeding. For example, to transfer and incorporate a genetically controlled trait from a donor to a recipient will require that the breeder makes a cross. For this, she/he needs to know the various steps complete with their sequence and duration that the plant goes through when it reproduces in nature. For a naturally self pollinated plant, the breeder emasculates the female parent and artificially pollinates it with the desired pollen to obtain the hybrid. For successful emasculation and imposed pollination prior knowledge of crucial floral biology aspects such as time of anthesis and period of stigma receptivity of the plant under consideration are essential. Other aspects for which prior knowledge will be very helpful include flower morphology; hermaphrodity and unisexuality; type of pollination and mechanisms promoting self and cross pollination; self incompatibility etc.

Germplasm

The enormous diversity of plant kingdom offers rich material for studying a number of complex phenomena of the biological systems. For a diversity rich country like India, the scientific temptation of probing different aspects of biodiversity is natural and academically justifiable. It is important to maintain clarity, however, on whether the programme relating to germplasm at an institute is being operated as an academic activity for enriching literature on the subject or it is aimed at achieving clearly defined and tangible goals for promoting the crop. This distinction is unfortunately missing currently in most institutes for programmes on plants of medicinal and nutraceutical relevance and deserves attention from the management for rationalisation and rectification.

Having said that, relevant germplasm is an important requirement, firstly as a starting point for successful introduction of a new plant to commerce and later for upgrading the plant for productivity and quality attributes. The following examples indicate the kind of criteria one might use for transition from random to specifics:

For plants where seeds are the commodity of commerce, undomesticated types may present seed shattering as a major limitation. Looking for non-shattering germplasm would aid bringing the species under cultivation. Again, intensive cultivation will aggravate pest and pathogen problems compared to growth under the wild. Looking for resistance donors would be the right priority. Similarly, vegetatively propagated material will require special attention with respect to viruses, pathogens and pests. Proper assessment of diversity of natural populations will be essential before selecting a genotype for agronomic assessment. This may call for working out appropriate seed maintenance/rejuvenation plans.

Pattern of Inheritance of Key Traits and Approach to Designed Breeding

When the yield levels of the best locally available or introduced material become stagnant even with the most optimised set of agronomic practices, recourse to breeding for achieving higher productivity threshold becomes necessary. For achieving improvement by breeding, it is necessary to work out the mode of inheritance of key components that contribute to the quantity and quality of the harvestable crop yield. Mendelian genetic analysis is the best and inevitable approach for meeting this objective. By its very nature, genetic analysis of plant characteristics is a laborious and time demanding activity because the exercise requires data recording and analysis for variable number of generations over several crop seasons. Such genetic analysis has been a neglected activity at most institutes dealing with plants of medicinal relevance and this demands urgent attention for course correction. Knowledge of the nature and number of genes involved in controlling the desired traits is an essential prerequisite. This information serves as the basis for choosing the specific breeding approach that is most likely to succeed for bringing together the characteristics of donor parents into the desired improved variety. The underlying basis of character assembly is genetic recombination between the relevant genes and their aggregation in new formations by recombination, segregation and independent assortment. Information on ploidy status and their relationship among genomes of the species involved in phylogenetic evolution of the crop under study can also be important for designing strategies for upgrading specific attributes of critical importance. Characters such as free threshing grains and ear head compactness in wheat exemplify this argument.

It is important to keep in mind that most of the plant species that we are dealing with are largely undomesticated. The first step for bringing them under cultivation would, therefore, be their domestication through a well planned agenda. They are likely to exhibit variable durations of seed dormancy and their seed viability may vary because of non-synchronous flowering and perennial nature. Knowledge of these aspects would be an inevitable starting point for achieving acceptable crop stand and the required product uniformity.

Information on growth and development pattern (plant phenology) and favourable conducive environmental requirements such as season, rainfall, humidity and photoperiod would be a prerequisite for successful cultivation. Prior knowledge about diseases and pests prevalent in natural habitats would be of immense value for avoiding or restricting damage from diseases and pests when these plants are brought under cultivation. It is also important to keep in mind that these plants are valued economically because of their chemical constituents. These constituents are believed to be elicited and synthesised in response to specific stresses. It is necessary, therefore, to identify these critical environmental factors for maximizing the concentration of the desired constituent(s). It will be equally important to identify, any i) genotype x environment ii) genotype x plant growth stage or iii) genotype x environment x plant growth stage interactions that are relevant to synthesis of the desired chemical constituent.

In plant species of nutraceutical relevance, the concentration of economically important chemical constituent is more important than the gross yield of plant part in which the metabolite is expressed. This is because the cost of extraction of the targeted metabolite is an important issue for the processing industry. Therefore, breeding for higher yield of plant part, contrary to the requirement in case of field/horticultural crops, is not sufficient. A minimum threshold concentration of the desired chemical constituent is a non-negotiable requirement. Since produce containing higher constituent is likely to fetch higher price to the farmer, breeding for higher concentration of the chemical constituent should be the first and over-riding priority in breeding of these plants. This, however, requires rapid, accurate and reliable methods for quantitative estimation of economically important chemical constituent. Without availability of such economically viable methods not much progress can be expected. When such methods are available, mutation breeding (including mutant TILLING, site-targeted mutagenesis via gene targeting) can be adopted for self-pollinating species. For those species in which vegetative plant part is economically important (and the species can also be propagated by vegetative means), induced polyploidy can be evaluated for obtaining rapid improvement.

Special breeding procedures have to be developed when direct selection cannot be made for the character desired to be improved. A good example of this is where roots are the produce of economic value. Direct selection will result in destruction of the selected plant and result in dead end. To overcome this difficulty, correlation of root yield is worked out with an above the ground plant organ and indirect selection is made via the correlated surrogate character for achieving breeding gains.

Marker assisted selection combined with availability of rapid and reliable methods for estimation of active chemical constituent can be expected to yield rapid gains for improvement of these plants by scientifically designed breeding approach.

The Role and Relevance of Molecular Methodologies

'Omics' and CRISPR (Clustered Regularly Interspaced Short Palindromic Repeats) gene editing technologies are flavour of current scenario of research pursuit. These technologies are incisive, precise and powerful. They have captured the interest and imagination of the younger generation of research community. These are, however, not replacement at the present juncture for the time tested conventional technologies; especially when the need is of early realisation of tangible results. Experience from improvement effort with agricultural crops has shown that molecular technologies have contributed to success only when these methodologies are fully integrated with traditional approach of recombination breeding. Well deliberated, open minded and rational analysis is crucial for research agenda setting and choosing approach for yield and quality improvement in the current scenario of rapid scientific developments. For example, there are a number of perennial/tree species of medicinal relevance which are sourced from the wild. Bringing them into cultivation would be challenging both from the view point of devising agronomic practices and the requirement of selective propagation of plus trees (leave aside breeding improvements). Also, the site of metabolic activity where a desired compound is synthesised may be very different from where it is ultimately to be harvested. For example, nicotine in tobacco is synthesized in the roots and is commercially extracted from leaves where it accumulates in vacuoles of the leaf cells. Considering, that genomics information could be quite readily obtained with moderate cost; breeding of any plant species in coming years would invariably involve use of molecular tools. It is unfortunate in any case that good plant geneticists and breeders have increasingly been rendered marginalised partners among research teams charged with the task of developing improved crop varieties. This perhaps explains in part why we are faced with yield stagnation in increasing number of crops in our country. This short coming should be addressed urgently.

Holistic Agronomic Management

Best agronomic practices are then worked out for such parameters as date of planting, rate of seeding, distance between rows and from plant to plant, depth of planting, schedule for irrigation and fertiliser application, intercultural operations, stage and date of harvesting, appropriate processing and storage of the harvest.

For most agriculturally important crops, this exercise is uncomplicated because the parameters that promote optimal plant growth and development are also invariably the same as those that give the best yield and product quality. For medicinal plants however, where the desired end products of curative and commercial significance are secondary plant products, the situation is different. It is so because the concentration of secondary plant products are significantly enhanced when the plants are subjected to an imposed stress. This important conclusion has been empirically verified by numerous studies on different plant species when the imposed stress was drought or salinity. In most cases, though, the stress induced elevation of secondary product content does not fully compensate the corresponding decrease of harvestable biomass. Agronomic research for medicinal plants, therefore, presents obviously serious challenges but also opens up rewarding opportunities. These opportunities arise from rapid technological advances which have made proteomics a very precise and rewarding instrument of knowledge enhancement. Increase or decrease of different proteins produced under a stress can now be measured before actual appearance of the physiologically observable effects. Proteomics therefore provides a rapid and precise method of predicting how a cultivar will respond to a natural or an imposed stress. Such proteomic analysis has revealed that proteins involved in biosynthesis of alkaloids have a dominant role in production of secondary metabolites. And, secondary metabolites are the major contributors to the healing properties of medicinal plants and fragrance of the aromatics and spices.

Agronomic research on medicinal plants offers unusual opportunities but also critical challenges. To yield meaningful outcomes, research agenda will have to be thoughtfully crafted on multi-disciplinary principles and executed through interactive and well coordinated team work. It is only such holistic approach which can ensure that enhancement of harvestable bulk plant yield is accompanied with the high content of medicinally relevant constituents of desired chemical profile and pharmacological activity.

(B) CHEMISTRY, CHEMICAL BIOLOGY AND PHARMACOLOGY ASPECTS: (Ram A. Vishwakarma)

Natural Products Chemistry

Isolation, structural characterisation and synthesis of small molecule secondary metabolites (natural products) from medicinal and nutraceutical plants used in various traditional systems of medicine and nutrition have been pursued globally with great vigour leading to the discovery of a very large repertoire of the compounds with remarkable structural and biosynthetic diversity. These natural products have been catalogued in a number of compendia and digital databases such as the "Dictionary of Natural Products". There are three major issues which have rarely been addressed in the context of natural products chemistry; (a) dynamics of the biosynthetic production of natural products with reference to location and season (b) metabolic flux and tissue distribution of small molecules and (c) impact of symbiotic microbiome of individual plant of the ecosystem. The new tools of analytical chemistry (imaging mass spectrometry) and cell biology (flow cytometry) are providing new direction for research with direct impact on the quality of botanical material for product development.

Chemistry, Manufacturing and Controls (CMC)

One of the major hurdles in acceptability of medicinal (herbal/botanical/ phytopharmaceutical drugs) or nutritional products (nutraceutical/dietary supplements/ functional foods) derived from Indian medicinal/nutraceutical plants is lack of quality parameters and standards required to meet global regulatory standards. There are specific good manufacturing practices (GMP) guidelines for extraction and formulation of botanical products meeting required quality control and quality assurance (QC-QA) standards. The botanical material handling, processing and extraction is to be done in cGMP facilities, inspected and approved by country specific regulatory bodies e.g. US-FDA, EMA and DCGI. Besides, the QC-QA is also mandated to be done by nationally accredited facilities, where a number of specific parameters have to be defined and approved for the individual plant based product. From the regulatory perspective for botanical herbal products, a far greater emphasis is placed on safety and label claim. The key quality parameters include concentration ranges of chemical markers to establish integrity of the raw material, microbial contamination, residual herbicides, pesticides, antibiotics, xenobiotics, heavy metals and know toxic natural products such as aflatoxins and other mycotoxins. These CMC guidelines have been laid down by US-FDA and DCGI/FSSAI. There are not many government and industrial laboratories equipped and accredited for CMC and Qc-QA of botanical medicinal and nutraceutical products and consequently a majority of the products in Indian market are sub-standard posing health hazard to the gullible population. These issues are critical for successful global approval of Indian plant derived medicinal, nutraceutical and food products.

Natural Products Pharmacology

Although a large number of natural products have been isolated from medicinal plants, only a tiny minority of these compounds have been rigorously evaluated and validated for pharmacological activities either in pharmaceutical industry or academic laboratories. In spite of this, the success rate of discovery of new drugs from natural products has been pretty remarkable, compared to synthetic small molecules; almost 50% of US-FDA approved drugs have their roots and inspiration in the natural products scaffolds, testifying the potential of this vital biodiversity resource. However, the trajectory of discovery of the pharmacologically active compounds from biodiversity has so far followed the reductionist paradigm – a single pure natural product isolated from plant targeting a protein target involved in disease biology. The advances in synthetic medicinal chemistry, pharmacokinetics and pharmacology enabled the development of several such natural products derived drugs which are currently used in clinic. There is slight shift from this "reductionist" approach due to appreciation of the limitations of this approach. Managing chronic, complex conditions such as cardiovascular diseases, cancer, diabetes and metabolic syndrome, and neurodegenerative disorders (Alzheimers) is posing serious challenges to current drug discovery paradigm, and modern science and medicine are looking for age-old traditional systems of medicine (Ayurveda and TSM) for new solutions and these arguments are being encouraged by new discoveries in so-called "network pharmacology" and the role of minor food constituents in health homeostasis and disease progression. There is now greater realization that accumulated wisdom, materials (plants) and concepts of indigenous medicine can be harnessed with the application of new molecular pharmacology. There is also a new evidence based thinking that body's own resources and extraordinary ability to repair and rebuild tissues and organs to prevent or reverse illnesses is possible by approaches driven by systems biology and network pharmacology. In this new approach to address complex multi-factor diseases (diabetes and neurodegenerative disorders) driven by network pharmacology, it is now possible to engage a number of protein targets in a given signalling pathway through a number of low-affinity ligands (natural products) and derive overall aggregate high affinity (nano-molar) response enough to impact the disease pathology and progression. This poly-pharmacology based approach appears to be the way forward for treating diseases of multiple aetiologies. In fact, many of the natural products of nutraceutical plants must be relooked from this perspective to tailor person-specific diets and medications. A recent example of linking gut microbiome, food ingredients and glycaemic control in type-2 diabetes patients is indicating towards this horizon. Hence, the next generation questions on medicinal and nutraceutical plants and their compounds from India will have to be asked from a new vantage point. Although new science in this direction is moving positively in Western countries and particularly in China, the situation in India is not encouraging.

Chemical Biology

The diversity of secondary metabolites produced by medicinal and nutraceutical plants is mind-boggling and building that level of structural, biosynthetic and genetic diversity must have been a herculean enterprise for the evolution over thousands of years! Over the last several decades, we have deciphered in reasonable details the biosynthetic machinery (metabolic intermediates, proteins and genes) involved in building plant secondary metabolites, however our understanding on why these compounds are made by the plants is rudimentary and confusing. The entire focus so far has been harnessing such substances for treating human diseases without giving due consideration to understand why plants make them. Obviously, the plants do not make secondary metabolites for treatment of human diseases, which clearly is a fortuitous serendipity. Therefore, systematic chemical-biology driven efforts are required to decipher the functional and physiological role of natural products in plant *per se*. These questions are even more interesting considering the fact that some natural products are widely distributed across the genus and species while the others are restricted only to a single species (for example artemisinin and forskolin). So there is whole new frontier waiting to be discovered in chemical biology of plants.

Chemical Ecology

Chemical ecology is a new discipline driven by the recognition that organisms (plants, microbes, insects and mammals) invariably use small-molecule natural products to communicate with each other in diverse ways (smell, taste, vision, mating, symbiosis, deterrence, pathology and pharmacology). Since the vocabulary of living things is overwhelmingly chemical, it is believed that the understanding of the information content of the mediating molecules are going to radically change the ways we look at mechanism of biological communication and species survival. Despite ideal biodiversity and ecosystems, surprisingly no efforts have been made in India in this area. These studies may have far reaching consequences for the questions such as: how do natural products (mostly chiral) arise in evolution and adapt to diverse functions? How they are biosynthesized and how the rate and timing of their production controlled? Eventually some of these discoveries will be translated to new medicines. Chemical ecology driven approaches for new drug discovery from medicinal plants offer new opportunity to decipher functional role of secondary metabolites.

Natural Products in Regenerative Medicine

Treatment of chronic diseases (cardiovascular, neurodegenerative, musculoskeletal, diabetes, cancer) depends on our ability to control cell fate, allogenic rejection and access of selectively differentiated stem cells. Cell-based phenotypic and pathway specific screening of natural products and their synthetic analogues provide great new opportunity to selectively control stem cell proliferation, differentiation and de-differentiation. Recent studies show great promise in discovering effective medicines for tissue repair and regeneration. A recent example where the key transcription factors (the Yamanaka factors; *Oct4*, *Sox2*, *Klf4* and *c-Myc*) could be replaced with a cocktail of eight small organic molecules (including Indian natural product Forskolin) for reprogramming of pluripotent stem cells from somatic cells indicate the future for regenerative medicine. If the natural resources form Indian medicinal and nutraceutical plants are coupled with stem cell biology, very exciting science and new therapeutics will emerge. Several studies globally are hinting towards a new medicine paradigm where ideas from traditional systems of medicine will be harnessed for regenerative medicine through stem cell biology and pharmacology.

In the conclusion, there is now a far greater appreciation worldwide of the potential of natural products where nature is looked at the ultimate source for inspiration and solution to the problems of health and disease, with integration of indigenous systems with modern medicine. In this regard, the decisions by US-FDA (US Food and Drug Administration) and recently by Indian DCGI (Drugs Controller General of India) of creating regulatory approval pathway for botanical and phytopharma-ceutical drugs and inclusion of a number of Indian medicinal plants for nutritional products in the priority list of FSSAI (Food Safety Standard Authority of India) are important milestones.

2

Panax ginseng

Ram Krishen Ogra

***Panax gingseng* C.A. May**
(A) Field view of *Panax ginseng*, **(B)** Whole plant, **(C)** Fruit bearing plant **(D)** Young rhizomes, **(E)** Crude drug material (under-ground parts)

INTRODUCTION

The term 'ginseng' is applied to a range of plants from family Araliaceae involving genus *Panax*. Out of the many species available in nature, Himalayas are also home to a commercially lesser known variety called as *Panax psuedoginseng* Wall (Nayar and Shastry 1990). In the interior temperate areas of the eastern Himalayas, the species is distributed across Sikkim, Bhutan, Nepal, and extends into China. Keeping in view its high commercial international demand, CSIR-Institute of Himalayan Bioresource Technology (CSIR-IHBT) Palampur has recently taken an initiative to introduce and cultivate Chinese ginseng (*Panax ginseng*) in the Lahaul valley of Himachal Pradesh, in India. This chapter addresses the different biological, chemical, medicinal and cultivation aspects of both the indigenous Himalayan variety (*P. psuedoginseng*) and the introduced Chinese species (*P. ginseng*) in particular.

BIOLOGICAL ASPECTS

Systemic Classification

Kingdom	:	Plantae
Division	:	Magnoliopsida
Class	:	Magnoliopsida
Order	:	Apiales
Family	:	Araliaceae
Genus	:	*Panax* L.
Species	:	*Panax ginseng* C.A. May (Chinese ginseng)
		Panax psuedoginseng Wall. (Indian ginseng)

Synonyms

***Panax ginseng* C.A. May.** Rep.Pharm.Prakt. Chem. Russ. 7:524, (1842), *Aralia ginseng* (C.A.Mey.) Baill, *A. quinquefolia* var. *ginseng* (C.A.Mey.) auct., *Panax chin-seng* Nees, *Panax verus* Oken

***Panax psuedoginseng* Wall.**, *Panax chin-seng* T. Nees var. *nepalensis*, *Panax japonicas* var. *bipinnatifidus. A. psuedoginseng* (Wall.) Benthex C.B. Clarke, *A. quinquefolia* var. *pseudoginseng. P. psuedoginseng* Wall. sub sp. *japonicus* Hara.

Common names

***P. ginseng*:** Asiatic ginseng, Oriental ginseng, Chinese ginseng, Korean ginseng, Ningin, Seng and Sang, Schinsent, Jintsam.

***P. psuedoginseng*:** Indian ginseng, Himalayan ginseng, Nepal ginseng, Psuedoginseng.

Diversity within genus (number of species; their broad distribution range)

***Panax* spp.:** Depending on the geographical origin, the ginseng plants are classified as Chinese, Korean, Japanese, Himalayan or American. The genus *Panax* includes species like *P. ginseng, P. psuedoginseng, P. quinquifolium, P. trifolium, P. japonicus, P. vietnamenses, P. wangianum* and *P. zingiberences.* Most of the species have an erect rhizome and fleshy roots, except for some sub-species, which may have adventitious roots instead of primary fleshy roots. The three main species/groups of ginseng having medicinal value are:

1. ***P. ginseng*** C.A.Mey (Chinese, Korean or oriental ginseng): The roots are thick, fleshy, fusiform, cream-yellow to buff-yellow in colour, persistent and irregularly branched. The cultivated plants are short thick and fleshy in contrast to somewhat elongated rhizomes in the wild ones (Baranov, 1966; Thompson, 1987). The rhizome part is unbranched, bearing leaf scars from which the age of a plant can be determined. The leaves are palmately compound, verticellate, petiolate and up to 6 in number depending upon age of the plant. Each leaf has 5 leaflets and occasionally 3-7 leaflets. The leaflets are serrate, cuneate,and acuminate with some stiff hairs along the marginal veins.The inflorescence is an umbel with 4-40 flowers on a 7-20cm long peduncle which may vary depending on the environment and age of the plant. The flowers that open in June-July are small, and each flower is 2-3 mm in dimension, has 5-toothed green calyx, 5 petals yellow-green entire petals, 5 short stamens with oblong anthers. The ovary is inferior with a bifid stigma. The fruit, of the size of a small pea, is initially green and turns to a red drupe on maturity. The seeds, also called as pyrenes, are white to buff in colour.

2. ***P. notoginseng*** (Burk.) F.H. Chen (*P. psuedoginseng* Wall., Sanchi ginseng): This includes sub-species *himalaicum* (Burk), sub-species *psuedoginseng* and several other sub-species of wild Asiatic origin. It is distributed in the Szechwan (Sichuan) and Kiangsu areas of China, North Vietnam and eastern Himalayas in India, Nepal, Bhutan and Burma. The aerial parts resemble that of *P. ginseng* in habit, but the roots are fleshy, short, cylindrical, 2-4 cm long, 2-4 cm in diameter, and smooth skinned and yellow-green to yellow-brown in colour. In India, *P. psuedoginseng* is reported from Sikkim, Arunachal Pradesh, Manipur and Meghalaya

3. ***P. quinquifolium* L.** (American ginseng): It is indigenous to the eastern temperate forest areas of North America covering Qubec to Minnesota in the north to Oklahoma, the Ozark Plateau and Georgia in the south

(Carpenter and Cottam, 1982; Thompson, 1987). The growth habit is similar to *P. ginseng.* The leaves are palmately compound with a 2-10 cm long petiole. The roots are fleshy, spindle shaped, somewhat forked in older plants. The rhizome is generally erect. The inflorescence is umbel and green in colour.

Geographical distribution

The genus *Panax* L. has 16-18 species with bicentric and disjunct distribution in North America and Eastern Asia (Wen and Zimmer, 1996; Wen, 2001). Most of the species of *Panax* are Asiatic (Hu 1976; Wen 2001) and only two species, *viz.*, *P. quinquefolius* L. and *P. trifolius* L. are found in the Eastern North America (Burkill, 1902; Graham, 1966; Hu *et al.*, 1980; Proctor and Bailey, 1987). All other species are distributed in the far-eastAsiatic countries, in southwestern and central provinces of China, Korea, and Japan. The wild populations of *Panax,* were distributed from central China to Siberia and Korea at one time, have now become scarce due to over exploitation. Currently, wild populations are restricted to mixed broad leaf and coniferous forests in Heilong Jiang, Jilin Kirin (Manchuria), Liaoning (near North Korea), Long White Mountain area of northeastern China and in Korea (Hu. 1976, 1978). Eastern Himalayas covering Nepal, Bhutan and India, and Korean peninsula along with Japan in the Far East, host maximum species diversity in genus *Panax* (Burkill, 1902; Hara, 1970; Yang, 1981; Wen and Zimmer, 1996; Yoo *et al.,* 2001). *P. notoginseng* (Burk.) F.H. Chen, *P. psuedoginseng* Wallach, Sanchi ginseng includes sub-species *himalaicum* (Burk), sub-species *psuedoginseng* and several other sub-species of wild Asiatic origin. There are various sub-species and races reported from Eastern Himalayas (Banerjee, 1968, Hara, 1970) as well. Areas in the central-western China and eastern Himalayas are the current centers of diversity of genus *Panax* (Wen and Zimmer, 1996).

Habitat and ecology

1. ***P. ginseng*:** It is endemic to Manchurian flora in China. The mixed and broad leave temperate forests are the natural habitat of *P. ginseng*. Earlier, in the beginning of 20th century, it used to grow naturally in the forests of north-east China and the northern Korean peninsula but now it is almost extinct in the natural habitats. *P. ginseng*, having relatively a narrow ecological amplitude, is a shade loving plant and its growth is ensured only under forest shade of mixed conifers like *Pinus, Abies*, *Picea* and broad leaved trees like *Quercus, Betula* forests. A 60-80 percent forest shade cover is most favourable for its development. *P. ginseng* prefers soils which are weakly acidic, well drained, but rich in moisture and humus.

For commercial cultivation, these conditions are tailored to provide suitable environment.

2. ***P. psuedoginseng*:** The species is distributed across an altitudinal range of 2900-4000 m, in China, Nepal, Bhutan, India, and Myanmar, and grows in specific niches containing deep humus rich soils of forests of temperate Himalayas, represented by mixed conifer-oak, Birch, Hemlock and Acer.

Taxonomic enumeration: Botanically, genus *Panax* is characterized by an underground rhizome and fleshy roots, an aerial shoot with a whorl of digitately compound serrate leaves,pinnate-serrate leaflets, terminal umbellate inflorescence, small flowers, inferior ovary and fleshy red or orange coloured fruits, that contain 2-5 pyrenes. Anatomically, it is characterized by occurrence of variable secretary canals in the cortex, phloem and medulla.

1. ***P. ginseng* C.A. Meyer** (Chinese, Korean or oriental ginseng): The roots are thick, fleshy, fusiform, cream yellow to yellow buff in colour, persistent and irregularly branched. The cultivated plants have short thick and fleshy rhizomes, in contrast to somewhat elongated rhizomes in the wild ones (Baranov, 1966; Thompson, 1987). The rhizome part is un-branched, bearing leaf scars from which the age of a plant can be determined. The leavesare up to 6 in number depending upon age of the plant, and are palmately compound, verticellate, petiolate. Each leaf has 5 leaflets and occasionally 3-7 leaflets. The leaflets are serrate, cuneate, acuminate, with some stiff hairs along the marginal veins. The inflorescence is a solitary umbel. Peduncle 7-20cm long with 4-40 flowers on it depending on the environment and age of the plant. Each flower has 5-toothed green calyx, 5 yellowgreen entire petals, 5 short stamens with oblong anthers. The ovary is inferior with a bifid stigma. The fruit is initially green and turns to a red drupe on maturity. The seeds, also called as pyrenes, are white to buff incolor.

2. ***P. pseudoginseng*:** It is a perennial herb having horizontal knotted root stock with 2-5 fleshy (2-4 cm long and about 1 cm in diameter) fusiform roots. Rhizomes are horizontal, with thick nodes and internodes. Aerial portion are up to 50 cm tall, leaves usually 4 in number, palmately compound, verticellate at apex of stem. Numerous lanceolate, stipule-like appendages occur at the base of petioles. Fruits are drupaceous, globose, 3-5 mm across, dull green to black. Flowering takes place in May-June and fruiting around July-October, depending upon the altitude.

The taxonomy of Indian ginseng is controversial (Banerjee,1968;Wen *et al.*, 2001; Pandey *et al.*, 2002, 2004, 2007). In India four different taxa of ginseng have been recognized, *viz, P. burkillianus* Ben. & Viswa., *P. sikkimensis* Ban., *P. pseudoginseng Wall.* and *P. bipinnatifidus* Seem. (Bennet and Viswanathan,1984). It was concluded by them that *P. burkillianus* isthe commonest Indian ginseng, having horizontal creeping rhizomes (largest being 35 cm long and 18 mm diameter), narrow lanceolate leaflets (not more than 2.5 cm wide), caudate apex and crimson red fruits. On the basis of nature of rhizomes, leaflets and size of the seeds, Bannerji (1968) classified all the Himalayan races of *Panax* under *P. pseudoginseng* Wall. Further *P. assamicus* and *P. sikkimensis* were treated as *P. pseudoginseng* var. *angustifolius* and var. *himalaicus*, respectively by Bennerji (1968). Wen and Zimmer (1996) reported that *P. bipinnatifidus*, falling under Himalayan ginseng has polyphyletic nature. A key to the identification of different Himalayan ginseng taxa has been given by Sharma and Pandit (2011).

Phenology

Ginseng emerges in late April from the underground perenating roots. The plant foliage increases in number and complexity depending upon the age of plant. The first year ginseng seedling has three small serrated leaflets joined at the top of a 2-4 inch stalk. During the first year, small carrot shaped white fleshy root is formed. At the onset of winter, the foliage dies and a small perenating bud is formed at the top of the rootlet. The perenating bud germinates next year into a two year old plant with two compound leaves each having 3-5 leaflets terminating a 4-7 inch stalk. In the subsequent years, both the height of stalk and number of compound leaves increase. Usually one compound leaf with 3-5 leaflets is added every year. Correspondingly, the size and weight of the root also increases but the growth is very slow. Under ideal conditions the fresh weight per root may reach upto 80-100 gm. after 6 years of cultivation. Flowering starts in late spring or early summer usually in the third year. The erect flowering stalk arises from the juncture of compound leaves terminating into an umbelliferous cluster of small greenish white flowers. The pollinated flowers develop into small green berries in late summer. These ripen into bright red berries by early autumn containing flattened round creamy yellow seeds with hard coat. The seeds are immature at this stage and need stratification at low temperatures for 1-2 years for the embryos to mature.

Cytology: The base chromosome number is 12. *P. ginseng, P. quinquifolium* and *P. japonicum* are tetraploid (2n=48), whereas *P. psuedoginseng, P. notoginseng* and *P. trifolium* are diploid (2n=24) (Thompson, 1987). Shaw and But (1995), on the basis of 18S ribosomal RNA, have shown that *P. ginseng* and *P. quinquifolium* are more closely related to each other than to *P. notoginseng*. Similar observations were made by Kwon *et al*. (2009) and Lee and Wen (2003) based on the sequencing of the chloroplast trnC-trnD region and the internal transcribed spacer (ITS) regions of nuclear ribosomal DNA. *P. trifolius* was observed to be least similar. The central western China and eastern Himalayas are the current centers of diversity in *Panax (*Wen and Zimmer, 1996).

B. REPRODUCTIVE BIOLOGY

Reproductive phenology

Panax spp. reproduce through production of seeds only and there is no clonal or vegetative propagation except for some reported clonal propagation in *P. psuedoginseng*. Flowering takes place third year onwards in the form of single umbels per plant, with rather small whitish-green flowers borne on a stalk. The flowers mature centripetally and thus the centrally located flowers mature in the last. Ginseng flowers are perfect, but maturity of male and female reproductive parts is separated temporally in each flower. Ginseng individuals are not obligate out crossers, and the flowers are visited by generalist pollinators. The reproduction is both autogamous and cross-pollinated (Carpenter and Cottam, 2011). The cross pollination takes place through general pollinators like syrphid flies and halictid bees.

Ginseng flowers mature into bright red berries containing one or two seeds. In the wild, the dispersal is primarily by gravity. Birds and wild animals also play some role in their dispersal. The flowering and seed production commences after 3rd to 4th year of growth and the life cycle is repeated. The shoots dry every year in the fall leaving a scar. The root forms a solitary bud which sprouts in the following spring. The roots are harvested after 5th year of growth in commercial production.

C. MEDICINAL ASPECTS

Plant material used as crude drug

Matured primary roots, following about 5 years of growth, are used as the crude drug in dried form. Dry roots appear wrinkled.These roots are fleshy, fusiform, cream-yellow or buff in colour, firm and compact in texture and weigh 30-60 g per plant on an average. The unbranched rhizome with leaf scars of wild ginseng also forms part of the crude drug.

Organoleptic properties

The fresh roots have a strong bitter sweet taste and a characteristic aroma that gradually get lost on storage. The dry roots become slightly sweet but yet have a bitter aftertaste.

Macroscopic characteristics

The dry roots are long 5-20 cm long, cylindrical to fusiform, usually branching from the middle to 2-5 rootlets. The root base can be 0.5-3cm in diameter. The root surface is light yellow brown to light grayish brown with longitudinal wrinkles. The root crown is somewhat constricted having short remains of rhizome. The fractured surface of root is light brown except in the vicinity of cambial area where the colour is brown.

Microscopic characteristics: Powdered ginseng shows parenchyma cells containing starch grains and occasionally gelatinized starch. The diameter of reticulate vessels is around 45µm while those of scalariform and spiral vessels range from 15-40 µm. Secretary cells have a mass of yellow glistened material of calcium oxalate crystals. Starch grains are 3-15 µm in diameter, and are found either singly or in aggregates of 2-4.

Powdered plant material: The powdered Ginseng is light yellowish-white to light yellowish-brown in colour with a characteristic odor that gets diminished with age.

Substitutes and adulterants: Due to its high commercial value, ginseng roots are frequently adulterated. The adulteration is further compounded with the taxonomic confusion of different land races, intermixed source plants and unrelated plants labeled as ginseng in the market. The ginseng adulteration can be divided into following five categories: 1. Plants misbranded as 'ginseng' in commerce but not belonging to genus *Panax* e.g., *Codonopsis pilosula* and *C. tangshen* (family Companulaceae). 2. Deliberate adulteration of one *P. ginseng* with other species having lower market value, for economic reasons. 3. *P. ginseng* is adulterated with plant materials of different genus/ species that may have a casual visual resemblance to a *Panax* species, or the use of pharmacologically inert bulking agents, particularly in powder form such as sawdust or di-calcium phosphate. 4. Substitution of various ginseng plant parts like leaves stem in the extract without declaring all parts on the label. The ginsenoside profile of ginseng leaf being different than that of ginseng root, mixing or substituting leaf for root/rhizome may affect product bioactivity 5. Intentional mixing of waste product left after extraction as ginseng root powder.

General identity tests

Purity tests

- Foreign matter : Not more than 2 percent
- Total ash : Not more than 4.2 percent
- Heavy metals : Not more than 15 ppm
- Arsenic : Not more than 2 ppm
- Acid- insoluble ash : Not less less than 1 percent
- Alcohol-soluble extractive : Not less than 14 percent
- Water saturated n-butanol extracts : Not less than 60 mg/g^3 (indicative of crude saponins qualitatively positive for ginsenosides)

The usual tests for ginseng species identification rely on botanical identification of fresh plant or root specimens and biochemical profiling of active and marker compounds e.g., ginsenosides. Presently, ginseng identification is based on: 1.DNA sequence analysis of the nuclear ribosomal internal transcribed spacer region. 2. Sequence analysis of chloroplast ribulose 1,5-bisphosphate carboxylase large subunit gene, and 3. Rapid amplification of polymorphic DNA for DNA fingerprinting. The results are also corroborated biochemically with HPLC. All these methods are helpful in distinguishing American from Korean ginseng types. Compared with other molecular techniques, Microsatellite marker technology has shown to be a robust, accurate, reproducible, reliable and sensitive tool to differentiate *P. ginseng* and *P. quinquefolius* with the resolution down to farm level, i.e., confirmation of its botanical identity and origin (Hon *et al*., 2003).

CHEMISTRY ASPECT

Chemical composition

The Korean ginseng is reported to have about 200 hundred chemical entities like ginsenosides, polysaccharides, polyacetylenes, peptides, and amino acids (Attele *et al*., 1999). Ginsenosides and polysaccharides are the main entities to which the pharmacological activities of Ginseng are attributed. Ginsenosides are the major and unique components of *P ginseng* (Attele *et al*., 1999). The main chemical constituents in oriental and American gingseng (*P. ginseng, P. notoginseng,* and *P. quinquefolius*) contain saponins, called ginsenosides (Fig. 1).

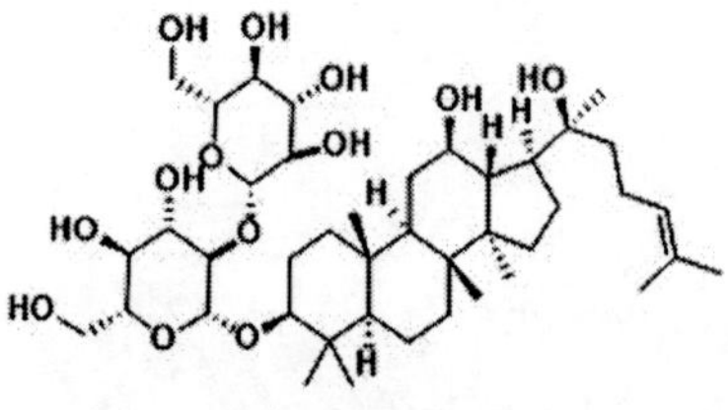

Ginsenoside Rg3

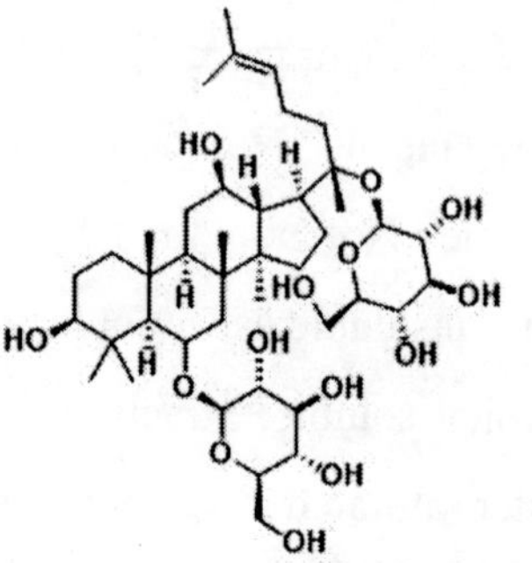

Ginsenoside Rg1

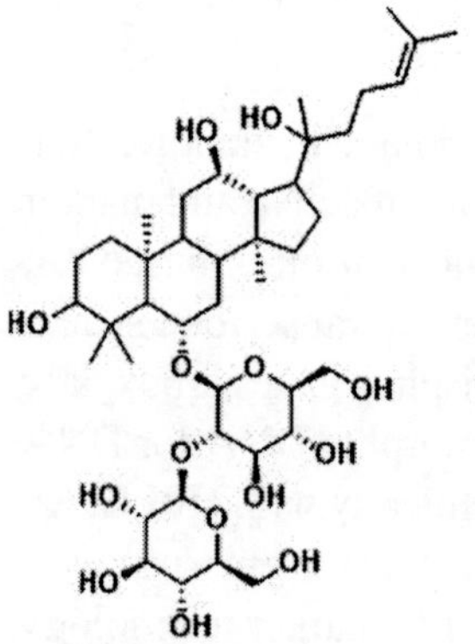

Ginsenoside Rf

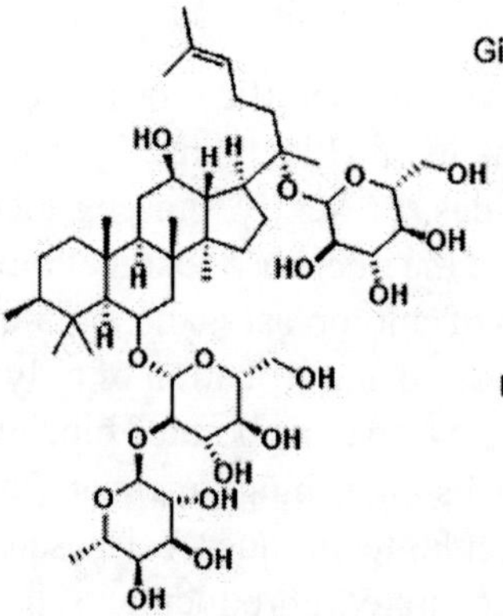

Ginsenoside Re

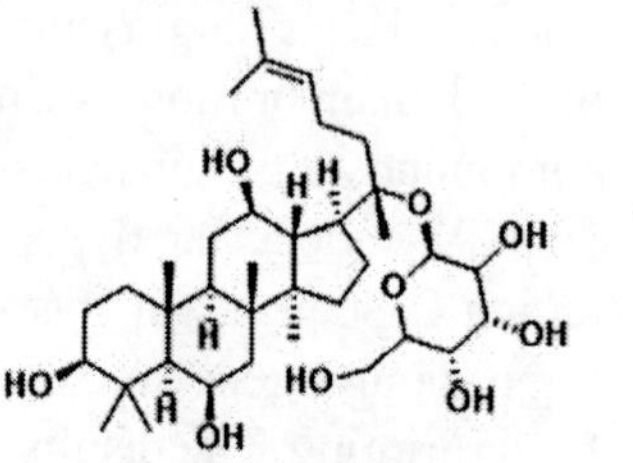

Ginsenoside F1

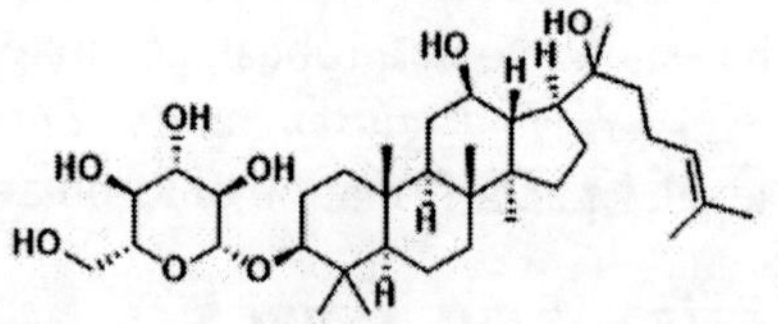

Ginsenoside Rh2

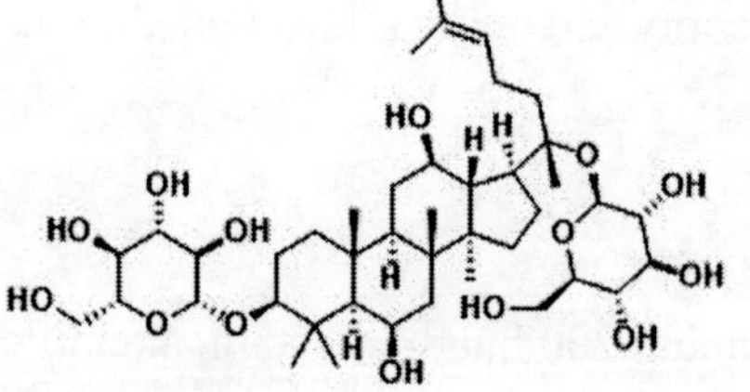

Ginsenoside F2

Chemical constituents of *Panax* species
(Chemical structures are contributed by Dinesh Kumar)

So far, about 289 saponins/ ginsenosides have been characterized from different *Panax* species (Yang *et al.*, 2014). The post-harvest processing of ginseng also leads to production of important biologically active derivatives. Freshly harvested air dried ginseng is called white ginseng, while red ginseng is prepared by steaming the fresh roots for a few hours, then drying it. The two differ in their chemical profile. Red ginseng has an altered and less polar ginsenoside profile than white ginseng of different biological activity (Kim, 2006; Sun *et al.*, 2011). The ginseng saponins are classified into three groups: 1) Dammarane-type. These saponins are further classified into (i) Protopanaxadiol having sugar moieties attached to OH at C-3 and or C-20 and (ii) Protopanaxatriol types. These have sugar moieties attached at C-3, C-6 and or C-20. 2) Cotillol-type having a five membered epoxy ring at C-20 and 3) Oleanane-type oligoglycosides, characterized by having a modified C-20 side chain.

The different ginsengs e.g., Korean/Chinese, American, Notoginseng including Himalayan ginseng can be differentiated chemically on the basis of specific ginsenosides or their relative predominance. For example, Chinese or Korean ginsengs and Notoginseng contain ginsenosides Rf and American ginseng can be distinguished by the presence of psuedoginsenoside F11 (Besso *et al.*, 1982).

Notoginsenoside R1 is found in two oriental ginsengs 'Korean' and 'Notoginseng' exclusively. The ratio of genocides Rg1/ Rb1 is also employed to differentiate American ginseng (less than 0.4) and a higher value in the oriental ginsengs.(Yuan *et al.,* 2010).

Polysaccharides

Many polysaccharides having immuno-potentiating properties are present in ginseng species. The *P. ginseng* has abundance of acidic polysaccharidesas compared to *P. quinquifolium* and *P. notoginseng*. The sepolysaccharides, also called ginsan, have a typical structure of pectin, constituting about 40 percent by weight of the dried roots(Ovodov and Solov'eva, 1966).

Wan *et al.* (2012) have isolated several water-soluble oligosaccharides from ginseng roots. Most of these like, α-Glcp-(1-6)-α-Glcp, α-Glcp-(1-6)-α- Glcp-(1-4)-α -Glcp, α-Glcp-(1-6)-α -Glcp-(1-6)-α-Glcp-(1-4)-α-Glcp, and besides six other malto-oligosaccharides (i.e., maltopentaose, maltohexaose, maltoheptaose, maltooctaose, maltononaose, maltodecaose) are well characterized.

Hypoglycemic glycans, such as panasans A-E, poanaxans F-H, panaxans I-L, panaxans M-P, and panaxans Q–U, have been isolated from the root of *P. ginseng* (Konno *et al*., 1984,1985)

Similarly, Tomoda *et. al.* (1993a) and Tomoda *et al.* (1993b) have reported immune-modulating glycans (ginsenan PA and ginsenan PB) from the root of the same species. The other immuno-modulationg glycans present are: D-galactose, L-rhamnose, D-galacturonic acid, and D-glucuronic acid, acidic lysaccharide ginsenan S-IAand ginsenan S-II A.

Other active compounds

Different flavonoids and polyacetylenes, such as kaempferol, trifolinand panasenoside, panaxynol, and panaxydol have been reported by Matsunaga *et al.* (1989). Also, sesquiterpene alcohols (pansinsanol A and panasinsanol B) and sesquiterpene hydrocarbons, (α-panasinsene, α-panasinsene, α-panasinsene, α-neoclovene, and α-panasinsene) have also been reported by Iwabuchi *et al.,*(1990). Two sesquitene alcohols, ginsenol and senecrassidiol, have been reported by Iwabuchi *et al.* (1988,1990). Several ginsenoynes (A, B, C, E, D, E, F, G, H, I, J and K) ginsenoynes (A, B, C, D, E, F, G, H, I, J and K) have been reported (Kim, 2012; Choi *et al.*, 2010). Polyacetylenes compounds (Liu *et al.*, 2007) and trilinolein (Chan *et al.*, 2002) have been isolated from *P. ginseng* leaves.

Red Korean ginseng (steamed *P. ginseng*), developed initially to enhance preservation of the product, has been in use for hundreds of years. The steaming and heating during the processing of red ginseng may lead to transformation of ginsenoside Rg3 to 20S-Rh2 20 R-Rh2. These ginsenosides get further chemically degraded to aglycone 20S-protopanaxadiol, 20R-protopanaxadiol, or 20-dehydroprotopanaxadiol (Hiromichi *et al.*,1982; Kasai *et al.*, 1983). Similarly, ginsenosides Rk1 and Rg5 are also transformed to products like Rk2 and Rh3. Rh1 which further get degraded into aglycones 20S-protopanaxatriol, 20-R protopanaxatriol and 20-dehydropropanaxatriol.

E. PHARMACOLOGICALASPECTS / Bioactivity

1. *P. ginseng*

Anti-aging/Antioxidant activity

Ginseng is considered a panacea for promoting longevity by ameliorating the oxidative stresses (Kim *et al.*, 2004). It has been shown to have anti-wrinkle effect (So *et al.*, 2008) and to help in the metabolic maintenance of the human body (Aryal, 2011), including homeostatic maintenance (Brekhman and Dardymov, 1969; Brekhman ,1957).

Ginseng extracts have been shown to improve learning and memory in normal, aged or brain damaged animals (Zhong *et al.*, 2000; Kennedy and Scholey, 2003; Jaenicke *et al.,* 1991). Ginsenosides Rb1 (Ryu *et al.*, 1997) and Rg1

(Han, 1987) have been shown to accelerate both memory acquisition and cognitive function in rats and mice (Mook *et al.,* 2001). Some phenolic compounds of ginseng (Kim *et al.,*2012) are also reported to have antioxidant activity.

The artificially induced axonal atrophy, and synaptic loss in a mouse model of Alzheimer disease has been shown to be ameliorated (Tohda *et al.,* 2004) by M1 a metabolite of ginsenoside Rb1 (Protopanaxatriol-type saponins). Similarly, ginsenosides Rg3(S), Rg5 and Rk1 have been shown to have neuro-protective properties and to reverse memory dysfunction induced by ethanol and scopolamine (Bao *et al.,* 2005).

Anti-diabetic activity

P. ginseng improves insulin sensitivity and glucose homeostasis (Sonnenborn and Proppert 1991; Yuan *et al*., 2011).The oral administration of ginseng root to diabetic KKAy mice for a month resulted in reduction of blood glucose levels and enhanced the insulin sensitivity (Chung *et al.,*2001). Similarly, ginsenoside Rb2 has been shown to be effective for streptozatocin-diabetic rats (Yokozawa *et al.,* 1985). Yun *et al.* (2004) reported improvement in insulin resistance index as well as decrease in obesity by administration of wild ginseng ethanol extract. The anti-diabetic activity of ginsenoside Rb2 is believed to be via AMP-activated protein kinase inhibiting palmitate-induced gluconeogenesis (Lee *et al*., 2011; Jung and Chung, 2011).

Immuno-regulatory activity

In Far East countries, *P. ginseng* has traditionally been used to increase stamina and strengthening the immune system. Out of the different polysaccharides which have been isolated, ginsan is a potent immune-modulator. This activity of ginsan is through elevation of (1.7-2 fold) heme-oxygenase activity, reduced hepatic cytochrome P-450 (CYP450) levels (by 20-34%), and prolongation of zoxazolamine-induced paralysis time (by 65-70%) without compromising the liver functions (Song *et al.,* 2004). It induces lymphoid cells to proliferate with the production of several cytokines (Song *et al.,* 2003). Ginsan improves γ radiation-induced immune-suppression by inducing mRNA expression of Th1 and Th2 type cytokines, and restores mRNA expression of INF-γ and Th1 cytokines (Han *et al.,* 2005). The acidic polysaccharide is helpful in treatment of chronic fatigue syndrome (Wang *et al.,* 2013). The red ginseng acidic polysaccharide and pidotimod (immuno-stimulant) is reported to have synergistic immune-stimulating activity against cyclophosphamide-induced immune-suppression (Du *et al.,* 2008).

Anti-cancer activity

P. ginseng is widely used for its strong anticancer activities by upregulating the anticancer enzymes (Yun, 2003). The cytotoxic activity of both saponins and other compounds from ginseng has been demonstrated against different kinds of cancer cell lines in culture such as L1210, L5187Y, Hela cells, Sarcoma 180 cells,A549, SK-OV-3, SK-Mel-2, P388, and K562 (Baek, 1995). Ginsenosides Rb1, Rc and Re act as weak phytoestrogens also. They bind with estrogen receptors and activating them at both at mRNA and protein levels in MCF-7 human breast cancer cells (Lee *et al.*,2003a; Lee*et al.*,2003b). Ginsenosides like Rg3 and Rh2 inhibit proliferation of prostate cancer cells and have anti-androgen activity (Kim *et al.,* 2004; Li, 2009). A ginsenoside metabolite, compound K, inhibits the growth of human monocytic leukemia cells U937 (Kang *et al.,* 2005). The inhibition of tumor cell proliferation and induction of cell apoptosis in mice with induced liver cancer has been reported by (Li *et al.,* 2005). Another ginsenoside Rk1 was found to induce apoptosis in SK-MEL-2 human melanoma (Kim *et al.*, 2012). Red ginseng has been suggested as an adjuvant therapy for the treatment of colorectal cancer (Fishbein *et al.*, 2009)

Neuro-regulation activity

Ginseng improves cognitive functions and prevents memory impairment (Jin *et al.,* 1996). *P ginseng* has been shown to have neuroprotective properties. It reduced cerebral inflammation and symptoms of experimental autoimmune encephalomyelitis. It has also been shown to improve motor function recovery after a spinal cord injury by reducing inflammation in mice (Bing *et al.*,2016). Korean red ginseng has also been shown to reduce the inattention and hyperactivity scores of children with 'Attention deficit hyperactivity disorder' (ADHD) symptoms. Ginseng total saponin (GTS) fraction supplementation leads to increase in the amount of norepinephrine and dopamine in mouse brain (Kim *et al.*, 1985). Besides, GTS is also reported to modulate dopaminergic activity at both presynaptic and post synaptic dopamine receptors (Kim *et al.*, 1998). The behavioral side effects induced by psychotropic agents were found to be ameliorated by administering GTS (Kim, 2005). Ginseng total saponins also modulate the methamphetamine-induced striatal dopaminergic neuronal systems by inhibiting dopamine increase (Oh *et al.*, 1997)

Ginsenoside Rh2 and compound K from red ginseng are helpful in treating ischemic brain injury in rats (Bae *et al.,* 2004). The red ginseng extract also modulates nerve growth factors in polycystic ovary murine model by decreasing the nerve growth factor proteins and the respective mRNA (Pak *et al.*, 2009).

Stress alleviation

Ginseng is helpful in improving both physical performance and psychological performance with anti-stress activity (Lee *et al.*, 2008). It increases ability to cope with stress and aging by regulating the hypothalamic-pituitary-adrenal axis, by increasing the plasma corticotrophin and corticosteroid levels (Bing *et al.*, 2016). Ginseng protects retinal cells against cellular oxidative stress caused by hydrogen peroxide. Ginseng therapy is used to increase performance and overall mood alleviation. It increased the score for well-being, and overall quality of life in patients with cancer-related fatigue (Bing *et al.*, 2016).

Lipid-regulating activity

P. ginseng saponins have been found to be very effective in regulating the lipid metabolism. The crude saponin extract from red ginseng decreased plasma cholesterol and triglyceride levels and inhibited aortic atheroma formation in animals fed with a diet containing high cholesterol and suffering from hypercholesterolemia (Moon *et al.,* 1984). The saponins were found to stimulate the absorption, metabolism and transport of lipids.

Antithrombotic activity

Yu *et al.*(2006) reported *in vivo* anti-platelet and antithrombotic activities of Korean redginseng, suggestive of its beneficial role to individuals with high risk of thrombotic and cardiovascular diseases.

Wound and ulcer healing activity

P. ginseng is known to increase the dermal fibroblasts proliferation and increased collagen synthesis in human dermal cells. It is considered to have beneficial effect in skin problems and wound healing. Ginsenoside Rb2(II) has been shown to enhance epidermal cell proliferation (Choi, 2002). Similarly, ginsenoside Rh3, a metabolic derivative of genocide R5, ameliorates chronic dermatitis or psoriasis (Shin *et al.*, 2006). Another ginsenoside, Rb1, has been shown to have anti-ulcer activity by increasing mucus secretion (Jeong, 2002; Jeong *et al.,* 2003).

Anti-bacterial activity

Ginseng extracts exhibit both bacteriostatic and bactericidal activity mainly through disruption of biofilms, inhibition of quorum-sensing and virulence factors, and altering motility. The less polar ginsenosides of red ginseng have higher antimicrobial efficacy than polar ginsenosides (Kachur and Suntres, 2016; Xue *et al.,* 2017). Ginseng extracts have also been shown to inhibit growth of several yeast and mold species.

These extracts have a very strong antiviral activity particularly against RNA viruses as demonstrated in cell culture and animal models and prevention of influenza virus. Quan (2007) and Cho *et al*. (2001) have reported the efficacy of ginseng for inhibiting acquired immune deficiency syndrome virus growth.

Anti-parasitic activity

Han *et.al*. (2011) reported that Protopanaxadiol-type ginsenosides suppress the parasite during early infection. It is considered that acidic ginseng polysaccharides significantly act against malaria parasite by stimulating the immune system. It is stipulated that ginseng polysaccharide ginsenosides have potential for use in the antimalarial treatments

Sexual health

Red ginseng is considered to be effective in improving sexual health by improving erectile dysfunction (Jang, 2008) and protection of sperms from chromatin dysfunction. It is also reported to have efficacy in improving female climacteric disorder (Tode and Kikuchi, 2003)

Anti-inflammatory/ anti-arthritic activity

Korean ginseng, particularly the red ginseng, has strong anti-inflammatory activity. It suppresses the pro-inflammatory genes IFNG, IL17A and TNF(Bing *et al*., 2016). Red ginseng is effective in reducing pro-inflammatory cytokine and fatigue levels in overweight patients with non-alcoholic fatty liver disease (NAFLD), besides improving the adiponectin levels (Hong *et al*., 2016).

Metabolic health

Ginseng is considered very useful in improving metabolic health and in weight management. Fermented red ginseng reduced body weight and improved glucose tolerance and insulin sensitivity in aged obese mice.

Other miscellaneous activities

Ginseng has gastro-protective (Park, 2005), hepatoprotective (Seong *et al*., 2005) and reno-protective (Kim *et al*., 2004) activities. Ginsengsaponins are shown to have other multifarious interactions in body metabolism. For example, they interact directly with Na+-K+-ATPase before disruption of membrane barriers of sarcolemmal vesicles (Lee *et al*., 1986), The saponins induce IP3-mediated Ca2+release from ERs for the activation of Ca2+-activated Cl- channel in *Xenopus* oocytes (Choi *et al*., 2001a, 2001b). It has been reported that GTS could modulate various cellular activities by inhibiting gap junction channel reconstitution (Hong *et al.,* 1996). Furthermore, it was found that CaM could modulate ginseng saponin-mediated Ca2+-activated Cl- channel activation (Lee *et al*., 2005).

2. *P. psuedoginseng*

P. psuedoginseng is mainly used as a hemostatic herbal drug and not as an adaptogen in case of *P. ginseng*. It is used to stop or slow down bleeding and is considered to build blood and also invigorate it. Other activities attributed to it are pain relief, anti-inflammatory, sore throat alleviation, regulation of cholesterol and blood pressure. Traditionally, it is used as a stimulant, aromatic bitter, stomachic and demulcent. The other uses attributed to it are as alliterative, carminative, tonic, expectorant, antipyretic and sedative (Selvam, 2012). In combination with several other herbs, *P. pseudoginseng* is used to treat prostate cancer. Dried powdered roots are widely used locally for ailments such as for diarrhea, dysentery, high blood pressure and for impotency (Jamir *et al*., 2012).

Threats and Conservation Aspects

P. ginseng had become extremely rare in Asia, mostly due to overharvest and deforestation (Taylor, 2006). In 1975, ginseng was included to the Appendix II of CITES (Convention on International Trade in Endangered Species), due to its over-extraction to meet the huge international trade and demand. *Ex-situ* conservation measures, including mandatory commercial cultivation to meet the industrial demands are suggested. However, the scope of its cultivation is restricted to limited suitable sites, since ginseng grows only in specific niches, with a consideration to factors such as slopes, aspects, elevations, and under an assortment of tree species, as discussed earlier. Climate change also poses a threat to the survival of ginseng, and may reduce its habitat.

Conservation strategies

Considering the excessive extraction of *P. pseudoginseng* from wild and large scale land use change of its natural habitats in India, effective conservation strategy becomes the first priority.

Identification of existing natural populations and its safeguarding from both human and animal damage is essential for protection of germplasm. The identified natural population sites can further be strengthened by planting mature seeds in the area to enrich the natural stocks. Since the ginseng seed requires stratification and germination asynchrony over a period of two to three years or more, protection of the designated area is necessary till the seeds germinate. Besides, both *in vivo* and *in vitro* conservation efforts are required to maintain the diversity of the races available in the Himalayan ginseng areas. Analysis of the genetic diversity and population structure of ginseng is essential for its conservation. Allozyme or RAPD based polymorphisms in wild ginseng populations are a low of about 7% (Zhuravlevb *et al.,* 1998,1999). This is against high (about 46%) polymorphism in cultivated *P. ginseng* (Mab *et al.,* 2000). Priority should be given to the conservation of diversity of the different races available in the wild without intermingling with the commercial cultivars.

CULTIVATION ASPECTS

Climate: The selection of site is crucial for success of ginseng cultivation (Li, 1995; Liu, 1988). The forested areas under canopy shade such as temperate Himalayas with adequate soil moisture offer cultivation opportunities. Water supply should be ensured during extreme dry spells. The ideal temperature for ginseng is 15-25^{0}C, which varies for different stages of growth. During perennation in winter months, roots do not tolerate very low soil temperature at sub-freezing temperatures. The northern aspect is favoured, though the southern aspects can also be considered for commercial cultivation by providing appropriate artificial shading infrastructure, if the soil profile is congenial to ginseng cultivation. Ginseng requires 70% shade for its optimal growth (Park, 1987). A few niches are available in Himachal where it can be grown among temperate broad leaved forests and side valleys. Deciduous broad leaved temperate forests areas with prevalence of trees like, Acer, Walnut, Oak, or Poplar are suitable for growing ginseng. Locations with coniferous stands or other shallow-rooted flora like ferns in the under story should be avoided, because they compete seriously for soil moisture and nutrients, besides posing adverse allelopathic effects on ginseng growth. Other than under canopy shade, *P. ginseng* can be commercially grown in open temperate valleys of inner Himalayas like Lahaul in Himachal Pradesh with 70% shade infrastructure created at the site of cultivation using shading nets.

Soil

Soil rich in humus, with proper balance of clay and sand is suitable and helps in maintaining proper soil moisture without either the need to irrigate frequently and avoid water logging or compaction of soil. The ideal soils are loams with high organic matter content. Heavy clay or light sandy soils should be avoided. Ginseng can grow well in slightly acidic pH (5 - 6.5). Amendments to adjust the pH levels by addition of lime or sulphur can be made as per requirement.

Nursery raising and propagation

Both *P. ginseng* and *P. pseudoginseng* are best propagated by seed. The seeds are recalcitrant and immature at the time of harvest. Consequently, these need to be kept moist and require special handling and stratification treatments till they are ready to germinate after about 18 months.

Seed collection

The phase of maturity of berries on inflorescence head is asynchronous. Only ripe berries, bright red in colour, which contain seeds, are picked to ensure high emergence (Liu, 1988). Thinning of green berries to about 10-20 in number is practiced for proper development of the seed. An average weight of 50g/100 seed is considered ideal for good germination and seedling vigor. Sometimes the seeds are loaded with contaminating flora on the surface due to in adequate washing of the berries and improper storage. Since the emerging seedlings are very delicate and highly susceptible to infections, particularly the damping off disease, the seeds should be thoroughly washed and surface sterilized with calcium per chlorate for 10 minutes, or alternatively given antifungal treatment of Bavistin (0.2%) for ten minutes, before storage or stratification.

Germination

The seeds of *P. ginseng* do not germinate immediately after harvest as the embryos are immature and require a period of 1 to 2 years to mature after harvest under cold, moist scarification, in specially designed containers placed under soil (Baranov, 1966). Since seeds are recalcitrant type, care is taken that the seeds do not lose moisture after harvest, and are stored for low temperature scarification. The seeds can be forced to crack to facilitate early germination (9-12 months)by subjecting them to alternating periods of warm and cold storage or stratification.During moist stratification, seeds may undergo changes in development of embryo and/or physiological ripening (Yu and Kim, 1992; Xiao *et al.,* 1987). Seeds which are surface sterilized can be mixed with sterile moist peat moss (1:2) and placed between two layers of peat moss in a pot, which are then sealed with cling film and stored for scarification at alternating temperatures of warm 15^{0}C and cold 4^{0}C for about 90 days each, till cracking of the seeds takes place. Most of the seeds germinate after 12 months of storage. The cracking may continue till 18 months, in asynchronous manner. However, depending upon the seed quality and storage conditions, seed cracking and germination may initiate after 6-9 months. Soon after the cracking, seeds need to be immediately transferred to field irrespective of the season. The stratified seeds can safely be transplanted anytime of the year till autumn and even if the temperatures are low. It is not advisable to delay the seeding of cracked seeds; in fact it should be done before the actual germination begins. It has been reported that GA3 treatments enhance emergence rate by 2-3 times compared to untreated seeds taken as control (Xiao *et al.,* 1987). Cracking and germination of seeds can be achieved initially under controlled conditions, for hastening the process of one year old nursery plants. Alternatively, the non-stratified (green seeds) can be sown immediately after harvest at favorable sites for cultivation.

The seeds stratify under natural conditions and germinate at proper time over a period of 18 months in nursery beds. The rootlets can later be transplanted after one year with proper spacing in the field.

Seed sowing

Depending on the site and soil conditions, seeds are sown at a depth of 1.25 to 2.5 cm. The seeds are then covered with sufficient soil or mulch so that they do not dry. It is essential to cover the seed beds with 2.5 to 5.0 cm of mulch. The seed should be sown at uniform depth and spacing to avoid uneven stands. Seeding rate differs in low density and high density planting in the first year. High density culture is usually practiced when green seeds are used for natural stratification in field. Furrows are prepared 15 cm apart along the length of the bed. Seeding is done 2-5 cm apart in each row. However, low density plantation is usually adapted when cracked seeds are available, and these seeds are sown 15-30 cm apart. The seeding rate is about 15 to 45 kg per acre, depending upon the seed density adopted in the first year of growth. The actual seeding rate and other cropping parameters related to ginseng cultivation need to be standardized as per requirements of the location.

Transplantation

In high density plantation, the rootlets produced in the first year, known as the planting stock, are transplanted to establish a ginseng field. Proper spacing needs to be ensured for good and disease free growth, till the harvest time, 4-5 years later. The usual spacing for rootlets is 10-15 cm plant to plant, in rows 15-20 cm apart. Transplantation also ensures uniform growth and harvesting for commercial production. The soil requirements and bed preparation for transplantation are same as for seed bed preparation. The rootlets are dug out during the dormancy period after the aerial portions die. The rootlets can either be transplanted immediately to the already prepared beds or may be stored in a cold room at 4^{0}C in moist peat moss, till next spring.

CROP MANAGEMENT

After the seed sowing and transplantation, the ginseng field requires specialized crop management, till the crop is ready for harvest after about 4-5 years.

1). Mulching: Organic materialis used for mulching of ginseng beds, immediately after sowing of the seeds, or else after the senescence of aerial parts. It prevents packing or hardening of the soil, helps in moisture conservation, and improves fertility and humus content of soil. The best mulching materials are leaves of deciduous broad leaved trees, wheat or oat straw. Care should be taken not to mulch with weed or wild grasses containing seeds. Well decomposed FYM may be used only after proper fumigation.

2). Shade management: Ginseng is a shade loving plant and requires about 70 % shading for ideal growth (Lee *et al.,* 1980), achieved by growing the crop under canopy shade or use of net under open habitat conditions. The ginseng beds are best prepared across the contour of land preferably on northern slopes. A shade-net structure is prepared on the area and a 70 % shading net open on the northern sides is placed so that no direct sunlight falls on the plants. The sides are left open in such a way, that no direct sunlight enters the field but ensuring at the same time that there is ample aeration and flow of air to minimize disease incidence. Traditionally, in Korea and China, every bed is covered with a separate thatched structure. The shade cover can be removed after the senescence of aerial parts and is replaced by organic mulch. The shade-net should be about 2.5 to 3.0 m high to allow cultural operations comfortably and to minimize the buildup of heat and humidity. The posts are planned in a grid formation 3.5 m apart and dug 30 to 60 cm deep to support the shading net.

Manuring: *P. ginseng* does not have specific requirement of fertilizers, except that the soil should be rich in humus sand porous but with adequate moisture. Addition of nitrogen can improve the growth of foliage with little effect on root biomass. A yearly dose of 25-30 kg/ha of nitrogen is suggested.The optimum growth is reported at 5.5 soil pH, although it can tolerate lower pH (4-5) also. Sometimes micronutrient supplementation may be required, if the soil analysis reports of area suggests so. Decomposed leaf litter without seeds of weeds is considered the best fertilizer for ginseng.

Irrigation: Ginseng crop requires adequate soil moisture, but not wet conditions, which may otherwise lead to root-rot. The ideal level of soil moisture is 40-50 % of the field capacity (Liu, 1988). Proper site selection, shading and appropriate mulching help in retaining moisture levels. Occasional irrigations may be required in commercial cultivation. Care should be taken not to go for sprinkler irrigation as it may lead to spread of spores of pathogens on leaves.

Guide for disease prevention measures

1. Select and prepare a site with good soil drainage. *Ginseng* is highly susceptible to water logging. The root-rot and damping-off of the plants usually occurs in wet and poorly drained soils.
2. Ginseng thrives well at sites with very good air circulation. On the contrary, the spread of foliar diseases is favored by stagnant and humid air. For commercial success, the beds should be free of any weeds and undergrowth.
3. Sites infested with root-knot nematodes should be avoided. Fumigate and treat the sites likely to suffer from nematode infestation. Destroy the pathogens of the site before going in for cultivation.

4. Use disease-free seeds and plants. Surface sterilizing the seeds before putting them for stratification helps in eliminating seed-borne diseases. The planting stock is delicate and needs to be handled with care to prevent damage.

5. Use proper spacing to avoid over-crowding in the beds. Dense planting may promote spread of diseases.

6. Upon appearance of disease symptoms remove and destroy diseased plants immediately.

7. It is better to plant ginseng in several small pockets than at one large consolidated plantation. This will prevent spread of disease and loss of entire crop.

8. Analyze the soil and profile of the site, particularly soil pH, soil nitrogen and organic matter/humus before plantation. Pure clay or sandy soils should be avoided.

9. It is better to rotate the crop and select a new site for ginseng cultivation, so that any infestation in the previous ginseng growing site is avoided.

Harvesting: Roots can be harvested fourth year onwards in the artificially shaded commercial fields, depending upon the growth of plants. Harvesting is carried out in September. Fruits start maturing by October. Rhizomes are washed and dried in plant driers at below 45^0 C (Nayar and Sastry, 1990). In case of naturally shaded crop, harvesting can be delayed because the growth is normally slow under these conditions. In any case, harvesting should be done after the berries ripen. A yield of about 4 t/ha of air dried roots is expected from a well-managed, artificially shaded, ginseng field. The roots are dug after removal of any mulch and vegetative aerial parts. Care is taken to dig out the roots with minimal damage. The roots are collected in baskets and washed carefully with clean water without scrubbing or damaging the skin. The roots are dried in a room or drying cabinet under dry hot air stream at 35°C after spreading them in a single layer preferably on a wire mesh. A room heater and a fan can be used for room drying. The roots should be dried slowly over 3-4 weeks' time to maintain quality of the roots. The average moisture content of dried roots should be about 10%. The drying process is complete when the roots break with a snap and can be stored for commercial disposal.

COMMERCIAL ASPECTS

Ginseng is amongst the top five highly traded medicinal plants in the world.

Trade data: Presently, ginseng grows in 35 countries, out of which four countries e.g., South Korea (27,480 tons), China (44,749 tons), Canada (6,486 tons), and the US (1,054 tons) are the biggest producers contributing 99% of the total of 80,080 tons of ginseng produced in the world (Baeg and So, 2013). However, a comprehensive and reliable statistical data about ginseng production and its market size in the world is still lacking. Ginseng is traded as fresh ginseng, dried ginseng, red ginseng and products like extracts, tea etc. The total world ginseng market is estimated to be worth $2,084 million, sold at the rate of 50-500 $ per pound in the international market. The annual ginseng imported in India during year 2009 was worth $ 48000 (Baeg and So 2013). With the growing interest in alternate medicine in India, the demand for ginseng is expected to further increase many fold.

Major users

Ginseng roots are the valuable ingredient in many products, such as dietary supplements, functional food, medicinal formulations, etc. It can be consumed in dried form, steamed, or as extract. With the progressive growth of consumer market of food supplements and alternative medicine, it offers great scope for ginseng consumption and demand. In 2009, the Codex Alimentarius Commission adopted an Asian regional standard to ensure the quality of ginseng products. The Commission has now approved its conversion into a worldwide standard. The standard applies to ginseng products used as food or food ingredient and does not apply to products used for medicinal purposes. This worldwide standard specifies the identity, essential composition and quality factors that ginseng products used as foods should comply to ensure fair practices in their trade. [ftp://ftp.fao.org/codex/reports/reports_2015/REP15_PFVe.pdf (paragraphs 78-87)]

Future Research

The Chinese or Korean ginseng has been fully domesticated, and the bulk supply of the world comes from commercial farming in China and Korea. Amongst the allied species, *P. ginseng* has been the best studied in terms of chemical composition, pharmacological activity, molecular characterization and gene expression, including efforts for its genetic improvement and management of diseases limiting plant productivity. The scope of ginseng cultivation within India remains restricted due to areas offering favorable climatic opportunities. It has recently been introduced in the high elevation areas in the state of Himachal Pradesh by CSIR-IHBT. Some of the major factors contributing to the success of its commercial cultivation are 1) collection of diversity within the species from different sources or countries, 2) identification of suitable sites, such as the temperate Himalayas, 3) evaluation of its relative performance through multi-location trials, 4) development of standardized package of practices for both forest areas and cultivated field, 5) region specific studies on pests and diseases and working out of appropriate packages of plant protection measures, 6) development of ginseng products and vigorous marketing strategies. It is concluded that inner temperate Himalayan valleys and mixed forests are highly suited for organic cultivation, an advantage associated with high commercial returns.

The species of *Panax* distributed in India is *P. pseudoginseng* in the north-eastern Himalayan states. While, many land races or sub-species are available in different isolated areas, the taxonomical aspects are not well resolved and remain controversial. This polytypic nature of Himalayan ginseng needs immediate attention for understanding its chemical characterization, genetic diversity, with concerted efforts for germplasm enhancement, selection, molecular characterization and disease-pests management, besides focus on conserving the available diversity. The other aspect is that despite being used in the local and folk medicine, the material from indigenous population of *P. ginseng* has not been put to any rigorous pharmacological screening so far. Well planned research therefore is needed to screen bioactive compounds, their activity and clinical effects. Also, seed biology and cultivation aspects of *P. pseudoginseng* are almost untouched areas, and work regarding improvement of cultivation methods for productivity enhancement remains yet another matter requiring attention. The modern tools of molecular breeding and biotechnology have yet not been applied for Himalayan ginseng for getting better insight into issues such as gene expression and gene regulation. Vast opportunities exists in the areas of pharmacological investigations, screening of bioactive compounds, clinical trials, biosynthesis of active ingredient/ compounds, to better achieve goals of developing technologies for improved plant growth, plant protection packages compatible with organic farming and generation of elite and uniform plant material.

PATENTS

1 Adamko D. J., Rosenthal K. L., Shan J., Sutherland S., Wu, Y. Activation of innate and adaptive immune responses by a ginseng extract. EP2268295A2, WO2009106975A2. 27.02.2009.

2 Brekhman II. *Panax ginseng*. Gosudaarst Isdat et Med Lit.; Leningrad: 1957.

3 An H.S., Lee T.S., Shin E.M., Kim S.H.. Method for mass production of wild ginseng roots by plant tissue culture .WO2005020672A1. 10.03.2005.

4 Bae K.G, Jung I.S., Song J.Y., Yun Y.S.. The hematopoitic,myeloprotecting, antitumor immune cells generating and radiosensititizing polysaccharide isolated from *Panax ginseng*. EP 1124982 A1, DE60035406T2. 16.08.2007.

5 Bombardelli E., Corti F.. Compositions for treatment of cancer-related fatigue US9566309B2. 14.02.2017.

6 Cao J, Zhang X. Qu, F., Guo Z., Zhao Y. 2015. Dammarane triterpenoids for pharmaceutical use: a patent review . (2005 – 2014) *Expert Opinion on Therapeutic Patents*. 25: 7.805-17.

7 Choi, S, Kim S Il , Park N Y. Method for manufacturing ginsenoside Rg3 and Rh2 enriched red ginseng concentrate WO2013012180A3, WO 2013012180A2. 24. 01. 2013.

8 Da C. S. R., Fast D. J., Gellenbeck K. W., Krempin D. W., Lin Y., Murray M. A., Rana J., Rebhun J. F., Schmidt H. R., Wilkins L. M.. Anti-resorptive and bone building dietary supplements and methods of use. WO2008051594A3, 09.10.2008.

9 Eraslan M., Tekin M.. *Tribulus terrestris, Avena sativa* and *Panax ginseng* extract combination US 20140205687A1 . WO2013028140A1. 24. 07. 2014.

10 Gianesello V, Soldati F., Vignutelli A, Peters M.. Composition comprising panax ginseng and paullinia cupana extracts .EP1545569B1. 14.12.2016.

11 Gow R., Sypert G., Li D., Ya X. Methods and compositions comprising Panax species US20070065526A1, CA2623030A1. 22.03.2007.

12 Hashimoto K., Nakata K., Sakanaka M., Tanaka J.. Skin tissue regeneration promoters comprising ginsenoside Rb1. EP1295893A1. 26.03.2003.

13 Hong S. M., Lee E. K., Jin Y. W.. Composition for preventing or treating aids containing plant stem cell line derived from cambium of *Panax ginseng* including wild ginseng or ginseng as active ingredient. EP2363137A2, N102170891A. 08.04.2010.

14 Huh G. H., Kim Y. H., Kim B. M.. Promoter for the high level expression in plant-tissue culture and vector using the same. US20100100986A1. WO2009022845A2, 22.04.2010.

15 Hwang W. I., Kim D. C., Lee S. DChae, H. J. , In M J., Choi S Y., Lee K. S., Lee E. S., Kim J. H. Exhaustive extraction method for ginseng. WO2010114191A1, 07.10. 2010.

16 Jang M. O., Lim M. J., Oh I S., Lee D. H., Lee E K., Jin Y. W.. Composition for anti-aging or antioxidant composition containing plant stem cell line derived from cambium or panax ginseng including wild ginseng and ginseng as active components. US20130302287A1. 14.11.2013.

17 Jang M. O., Lim., M. J., Oh I. S, Lee D. H.,. Lee E. K, Jin Y. W. Composition for anti-aging or antioxidant composition containing plant stem cell line derived from cambium of *Panax ginseng* including wild ginseng and ginseng as active components. US9095532B2. 04.08.2015.

18 Jin Y. W., Lee E. K.. Composition for Cancer Prevention or Treatment Containing as Active Ingredient Plant Stem Cell Line Dervied from Cambium of *Panax ginseng* Including Wild Ginseng or Ginseng. US20110229443A1. 22.09.2011.

19 Jin Y. W., Lee E. K Lim., M. J.. Composition for preventing or treating liver diseases, containing plant stem cell lines derived from the cambium of *Panax ginseng* including mountain ginseng or ginseng as active ingredient. EP2335712A2, CA2734101A1, 22.06.2011.

20 Jin Y. W., Lee E. K.. Composition for cancer prevention or treatment containing as active ingredient plant stem cell line dervied from cambium of *Panax ginseng* including wild ginseng or ginseng .US 9314492B2. CA2742950A1. 19.04.2016.

21 Kang S. S., Jang Y. P., Jee E. H. A method for extracting ginsenosides-abundant extract from ginseng by using weak alkali water WO 2011062332A1. 26.05.2011.

22 Kwak T. H., Shin M. S., Kim J.Y., Park J.K. Active fraction having anti-cancer and anti-metastasis isolated from leaves and stems of ginseng US7901716B2. CA2486915A1. 08.03.2011.

23 Lee S. H., Yu H. J., Cho N. S., Park J. H., Kim T. H., Kim K. H., Lee S. K., A method for preparing the inclusion complex of ginseng extract with gamma-cyclodextrin, and the composition comprising the same.WO2008127063 A1. 23.10.2008.

24 Lim M. J., Jin Y. W., Lee E. K.. Composition for enhancing immunity containing plant stem cell line derived from cambium of *Panax ginseng* including wild ginseng or ginseng

25 Lim M. J., Jin Y. W., Lee E. K.. Composition for enhancing immunity containing plant stem cell line derived from cambium of *Panax ginseng* including wild ginseng or ginseng as an active ingredient. EP2399596A2, CN102325538A. 28.12.2011.

26 Lim M. J., Jin Y. W., Lee E. K.. Composition for enhancing immunity containing plant stem cell line derived from cambium of *Panax ginseng* including wild ginseng or ginseng as an active ingredient. EP2399596A4, CN102325538A. 29.08.2012.

27 Maeda N., Nakata K., Sakanaka M., Tanaka J.. Brain cell or nerve cell protecting agents comprising ginseng. EP1213026A1. 12.06.2002.

28 Maeda N., Nakata K., Sakanaka M., Tanaka J.. Brain cell or nerve cell protecting agents comprising ginseng. EP1213026A1. 12.06.2002

29 Paek K.Y.. Method for the mass propagation of adventitious roots of ginseng, camphor ginseng and wild ginseng by tissue culture and the improvement of their saponin content. US6713303B2. 30.03.2004.

30 Ryu B.. Composition for brain-neuron protection and brain-disease prevention, alleviation or treatment comprising muskrat musk. US20150238539A1. 27.08.2015

31 Sakanaka M., Maeda N., Tanaka J., Nakata K.. Brain cell or nerve cell-protecting agents comprising medicinal ginseng. US7235267B1. CN1630528A, 26.07.2007'

32 Sakanaka M., Maeda N., Tanaka J., Nakata K.. Brain cell or nerve cell protecting agents comprising ginseng. EP1213026A4. WO2001015717A1. 16.06.2004.

33 Sakanaka M., Sato K., Suite J.. 115 Ehime University TANAKA Cerebrovascular regeneration/reconstruction promoters and nerve tissue secondary degeneration inhibitors comprising ginsenoside rb 1? EP 1170012 A1. 09.01.2002.

34 Sengupta S., Fan T., Toh S. A. E. S., Leung H. W., Wong N. S. R., Yeung H. W.. Pharmaceutically effective ginsenosides and their use WO2002069980A2. 12.09.2002.

35 Shengbo Q , (Shanghai, CN). Chinese medicinal compositions useful as anti-fatigue, anti-aging and gonadotrophic agent and processes for preparation thereof US8974841. 03.10.2015.

36 Shim H.J., Kang M.Y., Shin Y.D.. Method and system for sterilization of the soil for growing Korean ginseng by using an electron beam and system of growing Korean ginseng using an electron beam sterilization. US8590206B2. CN102552955A, 26.11.2013.

37 Sung J. H., Huh J. D., Hasegawa H., Matsumiya S., Uchiyama M.. A process for the preparation of metabolites of ginseng saponins. CA2223234C. 15.07.2003.

38 Sung K, Ginseng preparation using vinegar and process for thereof US 20060198908A1. 07.09.2006.

39 Tuttle B. D.. Dietary supplement for increasing energy, strength, and immune function. US6465018B1, CA2416575A1, EP1190628A1.15.10.2002

40 Vascoe J. M., Merrill R. E.. Composition of matter for sexual dysfunction US9572848B1. 21.02.2017.

41 Veninga, L., 1973. The Ginseng Book. Felton, CA: Big Tree Press.

42 Vignutelli A., Cerny J. Composition for the activation of the immune system. WO2005039320A2. CA2541109A1. 06.05.2005.

43 Vignutelli A., Cerny J., Composition for the activation of the immune system. WO2005039320A2. CA2541109A1. 06.05.2005.

44 Wuh H. C. K., Trant A. S.. Method and composition for improving sexual fitness. US6544563B2, 08.04.2003.

45 Y. W. Jin. Stability of secondary metabolite mass production through syncronized plant cell cultures. WO2007052876A1, CA2627533A1, 10.03. 2007.

46 Yun Y.S., Song J.Y. Jung I.S. Composition comprising polysaccharide extracted from panax ginseng preventing and treating liver diseases EP2273999A4. 24.07.2013.

47 Yun Y.S., Song J.Y., Jung I. S.. Composition comprising polysaccharide extracted from panax ginseng preventing and treating liver diseases.WO2009125964 A2, EP2273999A2, EP2273999A4, 24.07.2013.

48 Yun Y.S., Song J.Y., Jung I.S.. Composition Comprising Polysaccharide Extracted from *Panax ginseng* Preventing and Treating Liver Diseases. US20110046086A1. 24.02.2011.

REFERENCES

1. Aryal B, Maskey D, Kim M.J, Yang J.W., Kim H.G. 2011. Effect of ginseng on calretinin expression in mouse hippocampus following exposure to 835 MHz radiofrequency. *J Ginseng Res*; 35:138-148.
2. Attele A.S., Wu J.A., Yuan C.S.1999. Ginseng pharmacology: multiple constituents and multiple actions. *Biochem Pharmacol*. 58:1685-1693.
3. Bae E.A., Han M.J., Kim E.J., Kim D.H. 2004.Transformation of ginseng saponins to ginsenoside Rh2 by acids and human intestinal bacteria and biological activities of their transformants. *Arch Pharm Res*: 27:61-67.
4. Bae E.A., Hyun Y.J., Choo M..K., Oh J.K., Ryu J.H., Kim D.H. 2004,Protective effect of fermented red ginseng on transient focal ischemic rats. *Arch Pharm Res*.; 27:1136-1140.
5. Baeg I.H and So S.H. 2013. The world ginseng market and the ginseng (Korea). *J Ginseng R*es Vol. 37, No. 1, 1-7.
6. Baek N .I., Kim J .M., Park J. H., Ryu J. H., Kim D. S., Lee Y. H., Park J. D., Kim S. I. 1997 Ginsenoside Rs(3), a genuine dammarane-glycoside from Korean red ginseng. *Arch Pharm Res*. 20:280-282.
7. Baek N. I., Kim D. S., Lee Y. H, Park J. D., Lee C. B., Kim S. I. 1996 Ginsenoside Rh4, a genuine dammarane glycoside from Korean red ginseng. *Planta Med*. 62:86-87.
8. Baek N. I., Kim D. S., Lee Y. H., Park J. D., Lee C. B. and Kim S. I. 1995. Cytotoxicities of ginseng saponins and their degradation products against some cancer cell lines. *Arch Pharm Res*. 18:164-168.
9. Baek N. I., Kim D..S., Lee Y. H., Park J. D., Lee C. B., Kim S. I. 1996 Ginsenoside Rh4, a genuine dammarane glycoside from Korean red ginseng. *Planta Med.* 62:86-87.
10. Baek N. I., Kim J. M., Park J. H., Ryu J. H., Kim D. S., Lee Y. H., Park J. D., Kim S. I. 1997. Ginsenoside Rs(3), a genuine dammarane-glycoside from Korean red ginseng. *Arch Pharm Res*. 20:280-282.
11. Baek S.H., Lee J.G., Park S.Y., Bae O.N., Kim D.H., Park J.H. 2010. Pectic polysaccharides from *Panax ginseng* as the anti-rotavirus principals in ginseng. *Biomacromolecules*, 11:2044-2052.
12. Banerjee R.N.1968. A taxonomic revision of Indian *Panax* L. (Araliaceae). *Bull Bot Surv Ind* 40:20–27.
13. Bao H.Y., Zhang J., Yeo S.J., Myung C.S., Kim H.M., Kim J.M., Kang J.S. 2005. Memory enhancing and neuroprotective effects of selected ginsenosides. *Arch Pharm Res*. 28:335-342.
14. Baranov A. 1966.Recent advances in our knowledge of the morphology, cultivation and uses of ginseng (C.A. Mayer).*Econ. Bot.,* 20.403-406.
15. Bennet S.S.R., Viswanathan M.V. (1984). The common Indian ginseng. Ind For 110:1049–1052
16. Besso H., Kasai R., Saruwatari Y. 1982. Ginsenoside-Ra1 and Ginsenoside-Ra2, new dammarane-saponins of ginseng roots. *Chem Pharm Bull*.; 30:2380-2385.
17. Bing S.J., Ha D., Hwang I., Park E., Ahn G., Song J. Y., Jee Y. 2016. Protective Effects on Central Nervous System by Acidic Polysaccharide of *Panax ginseng* in Relapse-Remitting Experimental Autoimmune Encephalomyelitis-Induced SJL/J Mice. *Am J Chin Med.* 44(6):1099-1110..
18. Bittles A.H., Fulder S.J., Grant E.C., Nichills M.R. 1979. The effect of ginseng on lifespan and stress response in mice. *Gerontology*. 25: 125.
19. Brekhman II, Dardymov IV. 1969. New substances of plant origin which increase nonspecific resistance. Annu Rev Pharmacol. 9:419-430.
20. Brekhman II. 1957 *Panax ginseng*. Leningrad: GosudaarstIsdatet *Med Lit*.
21. Burkill I.H.1902 Ginseng in China. *Kew Bull* 4:4–11

22. Canon J.F.M.1979 *Panax psuedoginseng* Wall. In: Hara H, William LJH (eds) An enumeration of the flowering plants of Nepal, vol. 2. British Museum, London, p 192.
23. Carpenter, G. C., Grant C. 2011 Growth and reproduction of American ginseng (*Panax quinquefolius*) in Wisconsin, U.S.A.*Canadian J. Bot* 60(12):2692-2696
24. Chan P., Thomas G.N., Tomlinson B. 2002. Protective effects of trilinolein extracted from *Panax notoginseng* against cardiovascular disease. *Acta Pharmacol Sin*. 23:1157- 1162.
25. Cho C.W., Kim Y.C., Rhee Y.K., Lee Y.C and Kim K.T. 2014. Chemical composition characteristics of Korean straight ginseng products. *J. Ethn. Foods*. 1, 24-28
26. Cho Y.K., Sung H., Lee H.J., Joo C.H., Cho G.J. 2001.Long-term intake of Korean red ginseng in HIV-1-infected patients: development of resistance mutation to zidovudine is delayed.*Int Immunopharmacol*.1:1295-1305.
27. Choi H.K., Seong D.H., Rha K.H., Evaluation of clinical efficacy of Korean red ginseng for erectile dysfunction by international index of erectile function (IIEF). *Int J Ginseng Res* 2001; 25: 112–7.
28. Choi H.K., Wen J. 2000 A phylogenetic analysis of Panax (Araliaceae): integrating evidence of chloroplast DNA and the ITS sequences of DNA. *Pl Syst Evol* 224:109–120.
29. Choi K.T., Botanical characteristics, pharmacological effects and medicinal components of Korean *Panax ginseng* C A Meyer. *Acta Pharmacol Sin*. 2008; 29:1109-1118.
30. Choi K.T.,2008.Botanical characteristics, pharmacological effects and medicinal components of Korean *Panax ginseng* C. A. Meyer. *Acta Pharmacol Sin* 29:1109-1118.
31. Choi RC, Zhu JT, Leung KW, Chu GK, Xie HQ, Chen
32. Choi S. 2002 Epidermis proliferative effect of the Panax ginseng Ginsenoside Rb2. Arch Pharm Res; 25:7176.
33. Choi S., Kim H.J., Ko Y.S., Jeong S.W., Kim Y.I., Simonds W.F., Nah S.Y.2001b. Gáq/11 coupled to mammalian phospholipase C â3-like enzyme mediates the ginsenoside effect on Ca2+-activated Cl" current in the Xenopus oocyte. *J Biol Chem*. 276:48797-48802.
34. Choi S., Rho S.H., Jung S.Y., Kim S.C, Park C.S, Nah S.Y. 2001a A novel activation of Ca2+-activated Cl" channel in Xenopus oocytes by ginseng saponins: Evidence for the involvement of phospholipase C and intracellular Ca2+ mobilization. *Br J Pharmacol*.132:641-648.
35. Choi S.J., Sohn H.O., Shin H.J., Hyun H.C., Lee D.W., Song Y.B, Lee S.H., Lim H.S., Lee C.W., Moon J.Y., 2008. Protective effects of Korean Panax ginseng extracts against TCDD induced toxicities in rat. *J Ginseng Res*. 32:382-389.
36. Choi, J.H., Oh S. K. 1985.Studies on the anti-aging actioorean ginseng. *Han'guk Sikp'um Kwahakhoechi*. 17:506-515.
37. Chung S.H., Choi C.G., Park S.H. 2001 Comparisons between white ginseng radix and rootlet for antidiabetic activity and mechanism in KKAy mice. *Arch Pharm Res*. 24:214-218.
38. Du X.F., Jiang C.Z., Wu C.F., Won E.K., Choung S.Y. 2008;. Synergistic immunostimulating activity of pidotimod and red ginseng acidic polysaccharide against cyclophosphamide-induced immunosuppression. *Arch Pharm Res*. 31:1153-1159.
39. Fishbein A.B., Wang C.Z., Li X..L., MehendaleS..R., Sun S.,Aung H.H., Yuan C.S. 2009. Asian ginseng enhances the antiproliferative effect of 5-fluorouracil on human colorectal cancer: Comparison between white and red ginseng..*Arch Pharm Res*.; 32:505-513.
40. *FulderS. 1996 Ginseng* Book: Nature's Ancient Healer book. *Diane Pub Co*: Pp.109. ISBN-13: 978-0756759513
41. Graham S. A. 1966. The genera of Araliaceae in southern United States. *J Arnold Arbor* 47:126–136.
42. Han H., Chen Y., Bi H., Yu L., Sun C., Li S, Oumar S.A., Zhou Y. 2011. In vivo antimalarial activity of ginseng extracts. Pharm Biol. 49 (3):283-89.

43. Han S.K., Song J.Y., Yun Y.S., Yi S.Y. 2005 'Ginsan improved Th1 immune response inhibited by gamma radiation.*Arch Pharm Res*. 28:343-350. 6
44. Han Y.N., Ryu S.Y., Han B.H., Woo L.K. 1987. Spinacine from *Panax ginseng*. *Arch Pharm Res*. 10:258-259
45. Hara H. 1970 . On the Asiatic species of the genus Panax. J Jap Bot 45:197–212
46. Harding, A.R. 1972. Ginseng and other medicinal plants.Arthur Richard Harding, Columbus, Ohio, USA.
47. Haruyo K., Shuichi S., Yoshiteru I., Junzo S. 1982.Studies on the saponins of ginseng. IV. On the structure and enzymatic hydrolysis of ginsenoside Ra1. *Chem Pharm Bull*. 30:2393-2398.
48. Haruyo K., Shuichi S., Yoshiteru I., Junzo S., 1982. Studies on the saponins of ginseng. IV. On the structure and enzymatic hydrolysis of ginsenoside Ra1. *Chem Pharm Bull* 30:2393-2398.
49. Hiromichi B., Ryoji K., Yuhichiro S., Tohru F. 1982 Ginsenoside Ra1 and ginsenoside Ra2, new dammarane-saponins of ginseng roots. *Chem Pharm Bull*. 30:2380-2385.
50. Hon C. C., Chow Y. C., Zeng F .Y., Leung F. C. C. 2003 Genetic authentication of ginseng and other traditional Chinese medicine. *Acta Pharmacol Sin* 24 (9): 841-846
51. Hong E.J., Huh K., Rhee S.K. 1996. Effect of ginseng saponin on gap junction channel reconstituted with connexin *Arch Pharm Res*. 19:264-268.
52. Hong M., Lee Y. H., Kim S., Suk K.T., CS. Bang, Yoon J H,Baik G H, Kim D J, Kim.2016. Anti-inflammatory and anti-fatigue effect of Korean Red Ginseng in patients with nonalcoholic fatty liver disease.*J Ginseng Res*; 40:203-210
53. Hu S.Y., Rüdenberg L, Tredici P.D. (1980) Studies of American ginsengs. *Rhodora*, **82**, 627–636.
54. Hu S.Y.1976 The genus *Panax* (ginseng) in Chinese medicine. *Eco Bot* 30:11–28
55. Iwabuchi H, Yoshikura M, Kamisako W. 1990. Studies on the sesquiterpenoids of *Panax ginseng* C. A. Meyer. II. Isolation and structure determination of ginsenol, a novel sesquiterpene alcohol. *Chem Pharm Bull* (Tokyo);36:2447-2451
56. Iwabuchi H., Kato N., Yoshikura M. 1988. Studies on the sesquiterpenoids of *Panax ginseng* C. A. Meyer. IV. *Chem Pharm Bull* (Tokyo). 38:1405-1407.
57. Iwabuchi H., Yoshikura M., Ikawa Y., Kamisako W.1987. Studies on the sesquiterpenoids of *Panax ginseng* C. A. Meyer. Isolation and structure determination of sesquiterpene alcohols, panasinsanols A and B. *Chem Pharm Bull* (Tokyo): 35:1975-1981.
58. Jaenicke B., Kim E.J., Ahn J.W., Lee H.S. 1991. Effect of *Panax ginseng* extract on passive avoidance retention in old rats. *Arch Pharm Res*. 14:25-29.
59. Jamir S. L., Deb C. R., Jamir N. S 2016. Macropropagation and Production of Clonal Planting Materials of *Panax pseudoginseng* Wall. *Open Journal of Forestry*, 2016, 6, 135-141
60. Jamir, N.S., Lanusunep, and Pongener, N. (2012). Medico-Herbal Medicine Practiced by the Naga Tribes in the State of Nagaland (India). *Indian J. Fundamental Applied Life Sci*., 2, 328-333.
61. Jang, D.J., Lee, M.S., Shin B.C., Lee,Y. C., Ernst E. 2008.Red ginseng for treating erectile dysfunction: a systematic review. Br J.Clin Pharmacol.66 (4): 444–450.
62. Jeong, C.S., Hyun, J.E., Kim, Y.S., Lee, E.S. 2003. Ginsenoside Rb1 the anti-ulcer constituent from the head of *Panax ginseng*. Arch Pharm Res.; 26:906-911.
63. Jeong, C.S. 2002. Effect of butanol fraction of *Panax ginseng* head on gastric lesion and ulcer. *Arch Pharm Res*; 25:61-66.
64. Jin S.H., Kyung J.S., Kim S.C, Nam K.Y. 1996. Effects of red ginseng saponin on normal and scopolarmine-induced memory impairment of mice in passive avoidance task.*Korean J Ginseng Sci*. 20:7-14.

65. Jung M.S., Chung S.H. 2011. AMP-activated protein kinase: A potential target for ginsenosides. *Arch Pharm Res*. 34:1037-1040
66. Kachur K and Suntres.Z. E.2016.The antimicrobial properties of ginseng and ginseng extracts. *Expert Review of Anti-infective Therapy* Vol. 14 (1), 81-94
67. Kang K.A., Kim Y.W., Kim S.U., Chae S., Koh Y.S., Kim H.S., Hyun J.W.2005. G1 phase arrest of the cell cycle by a ginseng metabolite, compound K, in U937 human monocyticleukemia cells. *Arch Pharm Res*. 28:685-690.
68. Kasai R., Besso H., Tanaka O., Saruwatani Y., Fuwa T. 1983. Saponins of red ginseng.*Chem Pharm Bull*. 31:2120-2125.
69. Kennedy D.O., Scholey A.B. 2003 Ginseng: Potential for the enhancement of cognitive performance and mood. *Pharmacol Biochem Behav*. 75:687-700.
70. Kim B.H, Lee SY, Cho HJ, You SN, Kim YJ, Park YM, Lee JK, Baik MY, Park CS, Ahn SC. 2006. Biotransformation of Korean *Panax ginseng* by pectinex. *Biol Pharm Bull* 29:2472-2478
71. Kim D. H. 2012. Chemical Diversity of *Panax ginseng, Panax quinquifolium*, and *Panax notoginseng*. *J. Ginseng Res*. Vol. 36, No. 1, 1-15.
72. Kim D.J, Seong K.S., Kim D.W., Ko S.R, Chang C.C. 2004. Antioxidant effects of red ginseng saponins on paraquat-inducedoxidative stress. *J Ginseng Res*. 28:5-10.
73. Kim D.S., Chang Y.J., Zedk U., Zhao P., Liu Y. Q., Yang C.R. 1995. Dammarane saponins from *Panax ginseng*. *Phytochemistry*. 40:1493-1497
74. Kim H. D.2012. Chemical Diversity of *Panax ginseng, Panax quinquifolium* and*Panax notoginseng .J. Ginseng Res*. 36(1) 1-15
75. Kim H.C., Shin E..J., Jang C.G., Lee M.K., Eun J.S., Hong J.T., Oh K.W. 2005. Pharmacological action of Panax Ginseng onthe behavioral toxicities induced by psychotropic agents. *Arch Pharm Res*. 28:995-1001.
76. Kim H.S., Lee E.H., Ko S.R., Choi K.J., Park J.H., Im D.S. 2004. Effects of ginsenosides Rg3 and Rh2 on the proliferation of prostate cancer cells.*Arch Pharm Res*. 27:429-435.
77. Kim H.S., Zhang Y.H., Fang L.H., Lee M..K. 1998, Effect of ginseng total saponin on bovine adrenal tyrosine hydroxylase. *Arch Pharm Res*.; 21:782-784.
78. Kim J.H., Yi S.M., Choi J.E., Son S.W. 2009. Study of the efficacy of Korean red ginseng in the treatment of androgenic alopecia. *J Ginseng Res*. 33:223-228.
79. Kim J.S,,Joo E.J., Chun J., Ha Y,W., Lee J., Han Y., Kim YS. 2012. Induction of apoptosis by ginsenoside Rk1 in SKMEL-2-human melanoma. *Arch Pharm Res*. 35:717-722.
80. Kim W. Y., Kim J. M., Han S. B., Lee S. K., Kim N. D., Park M. K., Kim C. K., Park J. H. 2000. Steaming of ginseng at high temperature enhances biological activity. *J Nat Prod*. 63:1702-4.
81. Kim Y. C., Lee J. H.,Kim M. S., Lee N. G. 1985. Effect of the saponin fraction of *Panax ginseng* on catecholamines in mouse brain *Archives of Pharmacol Res*. 8(1): 45-48.
82. Kim Y.L and Chung S.H. 2004. Renoprotective effects of Koreanred ginseng. *J Ginseng Res* 28:111-119.
83. Kitagawa I, Taniyama T, Shibuya H, Noda T, Yoshikawa M.1987. Chemical studies on crude drug processing. V. On the constituents of ginseng radix rubra (2): comparison of the constituents of white ginseng and red ginseng prepared from the same *Panax ginseng* root. *Yakugaku Zasshi*: 107:495-505.
84. Kitagawa I., Yoshikawa M., Yoshihara M., Hayashi T., Taniyama T.1983.Chemical studies of crude drugs (1). Constituents of *ginseng radix* rubra. *Yakugaku Zasshi*: 103:612- 622.
85. Konno C, Murakami M, Oshima Y, Hikino H. 1985. Isolation and hypoglycemic activity of panaxans Q, R, S, T and U, glycans of *Panax ginseng* roots. *J Ethnopharmacol*. 14:69-74

86. Konno C, Sugiyama K, Kano M, Takahashi M, Hikino H. 1984. Isolation and hypoglycaemic activity of panaxans A, B,C, D and E, glycans of *Panax ginseng* roots. *Planta Med.* 50:434-436.
87. Kwang-tae CHOI1 2008 'Botanical characteristics, pharmacological effects and medicinal components of Korean *Panax ginseng* C A Meyer' *Acta Pharmacol Sin*. 29 (9): 1109–11
88. Kwon H.K., Ahn C.H, Choi.Y.E. 2009. Molecular authentication of *Panax notoginseng* by specific AFLP- derived SCAR marker. *Journal of Medicinal Plants Research* 3(11) :957-966.
89. Lee C.,Wen J. 2004. Phylogeny of Panax using chloroplast trnC–trnDintergenic region and the utility of trnC–trnD in interspecific studies of plants.*Mol. Phylogen.Evol.* 31 (3). 894-903.
90. Lee J.H., Jeong S.M., Lee B.H., Kim J.H., Ko S.R., Kim S.H., Nah S.Y. 2005. Effect of calmodulin on ginseng saponininduced Ca2+-Activated CI-channel activation in Xenopuslaevis oocytes. *Arch Pharm Res*. 28:413420. 98
91. Lee K.T., Jung T.W., Lee H.J., Kim S.G., Shin Y.S., Whang W.K. 2011. The antidiabetic effect of ginsenoside Rb2 via activation of AMPK. *Arch Pharm Res*. 34:1201- 1208.
92. Lee M.J., Kim E.H., Rhee D.K,. 2008. Effects of *Panax ginseng* on stress. J Ginseng Res. 32:8-14.
93. Lee S.T., Chu K., Kim J.M., Park H.J., Kim M. 2007. Cognitive improvement by ginseng in Alzheimer's disease. *J Ginseng Res*. 31:51-53.
94. Lee S.W, Lee J.S., Kim Y.H., Jin K.D. 1986 Effect of ginseng saponin on the Na+, K+-ATPase of dog cardiac sarcolemma. *Arch Pharm Res*.9:29-38
95. Lee Y., Jin Y., Lim W., Ji S., Choi S., Jang S., Lee S. 2003a .A ginsenoside-Rh1, a component of ginseng saponin, activates estrogen receptor in human breast carcinoma MCF7 cells. *J Steroid Biochem Mol Biol*. 84:463-468.
96. Lee Y.J., Jin Y.R., Lim W.C., Park W.K., Cho J.Y., Jang S., Lee S.K.. 2003b. Ginsenoside-Rb1 acts as a weak phytoestrogen in MCF-7 human breast cancer cells. *Arch Pharm Res*.; 26:58-63.
97. Lee, J.C., S.K. Cheon, Y.T. Kim, Jo, J.S. 1980. Studies on the effect of shading materials on the temperature, light intensity, photosynthesis,and the root growth of Korean ginseng (*Panax ginseng* C.A. Meyer). *J. Korean Soc. Crop Sci.* 2.5:91-98.
98. Li W., Liu Y., Zhang J.W., Ai C.Z., Xiang N., Liu H.X., Yang L. 2005 Anti-androgen-independent prostate cancer effects of ginsenoside metabolites *in vitro*: Mechanism and possible structure-activity relationship investigation.*Arch Pharm Res*. 32:49-57.
99. Li X., Guan Y.S., Zhou X.P., Sun L., Liu Y., He Q., Mao Y.Q. 2005. Anticarcinogenic effect of 20 (R)-ginsenoside Rg3 oninduced hepatocellular carcinoma in rats. *J Sichuan Univ*.(Med Sci Edi). 36:217-220.
100. Li, T.S.C. 1995. Asian and american ginseng—A-review. *HortTechnology* 5:27–34.
101. Liu J.H, Lee C.S., Leung K.M., Yan Z.K., Shen B.H., Zhao Z.Z., Jiang Z.H. 2007. Quantification of two polyacetylenes in Radix Ginseng and roots of related *Panax* species using a gas chromatography-mass spectrometric method. *J Agric Food Chem*. 55:8830-8835.
102. Liu, C.N. 1988. Cultural methods of ginseng. Wu-Chou, Taiwan.
103. Lu J.M., Yao Q., Chen C. 2009. Ginseng compounds: an update on their molecular mechanisms and medical applications. *Curr. Vasc. Pharmacol*. 7:293-302.
104. Ma X..J., Wang X.Q., Xu Z.X., Xiao P.G., Hong D.Y. 2000. RAPD variation within and among populations of ginseng cultivars. *Acta Bot Sinica*: 42:587–590.
105. Matsunaga H, Katano M, Yamamoto H, Mori M, Takata K. 1989. Studies on the panaxytriol of *Panax ginseng* C. A. Meyer. Isolation, determination and antitumor activity. *Chem Pharm Bull* (Tokyo) 37:1279-1281.

106. Mook J.I., Hong H.S., Boo J.H., Lee K.H., Yun S.H., Cheong M.Y., Joo I., Huh K., Jung M.W. 2001 Ginsenoside Rb1 and Rg1 improve spatial learning and increase hippocampal synaptophysin level in mice. *J Neurosci Res*. 63:509-515.
107. Moon C.K., Kang N.Y., Yun Y.P., Lee S.H., Lee H.A., Kang T.L. 1984. Effects of red ginseng-crude saponin on plasma lipid levels in rats fed on a diet high in cholesterol and triglyceride. *Arch Pharm Res*.; 7:41-45.
108. Nayar, M.P. & A.R.K. Sastry (1990). Red Data Book on Indian Plants. Vol. 3, Pp. 28-29.Official monographs for part II (Ginseng) Codex Stan 321-2015. Official Monographs for Part II 927-929.
109. Oh K.W., Kim H.S., Wagner G.C.1997. Inhibitory effects of ginseng total saponin on methamphetamine-induced striatal dopamine increase in mice. *Arch Pharm Res*.20:516-518.
110. OvodovY.S, and Solov'eva T.F. 1966. Polysaccharides of *Panax ginseng*. *Chem Natural Compounds*. 2:243-245.
111. Pak S.C., Kim S.E., Oh D.M., Shim K.M., Jeong M.J., Lim S.C., Bae C.S. 2009; Effect of Korean red ginseng extract in a steroid-induced polycystic ovary murine model. *Arch Pharm Res*. 32:347-352.
112. PandeyA.K., Ali M.A. and Mao, A.A.2007. Genus Panax L. (Araliaceae) in India. *Pleione.* 1(2): 46 - 54. 2007.
113. Pandey A.K., Lee, C. & Wen, J. 2002. A molecular systematic study of *Aralia* L. and *Panax* L. (Araliaceae) inIndia and its taxonomic implications. *Rheedea.* 12(2), 89 – 99.
114. Pandey A.K., Wen J. & Pathak, M.K. 2004. Relationships among Indian Araliaceae inferred from Internal transcribed spacer sequences of nuclear ribosomal DNA. *Rheedea* 14: 1 – 8.
115. Park J.K., Nam K.Y., Hyun H.C., Jin S.H., Chepunov S.A., Chepunov N.E.1994. Effect of ginseng triol saponin fractions on the spatial memory functions studied with 12-arm radial maze. *Korean J Ginseng Sci*. 18: 32–8.
116. Park J.S, Hwang S.Y., Lee W.S, Yu K.W., Paek K.Y., Hwang B.Y., Han K.2006. The therapeutic effect of tissue cultured root of wild *Panax ginseng* CA Mayer on spermatogenetic disorder. *Arch Pharm Res*.; 29:800-807.
117. Park S., Yeo M., Jin J.H., Lee K.M., Jung J.Y., Choue R., Cho S.W., Hahm K.B. 2005. Rescue of Helicobacter pylori-induced cytotoxicity by red ginseng. *Dig Dis Sci*. 50:1218-1227.
118. Park, H. 1987. Effect of light and planting density on yield and quality of *Panax ginseng*. *Korean J. Crop Sci*. 32:386–391
119. Proctor J.T.A., Bailey W.G., 1987. Ginseng: industry, botany, and culture. *Hort Rev.* 9:187–236
120. Quan FS, Compans RW, Cho YK, Kang SM. 2007. Ginseng and Salviae herbs play a role as immune activators and modulate immune responses during influenza virus infection. *Vaccine*; 25:272-282.
121. Ru W., Wang D., Xu Y., He X., Sun Y E., Qian L., Zhou X., , Drug Y.Q. 2015 Chemical constituents and bioactivities of *Panax ginseng* (C. A. Mey.). *Discoveries & Therapeutics*. 9(1):23-32.
122. Ruan C.C., Liu Z., Li X., Liu X., Wang L.J., Pan H.Y., ZhengY.N., Sun G.Z., Zhang Y.S., Zhang L.X.. 2010. Isolation and characterization of a new ginsenoside from the fresh root of *Panax ginseng*. *Molecules*. 15:2319-2325.
123. Ryu J H, Park J H, Kim T H, Sohn D H, Kim J M, Park J H. 1996 'A genuine dammarane glycoside, (20E)-ginsenoside F4 from Korean red ginseng. *Arch Pharm Res*.19:335-336.
124. Ryu J.H., Park J.H., Eun J.H., Jung J.H., Sohn D.H. 1997 A dammarane glycoside from Korean red ginseng. *Phytochem*. 44:931-933.

125. Ryu J.H., Park J.H., Kim T.H., Sohn D.H., Kim J.M., Park J.H. 1996. A genuine dammarane glycoside, (20E)-ginsenoside F4 from Korean red ginseng. *Arch Pharm Res.*19:335-336.
126. Ryu J.H., Park J.H., Kim T.H., Sohn D.H., Kim J.M., Park J.H. 1996. A genuine dammarane glycoside, (20E)-ginsenoside F4 from Korean red ginseng. *Arch Pharm Res.*19:335-336.
127. Selvam, A.B.D. 2012. Pharmacognosy of Negative Listed Plants. Botanical Survey of India ISBN 81-8177-038-2. Pp. 138-148.
128. Seong G.S., Chun S.G., Chang C.C. 2005.Hepatoprotective effects of white and red ginseng extracts on acetaminophen-induced hepatotoxicity in mice. *J Ginseng Res.* 29:131-137.
129. Sharma S K. and Pandit M K. 2011. Plant A morphometric analysis and taxonomic study of Panax bipinnatifidus Seem. (Araliaceae) species complex from Sikkim Himalaya, India. *Plant Syst Evol*. 297:87-98.
130. Shaw, P. C., But P. P. (1995). Authentication of *Panax* species and their adulterants by random-primed polymerase chain reaction. *Planta Med.* 61, 466–469.
131. Shibata S, Tanaka O, Soma K, AandoT, Iida Y, Nakamura H. 1965. Studies on saponins and sapogenins of *Ginseng*. The structure of panaxatriol. *Tetrahedron Lett*. 42:207-213.
132. Shibata S., Ando T., TanakaO., Meguro Y., Sôma K., Iida, Y.1965. Saponins and sapogenins of *Panax ginseng* C.A.Meyer and some other *Panax spp. Yakugaku Zasshi*; 85:753-755.
133. Shibata S., Fujita M., Itokawa H., Tanaka O., Ishii T. 1963.Studies on the constituents of Japanese and Chinese crude drugs. XI. Panaxadiol, a sapogenin of ginseng roots. *Chem Pharm Bull* (Tokyo). 11:759-761.
134. Shin B.K., Kwon., S. W., Park J. H. 2015. Chemical diversity of ginseng saponins from *Panax ginseng*. *J. Ginseng Res*. 39(4): 287-298
135. Shin K.S., Lee J. J., Jin Y.R., Yu J. Y., Park E.S., Im J.H., You S.H., Oh K.W., Lee M.K., Effect of Korean red ginseng extract on blood circulation in healthy volunteersrandomized, double-blind, placebo-controlled trial. *J Ginseng. Res*. 31:109-116.
136. Shin Y.W., Bae E.A., Kim D.H. 2006. Inhibitory effect of ginsenoside Rg5 and its metabolite ginsenoside Rh3 in an oxazolone-induced mouse chronic dermatitis model. *Arch Pharm Res*. 29:685-690
137. Shin, B., Kwon, S. W., & Park, J. H. (2015). Chemical Diversity of Ginseng Saponins from *Panax ginseng*. *J Ginseng. Res*. 39: 287-298.
138. So S.H., Lee S.K., Hwang E.I., Koo B.S., Han G.H., Lee M.J.,Chung J.H., Kim N.M. 2008. Mechanisms of Korean red ginsengand herb extracts (KTNG0345) for anti-wrinkle activity. *J. Ginseng Res*. 32:39-47.
139. Song J.Y., Akhalaia M., Platonov A., Kim H.D., Jung I.S., Han Y.S., Yun Y.S. 2004 Effects of polysaccharide ginsan from *Panax ginseng* on liver function. *Arch Pharm Res.* 27:531-538
140. Song J.Y., Han S.K., Bae K.G., Lim D.S., Son S.J., Jung I.S, Yi S.Y., Yun Y.S. 2003.' Radioprotective effects of ginseng, an immunomodulator. *Radiat Res*. 159:768-774.
141. Sonnenborn U., Proppert Y. 1991 Ginseng (*Panax ginseng* CA Meyer). *Brit J Phytother.* 2:3-14.
142. Stephen F.1996. The ginseng book: Publishing Group, nature's ancient healer. Avery Garden City Park, New York, USA.
143. Sun B. S., Pan F. Y, Sung C. K..2011. Repetitious steaming-induced chemical transformations and global quality of black ginseng derived from *Panax ginseng* by HPLC-ESI-MS/MSn based chemical profiling approach. *Biotechnol Bioproc E*.16:956-65.
144. Susan G. Carpenter, G. Cottam 1982. Growth and reproduction of American ginseng (*Panax quinquefolius*) in Wisconsin, U.S.A. *Can.J. Bot.* 60(12): 2692-2696.
145. Tanaka O., Morita T., Kasai R., Kinouchi J., Sanada S., Ida Y., Shoh J. 1985 Study on saponins of rhizomes of *Panax pseudoginseng* sub sp. Himalaicus collected at Tzatogang and Pari-la, Bhutan-Himalaya. *Chem. Pharm. Bull.* 33 (6) 2323-2330.

146. Taylor, D.A. 2006. Ginseng, the Divine Root: The Curious History of the Plant that Captivated the World. Algonquin Books of Chapel Hill. Chapel Hill.
147. Thompson, G.A., 1987. Botanical Characteristics of Ginseng. In L.E. Craker and J.E. Simon (eds). Herbs Spices and Medicinal Plants. Recent Advances in Botany, Horticulture and Pharmacology. Vol.2. Oryx Press. Pp. 111-136.
148. Tode T. Kikuchi Y 2003. Effect of Korean red ginseng on psychological functions in patients with severe climacteric syndromes: a comprehensive study from the view point of traditional KAMPO-medicine and western medicine. *J Ginseng Res;*27:110-114.
149. Tode T., Kikuchi Y., Hirata J., Kita T., Nakata H., Nagata I. 1999.Effect of Korean red ginseng on psychological functions in patients with severe climacteric syndromes. *Int J Gynaecol Obstet.*. 67(3):169-74.
150. Tohda C., Matsumoto N., Zou K., Meselhy M..R., Komatsu K. 2004. Â (25-35)-induced memory impairment, axonal atrophy, and synaptic loss are ameliorated by M1, A metabolite of protopanaxadiol-type saponins. *Neuropsychopharmacology.* 29:860-868.
151. Tomoda M., Hirabayashi K, Shimizu N., Gonda R., Ohara N., Takada K. 1993a. Characterization of two novel polysaccharides having immunological activities from the root of *Panax ginseng*. *Biol Pharm Bull*: 16:1087-1090.
152. Tomoda M., Takeda K., Shimizu N., Gonda R., Ohara N., Takada K, Hirabayashi K. 1993b. Characterization of two acidic polysaccharides having immunological activities from the root of *Panax ginseng*. *Biol Pharm Bull*. 16:22-25.
153. Tung, N.H., Song, G.Y., Park, Y.J., Kim, Y.H. 2009 .Two new Dammarane-type saponins from the leaves of *Panax ginseng*. *Chem. Pharm. Bull*., 57, 1412–1414.
154. Wan D., Jiao L., Yang H., Liu S. 2012. Structural characterization and immunological activities of the water-soluble oligosaccharides isolated from the *Panax ginseng* roots. *Planta.* 235:1289-1297).
155. Wang J., Sun C., Zheng Y., Pan H., Zhou Y., Fan Y.. 2014 The effective mechanism of the polysaccharides from *Panax ginseng* on chronic fatigue syndrome. *Arch Pharm Res.* 37:530-**538.**
156. Wang S., Chen X., Han F., Li R, Li G., Zhao Y., Xu Z., Zhang L. 2016. Genetic diversity and population structure of ginseng in China based on RAPD analysis, *Open Life Sci.* 11: 387–390.
157. Wei J.X., Wang LA, Du H, Li R. 1985.Isolation and identification of sanchinoside B1 and B2 from rootlets of *Panax notoginseng* (Burk.) F. H. Chen. *Yao Xue Xue Bao.* 20:288-293
158. Wen J, Zimmer EA (1996) Phylogeny and biogeography of *Panax* L. (the ginseng genus, Araliaceae): inferences from ITS sequences of nuclear ribosomal RNA. *Mol Phylogenet Evol* 6:167–177
159. Wen J. 2001. Species diversity, nomenclature, phylogeny, biogeography, and classiûcation of the ginseng genus (*Panax* L., Araliaceae). In: Punja ZK (ed) Proceedings of the International Ginseng Workshop: utilization of biotechnological, genetic and cultural approaches for North American and Asian ginseng. Simon Fraser University Press, Canada, pp. 67–88.
160. Wenwen R., Dongliang W., Yunpeng X., Xianxian H., Yang-En S., LQian, Xiangshan Zhou, Yufeng Qin 2015; Chemical constituents and bioactivities of *Panax ginseng* (C. A. Mey.) *Drug Discoveries & Therapeutics*. 2015; 9(1):23-32
161. Xiao, P.G., Z.Y. Zim, F.Q. Zhang, W.H. Zim, J.I. Chen, G.D. Zhang, and G.T. Liu. 1987. Ginseng research and cultivation. Agr. Publ. House, Beijing, China.
162. Xue P., Yao Y., Yang X., Feng J. and Ren G. 2017. Improved antimicrobial effect of ginseng extract by heat transformation. *J Ginseng Res*: 41(2): 180–187.
163. Yang D.Q. 1981. The cyto-taxonomic studies on some species of *Panax* L. *Acta Phytotax Sin* 19:298–303
164. Yang W. Z., Hu Y., Wu W.Y., Ye M., Guo D. A.2014. Saponins in the genus *Panax* L. (Araliaceae): a systematic review of their chemical diversity. *Phytochemistry*;106:7-24.

165. Yang W.Z., Hu Y., Wu W.Y., Ye M., Guo D.A. 2014. Saponins in the genus Panax L. (Araliaceae): a systematic review of their chemical diversity. *Phytochemistry*. 106:7-24
166. Yokozawa T., Kobayashi T., Oura H., Kawashima Y. 1985 Hyperlipemia-improving effects of ginsenoside-Rb2 in strepzotocin-diabetic rats. *Chem Pharm Bull*. 33:3893-3898
167. Yoo K.O., MallaK.J., Wen J. 2001. Chloroplast DNA variation of Panax (Araliaceae) in Nepal and its taxonomic implications. Brittonia 53:447–453
168. Yu J.Y., Jin Y.R., Lee J..J., Chung J.H, Noh J.Y. You S.H, YunY.P. . 2006; Antiplatelet and antithrombotic activities of Korean Red Ginseng. *Arch. Pharm. Res* 29:898-903.
169. Yu, S.C. and W.K. Kim. 1992. Structure changes and histochemical study of endosperm of Panax ginseng C.A. Meyer during embryo development. Korean J.Ginseng Sci. 16:37–43.
170. Yuan C.S, Wang C.Z., Wicks S.M., Qi L.W. 2010. Chemical and pharmacological studies of saponins with a focus on American ginseng'. *J Ginseng Res*. 34:160-167.
171. Yuan H.D., Quan H.Y., Jung M.S., Kim S.J., Huang B., Kim D.Y., Chung S.H. 2011. Anti-diabetic effect of pectinase-processed ginseng radix (GINST) in high fat diet-fed ICR mice. *J Ginseng Res*; 35:308-314.
172. Yun S.N., Moon S.J., Ko S.K., Im B.O., Chung S.H. 2004. Wild ginseng prevents the onset of high-fat diet induced hyperglycemia and obesity in ICR mice. *Arch Pharm Res.*; 27:790-796.
173. Yun T.K. 2003. Experimental and epidemiological evidence on non-organ specific cancer preventive effect of Korean ginseng and identification of active compounds. *Mutat. Res*. 523-524:63-74.
174. Zhong Y.M., Nishijo H., Uwano T., Tamura R., Kawanishi K., Ono T. 2000. Red ginseng ameliorated place navigation deficits in young rats with hippocampal lesions and aged rats. *Physiol Behav*. 69:511-525.
175. Zhu G. Y, Li Y. W., Hau D. K., Jiang Z.H., Yu Z.L., Fong W. F. 2011 Protopanaxatriol-type ginsenosides from the root of Panax ginseng. *J Agric Food Chem*. 59:200-205.
176. Zhuo D.H. 1982. Preventive geriatrics: An overview from traditional Chinese medicine. *Am J Chin Med*. 10: 32
177. Zhuravlev Y N., Koren O.G., Kozyrenko M.M,, Reunova G.D., Artyukova E.V., Krylach T.Y., Muzarok T.I. 1999. Use of molecular markers to design the reintroduction strategy for *Panax ginseng*; In: Biodiversity and Allelopathy: From Organism to Ecosystems in the Pacific. Chou C.H., Waller G.R., Reinhardt C, editor. Taipei: *Academia Sinica*. pp. 183–192.
178. Zhuravlev Y.N., Reunova G.D., Kozyrenko M.M., Artyukova E.V., Muzarok T.I. 1998. Genetic variation of wild ginseng populations (RAPD analysis) *Mol Biol* (Moscow). 32:910–914.

3

Polygonatum species

English: King Solomon's-seal or Solomon's seal
Hindi: *Mahameda, Devarigaala*

B.D. Sharma and J.C. Rana

***Polygonatum* species**
(A) Herbarium sheet (*Polygonatum cirrhifolium*), **(B)** Whole plant (*P. cirrhifolium*)
(C) Flowering stage (*P. cirrhifolium*), **(D)** Flowering stage (*P. geminiflorum*)
(E) Crude drug material (rhizomes and roots), **(F)** Fruits (*P. cirrhifolium*)

Introduction

Species of *Polygonatum*, traditional known for their therapeutic properties for treating different ailments,are fairly well represented in temperate Northern Hemisphere in Asia, America and Europe and find use. In India, *P. cirrhifolium*, and *P. verticillatum* are the two important constituting species of *Ashtavarga*, but have attained a threatened status due to overexploitation. Of the 60 odd known species of the genus, less than 10 have been explored phyto-chemically. Their status varies species wise, at regional and global level.

A. BIOLOGICAL ASPECTS

Systematic classification

Kingdom : Plantae

Clade : Angiosperms/ Monocots

Class : Cestodia/Equisetopsida

Order : Asparagales

Family : Asparagaceae

Subfamily : Nolinoideae

Tribe : Polygonatae

Genus : *Polygonatum* Miller.

• Synonyms

The genus *Polygonatum* is known to have several synonyms, viz., *Axillaria* Raf. *Salomonia* Heist. ex Fabr., *Evallaria* Neck., *Siphyalis* Raf., *Codomale* Raf., *Troxilanthes* Raf., *Campydorum* Salisb., *Sigillum* Montandon in F. Friche-Joset, *Periballanthus* Franch. & Sav.

• Common names (in different Indian languages)

English: King Solomon's-seal or Solomon's seal

Sanskrit: *Mahameda, Vasucchidra, Tridanti, Devama*

Hindi: *Mahameda, Devarigaala*

Kannada: *Mahamedha*

Malayalam: *Mahameda*

Tamil: *Mahameda*

Telugu: *Mahameda*

• Diversity within genus

Genus *Polygonatum* is a member of family Asparagaceae (subfamily Nolinoideae), under the APG III classification system consist of 63 species worldwide, with about 20 endemic species occurring in China. About 9 species are well explored so far (Khan and Rauf 2015). Previously, these species were placed in family Ruscaceae and Convallariaceae. However, four species, *Polygonatum hookeri, P. cirrhifolium, P. verticillatum,* and *P. multiflorum* are commonly found in western Himalaya.

• Geographical distribution

The species are widely distributed in the temperate regions of America, Europe and Asia, especially China, Japan, Korea, India, Pakistan, Afghanistan, Nepal, Bhutan, Russia. They are usually wild perennial rhizomatous herbswhich sometimes also occur in mild temperate climatic conditions such as hills of south and N-E India. *P. odoratum,* a rhizomatous, perennial herb native to temperate Asia has been cultivated in Europe and haswidely naturalised outside this range (USDA, ARS, National Genetic Resources Program 2012), into woods, rocky habitats, generally growing on calcareous soils. This species has an extensive range in the northern hemisphere from Europe to the Himalayas up to Siberia.

• Habitat and ecology

Listed in the British Iris Flora 2012, *P. odoratum*, is stated to occupy rocky habitat and limestone outcrops. In Europe (Scandanavia south and east to Spain), *P. multiflorum* is found in woodland, usually on limestone. In Himalayas, some of these plant species are commonly found in forests, shrubberies and open slopes with altitudinal variations of 1500-3700 m, well distributed in Pakistan, India and southeastern Tibet. In India, fifteen species are reported to occur, which are distributed in the Himalaya, N-E region, temperate hills of southern India and other states (Devi *et al.,* 2014). They grow in moist shady and semi-shady conditions in forest or grassland areas on hill slopes rich in humus and loam soils (Sharma *et al*., 2014).

• Taxonomic enumeration

The genus *Polygonatum* has a controversial taxonomic status and some molecular studies have been done to seek clarity (Kim *et al.*, 2010). Taxonomists have repeatedlyrevised the same. The World Checklist in 2014 accepted 74 species and hybrids under *Polygonatum* Mill. *P. verticillatum* is often confused with *P. biflorum*, and exist in different forms such as a broader leaf form, a dwarf form, larger form, and colour of flower ranging from white to pink.

• Taxonomic Status: (Kumari *et al.*, 2014)

***P. cirrhifolium* (Wall.) Royle:** Tall, erect, weak herbs, with stout, creeping rhizomes; Stem is 60-120 cm high, terete, grooved. Leaves in whorls of 3-6, linear to narrowly lanceolate, 6-15 cm long, margins enrolled, apex coiled, tendril-like. Flowers white, tinged purple or green, in short stalked clusters of 2-4, from the leaf axils. Perianth is 6-parted, somewhat reflexed.

***P. verticillatum* (L) Allioni:** Annual-perennial, tall and erect herbs, 80-150 cm high; stem angled or grooved. Leaves are sessile, in whorls of 4-8, linear or narrowly lanceolate, 8-15 cm × 0.5-3 cm, acuminate or acute. Flowers are white, arranged in the terminal, whorls of leafy racemes. Perianth is oblong, 4-6 mm long, 6-parted, segments spreading. Berries are globose, 5 mm across, purple-black when ripe.

• Phenological period

Active growth resumes after winter dormancy; flowering occurs from May-June.

• Cytotypes

Therman (1950, 1952, 1953) has been one of the earliest worker to systematically investigate about 100 Asiatic, European and American accessions of *Polygonatum* for cytological details. A number of workers followed to add to the variation in the chromosome number (2n = 16, 18, 20, 21 22, 24, 26, 27, 28, 30, 31, 36, 38, 40, 42, 46, 59, 56, 60, 62, 64, 66, 84, 86– 91, 90) (Tamura, 1993; Yang *et al.*, 1988; Chen, 1989; Lattoo *et al.*, 2005; Deng *et al.*, 2009).

REPRODUCTIVE BIOLOGY

Reproductive phenology

- *Polygonatum* species is believed to be sensitive to environmental conditions especially temperature and moisture. This feature, resulting in evolution of unique aspects of reproductive biology,makes phenology of *Polygonatum* species interesting. In addition tospecies' perennation through rhizomes, multiplication and distribution generally happens through seeds. The seed exhibits dormancy, which is deep epicotyl morpho-physiological in nature (Sharma *et al.*, 2014). This type of dormancy breaks under natural conditions over a period varying from 45 days to over two years. This facilitates their germination under favourable conditions over a long span. After germination, warm temperature is required for further development. After radical emergence, tiny bulblets appear by July-August. The seeds, first have emergence of hypocotyls, which subsequently elongates up to 2-3 cm and then the epicotyls emerges as a spherical structure or a tiny bulblet at the proximal end of the hypocotyls some 35 to 45 days after the commencement of the germination. After two to three months, radical grows from the hypocotyl region and produces root hairs. With emergence of epicotyls by April in the following year, 2-3 leaves develop. Five to six leaves emerge by fourth year, and it is only during 7-8 year that flowering takes place, usually from June to July. Fruiting and seed setting takes place in autumn (Lattoo *et al.*, 2001; Sharma *et al.*, 2014, 2015).

Pollination and pollinators

The species of this genus have developed a system of gametophytic self-incompatibility to promote cross- and self-pollination. Pollination takes place through bees and some insects. Dormancy breaking experiments involving chemicals like gibberellic acid, hypochlorite have given variable results for species such as *P. cirrhifolium* and *P. verticillatum*.

C. MEDICINAL ASPECTS

• Plant material used as crude drug

The dried roots and rhizomes or underground plant parts (bulbs, pseudobulbs and tubers) constitute the crude drug. Leaves of *P. cirrhifolium* are also consumed as vegetable, while other species of the genus find listing as famine food (Wujisguleng *et al.*, 2012).

• General appearance

These are herbs which may grow tall (stem height of 60-120 cm), erect, weak in appearance and creeping rhizomes. Leaves are in whorls of3 to 6 and are linear to narrowly lanceolate, 6 to 15cm Long. Large morphological diversity has been observed in genus *Polygonatum.* Leaf arrangement varies greatly, such that some species may possess opposite leaves on the proximal part of the stem or three whorled leaves on the apical part of the stem. Flower colour may vary from white, tinged purple to green. Flowers appear in short stalked clusters of 2 to 4, arising from the leaf axils.

• Organoleptic properties

Sweet, astringent in taste and cold in potency as per *Ayurvedic* concept, specified to pacify *vata* and *pitta*.

• Microscopic characteristics

Chinmay *et al.* (2009) compared microscopic characteristics of *P. verticillatum* and *P. cirrifolium* to show a large number of mucilage canals and scattered vascular bundles in rhizomatic ground tissue of both the species, but reported a wide cortex, endodermal cells with thickening on radial and inner tangential walls and a polyarch stele in the adventitious root of the former only.The anatomical structures of *P. verticillatum* showed 6-9 layers of cork cells, large mucilaginous canals surrounding wide parenchymatous ground tissue, large number of vascular bundles mostly enclosed by bundle sheath,while vessels and tracheids greatly varied inshape and sizes with scalari form thickening present in rhizome. However, in *P. cirrifolium,* cork cells ranged from 5-10 layers, large mucilage canals surrounded by 7-8 epithelial cells and bundles of calcium oxalate raphides, with several scattered vascular bundles and absence of fibers in xylem.

• Powdered plant material

Rhizome powdered material of *P. cirrhifolium* was reported to be smooth surfaced, 0.4-0.8 cm broad and of cream color. It was orange-brownin color, rough or sandy surfaced and 1-2 cm broad in case of *P. verticillatum.* Histological examination revealed sclerenchymatous central zone in roots of *P. verticillatum,* which was absent in *P. cirrhifolium* (Pandey *et al.*, 2006).

According to The Ayurvedic Pharmacopoeia of India (2016), root powder of *P. cirrhifolium* is described as dark brown, and its microscopic examination shows epidermal cells with actinocytic stomataand cortical cells in surface view; the starch grains are ovoid with concentric striation, eithersingly or in groups, with presence of raphides and druses. The tracheids are elongated with pointed ends, wall slightly wavy towards tips, and thickenings to be reticulate; vessels with simple, cross wall perforation.

• Major Chemical constituent

Phenolics (lignins, flavonoids, coumarins), steroids (sterols, steroidal saponins, glycosides), triterpenes, tannins, alkaloid saponins, phyto-hormones, glycosides, flavonoids and polysaccharides are the major compounds isolated from *Polygonatum* (Alluri *et al*, 2006; Khan and Rauf, 2015). The rhizome of *P. odoratum* contains asparagine, mucilage, a cardio-tonic glycoside, saponin, and quinine gluconate. It has been used for intestinal problems and pain, for rheumatism, gout, water retention, and as a diuretic (https://en.wikipedia.org/wiki/Polygonatum_odoratum). The rhizomes of *P. cirrhifolium* contain glucose, sucrose, steroidal saponins A & B, polysaccharides, steroidal, phenol, stannin, α-L-rhamnopyranosyl, β-D-glucopyranoside, dauvosterol, β-sitosterol, 6-

nonadecenoic acid, 6-stearic acid and one organic compound (Wang, 2008). The chemical analysis of rhizomes of *P. veticillatum* shows the presence of lysine, serine, aspartic acid, threonine (free amino acids), diosgenin, β-sitosterol, sucrose and glucose (Saini *et al.*, 2010).

• Substitutes and adulterants

Substitutes and adulterants are employed for meeting the higher demand of the raw drug material of the species and also to elevate earnings. For the most demanded two species, *P. cirrhifolium* and *P. verticillatum* in the Indian market, the substitutes and adulterants have been mentioned in the Nighantus, but for species used in the traditional Chinese medicines no adulterants are specifically mentioned in the literature. The substitutes and adulterants used are given below;

Medicinal species	Substitutes	Adulterants
1. Meda (*P. verticillatum*)	*Asparagus racemosus* *Sida veronicaefolia* *Paederia foetida*	*P. multiflorum* *P. oppositifolia*
2. Mahameda (*P. cirrhifolium*)	*Asparagus racemosus* *Eulophia campestris*	*P. verticillatum* *P .multiflorum*

• General identity tests

The general identity given for *P. cirrhifolium* by The Ayurvedic Pharmacopoeia of India (2016) is as given below.

- Foreign matter : Not more than 3 per cent,
- Total Ash : Not more than 3.5 per cent,
- Acid-insoluble ash : Not more than 1 per cent,
- Alcohol-soluble extractive : Not less than 4.5 per cent,
- Water-soluble extractive : Not less than 70 per cent,

• Medicinal usage

In the traditional system Ayurvedic medicine, two *Polygonatum* species, *P. cirrhifolium* (Mahameda) and *P. verticillatum* (Meda), are described as constituents of Ashtavarga plants (Sharma and Balkrishna, 2005). These are the first ever discovered and used for making the world's earliest nutraceutical or health food, which is popularly known as "Chayavanprash". Based on knowledge of Ayurveda, both the species are described as *meda* (fat) *dhatu* enhancer in the body. They are also used in the Amchi (Tibetan) medicines and in the Traditional Chinese System of Medicine.

Traditionally, species of *Polygonatum* have been used in different countries to treat various body disorders/ailments. The rhizomes are reported to have hypoglycemic antimicrobial, antitumor, cytotoxic as well as phytotoxic activities, and suggested to be helpful in various ailments such as malaria, respiratory diseases, stomach disorders, herpes, osteoporosis, and serve as energy booster (Singh *et al.*, 2006). The antibacterial and antifungal activities of *Polygonatum* have been reported (Alluri *et al.*, 2006), and are effective in reducing sugar level in type 2 diabetes by different mechanism including glucosidase inhibition and increased insulin sensitivity (Singh, 2006; Sharma and Pal, 2007).

• Ethnobotanical uses

The rhizomes of *Polygonatum* species, or so called Solomon's Seal, have been used in traditional herbal medicines since times immemorial. They are adaptogenic, cardio-tonic, demulcent, antioxidant, diuretic, vital energy enhancer, hypoglycemic, aphrodisiac and tonics. They are widely used in the treatment of dry coughs, tuberculosis, burning sensation of body, and pulmonary problems (Savita *et al.*, 2014, Devi *et al.*, 2014, Singh *et al.*, 2013). Among others, beneficial effects reported in ethnic use includes that on kidneys and liver, improving bones strength, prevent gray hair, vision problems, vertigo and ringworms (Khan *et al.*, 2012b). Young shoots are consumed after cooking and often used as a substitute for asparagus. The roots are also consumed in a cooked form.

In China, several species of *Polygonatum* (*P. sibiricum, P. odoratum, P. filipes, P. acuminatifolium, P. stenophyllum, P. humile, P. inflatum* and *P. involucratum*) are traditionally used as food, consumed in cooked or raw forms. Different parts of plans such as young leaves, tender shoots and rhizomes are consumed raw or after frying, stewing, roasting, steaming and as decoction. The sprouts *of P. odoratum* are eaten after adding salt during winter months. In the Mongolian region of China, rhizomes of *P. acuminatifolium, P. filipes, P. humile, P. inflatum, P. involucratum, P. odoratum, P. sibiricum, P. stenophyllum.* As cereal food, the rhizomes of three species, *P. nakaianum, P. odoratum, P. sibiricum* are used as emergency food. These species are sundried, ground to powder and mixed with flour to make cakes. The rhizomes of *P. sibiricum*, as well as of other species, are used to make sweet dishes. *P. cirrhifolium, P. kingianum, P. macropodium and P. verticillatum* are used both for medicine and food (Wujisguleng *et al.*, 2012). In India, *P. cirrhifolium* and *P. verticillatum* are used both for medicine and food (Rana *et al.*, 2010; Devi *et al.*, 2013).

D. CHEMISTRY ASPECT

• Chemical composition

Lignans, flavonoids, coumarins, steroids, terpenes, fatty acids and aliphatic long chain compounds have been reported in species of *Polygonatum* (Deng *et al.*,

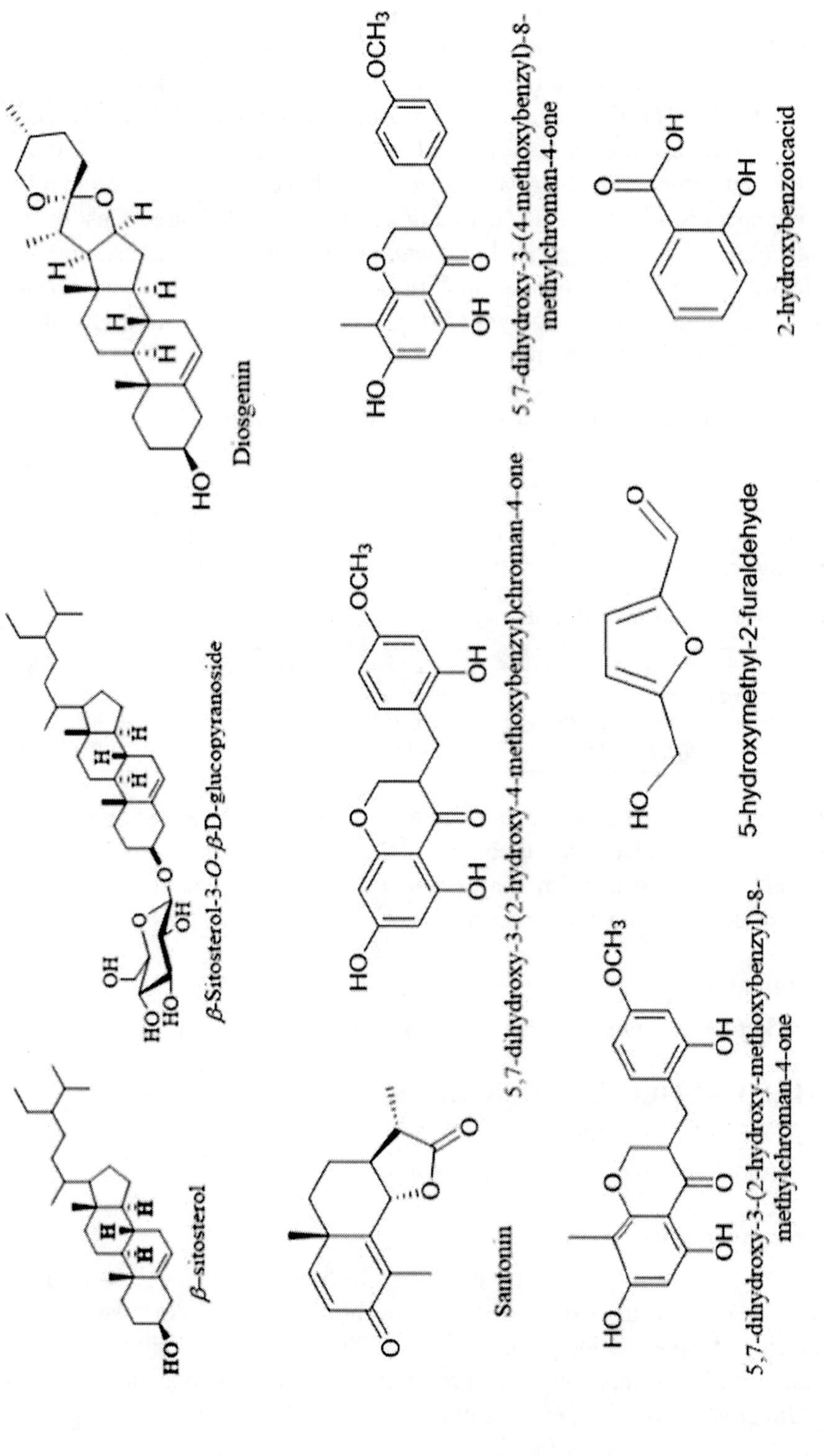

β-sitosterol

β-Sitosterol-3-*O*-β-D-glucopyranoside

Diosgenin

Santonin

5,7-dihydroxy-3-(2-hydroxy-4-methoxybenzyl)chroman-4-one

5,7-dihydroxy-3-(4-methoxybenzyl)-8-methylchroman-4-one

5,7-dihydroxy-3-(2-hydroxy-methoxybenzyl)-8-methylchroman-4-one

5-hydroxymethyl-2-furaldehyde

2-hydroxybenzoicacid

Chemical constituents of *Polygonatum* species

2012; Singh *et al.* 2013). Other phyto-constituents of importance identified are mucopolysaccharides, which are biologically active and considered as important bioactive components as antitumor, antidiabetes, and antioxidation agents (Tomoda *et al.*, 1971; Chen *et al.*, 2014). In addition, homoisoflavanones (Qian *et al.*, 2010), azetidine-2-carboxylic acid (Baek *et al.*, 2012; Kim *et al.*, 2006), and steroidal compounds (Sugiyama *et al.*, 1984) have also been reported from different species of the genus. Zhao *et al.* (2018) have attributed most of the pharmacological effects of genus *Polygonatum* to its polysaccharides, saponins and lectins. The nutritional composition of two species *P. cirrhifolium* (*Mahameda*) and *P. verticillatum* (*Meda*), collected from western Himalaya showed the following composition:

Major nutritional constituents (g/100g) of *Polygonatum*

Species	Moisture	Protein	Fat	Fibre	Carbohydrate	Ash	Energy Kcal/100g
P. cirrhifolium	81.56	20.27	1.23	8.33	16.08	8.00	121.70
P. verticillatum	84.53	16.20	0.46	12.33	17.07	7.45	108.23

Micronutrients status (mg/100g) of *Polygonatum*

Species	P	K	Na	Ca	Mg	Fe	Cu	Mn
P. cirrhifolium	100	20.75	14.42	1365.3	75.4	35.06	0.90	28.84
P. verticillatum	100	13.33	37.82	1338.3	90.1	23.64	0.21	28.64

On the basis of our and the earlier studies (Khan *et al.*, 2012c; Sharma *et al.*, 2014), it is concluded that rhizomes of the two *Polygonatum* species are excellent natural source of major and micronutrients such as protein, carbohydrates, calcium, magnesium, iron etc.

• Chemical markers

The marker compounds are 4-hydrobenzoic acid and trigonelline

E. PHARMACOLOGICAL ASPECTS

Bioactivity Plants of the genus, *Polygonatum* display a wide range of pharmacological activities which are summarized below;

• Anti-cancer activity

P. odoratum was reported to have secondary metabolite, 8-methyl-dihydrobenzopyrone that exhibited prominent anticancer activity in breast cancers by inducing phosphorylation of Bcl-2 (Rafi and Vastano, 2007). Similarly, saponins isolated from *P. sibiricum* have cytotoxic potential against human breast cancer cells (Jin *et al.*, 2004; Ahn *et al.*, 2006).

• Aphrodisiac activities

The aqueous extract of leaves of *P. verticillatum,* were evaluated for aphrodisiac activity in male rats. A significantly increased activity in sexually active and inactive male rats was observed (Tiwari, 2001; Kazmi *et al.*, 2012).

• Antioxidant/ antifatigue activities

Horng *et al.* (2014) reported rhizomes extract of *P. altelobatum* to increase endurance to exhaustion and attributed it to the presence of polyphenols, flavonoids and polysaccharides.

• Hypolipidemic activity

Chiang *et al.* (2013) demonstrated the hypolipidemic activity of fresh rhizomes of *P. altelobatum* on hamsters and observed a significant decrease of total cholesterol, low- density lipoprotein cholesterol (LDL) and very low density lipoprotein cholesterol (VLDL) but an increase in high-density lipoprotein cholesterol (HDL).

• Antidiabetic activity

Sharma and Pal (2007) showed a significant reduction in blood sugar level in non-insulin dependent diabetic rats with use of rhizome extract of *P. verticillatum.*

• Antimalarial activity

Khan *et al.* (2011) reported that crude extract of rhizomes of *P.verticillatum* possessed a significant anti-malarial activity which acted against *Plasmodium falciparum.*

•Antimicrobial activity

Broad spectrum antimicrobial activity was demonstrated by the crude extract of rhizomes of *P. verticillatum* against gram-negative bacteria (*Escherchia coli*, *Salmonella typhi*, *Shigella flexneri*) and gram-positive bacteria (*Styphlococcus aureus*) and for fungi, *Microsporum canis* and *Fusarium solani* due to presence of flavonoids and phenols (Khan *et al.*, 2012a; Wang *et al.*, 2006, 2007).

• Antiinflammatory activity

Debnath *et al.* (2013) reported, from a study conducted in South Korea, that water extract of rhizome of *P. sibiricum* acts as natural antioxidant and anti-inflammatory agents and has a wide scope for food and drug industries.

• Antiherpetic activities

Liu *et al.* (2011) reported that the chemically modified polysaccharide extracted from *P. cyrtonema* was effective to treat herpes simplex, a viral nervous disease.

• Aantipyretic activity

Khan *et al.* (2013) demonstrated that the methanol extract of *P. verticillatum* possessed a prominent antipyretic activity.

• Insecticidal activity

Saeed *et al.* (2010) recorded the insecticidal activity of aerial parts of the *P. verticillatum* against *Rhizopertha dominica* and *Leishmania major*.

• Tranquilizing activity

Gu *et al.* (2003) reported that anthraquinone derived from *P. multiflorum* possessed protective effect against brain disturbances induced by severe cerebral injury.

Clinical Studies

• Role in management of specific ailments

As an important constituent of *Chayavanprash*, *Polygonatum* has restorative role for mental vitality, general well being and for alleviating pain, fever, inflammation, allergy, and weakness. *P. verticillatum* is used in Ayurveda as an aphrodisiac.

• Toxicity related information

Generally, no serious toxicity is reported on its consumption, especially in quantity suggested in medicinal/ therapeutic usage. However, some members of this genus have poisonous fruits/ berries and seeds, and toxicity is attributed to presence of some anthraquinones (Lewis and Elvin-Lewis 2003). Common reaction after its ingestion is vomiting and diarrhea.

• Formulations

Among them, three species, *P. cirrhifolium, P. multiflorum* and *P. verticillatum*, have been mainly exploited for medicinal and other uses. Two species, *P. cirrhifolium* and *P. verticillatum* are the most demanded and *P. multiflorum* is used as an adulterant for the former species (Prakash *et al.*, 2013; Sagar *et al.*, 2014; Rana *et al.*, 2012). The former two species are used as ingredients of "Chayavanprash" and also needed for various other Ayurvedic formulations, like *Asoka Ghrta, Sivagutika, Amrtaprasa Ghrta, Dasam, Ularista,*

Dhanvantara Taila, Brhatmasa Taila, Mahanarayana Taila, Vasacandanadi Taila, etc.

F. THREAT AND CONSERVATION ASPECT

Population status

The status of any plant species actively sought in the pharmaceutical industry is a subject of variation with respect to time period of reporting and place of occurrence. This is the reason that different workers have reported different status to species of *Polygonatum* in different parts of Himalayas. Saboon *et al*. (2016) have recorded more than 20 studies where *P. verticillatum* has been listed as rare, vulnerable, threatened, endangered or critically endangered, in western Himalayas. *P. verticillatum* was listed as vulnerable in Jammu & Kashmir, Himachal Pradesh, Uttarakhand and Arunachal Pradesh while *P. cirrhifolium* was termed endangered in Himachal Pradesh but vulnerable in Uttarakhand (Ved *et al.*, 2003; Bist *et al.*, 2011; Lohani *et al.*, 2013). In Europe, *P. odoratum* is listed as Near Threatened in Luxembourg (Colling, 2005) and of least Concern in Denmark (NERI 2007), Norway (Artsdatabanken 2010), Switzerland (Moser *et al.* 2002) and the United Kingdom (Cheffings and Farrell, 2005).

Threats

Habitat loss/degradation/fragmentation

The Himalayan ecosystem has been known to be fragile, and prone to rapid degradation process due to high population and anthropogenic activities. The recent time has witnessed considerable loss of forest land due to increasing biotic pressure and exploitation for many valuable medicinal plants. In a study in central Himalaya (Kala *et al.*, 2006) the factors associated with declining status of medicinal plants are their habitat specificity, narrow range of distribution, increasing land-use disturbances, habitat alteration, heavy livestock grazing, explosion of human density, fragmentation and degradation of plant density.

In Europe the status of *Polygonatum* (*P. odoratum*) has declined in recent times due to destruction of limestone pavement, and it is reported to be becoming scarce (Khela, 2012; Online Atlas of the British and Irish Flora, 2012).

• Over-exploitation

Overexploitation of roots and rhizomes of *Polygonatum* has been reported by different workers in reference to Himalayas (Bist *et al.*, 2011; Lohani *et al.*, 2013; Bhatt *et al.*, 2014; Saboon *et al.*, 2016).

Conservation strategies

• *Ex-situ* conservation

For *ex-situ* conservation of *Polygonatum*, it is important to develop agro-technology protocols for large scale propagation, based on best of conventional (vegetative and seeds) methods. Lohani *et al.* (2012) studied survival, growth and yield parameters for *P. verticillatum* using different types of manures and fertilizers under field conditions. Supplementation with manure or organic compost yielded better results compared to control as reported in different studies (Butola *et al.,* 2011; Lohani *et al.*, 2012). Ex situ conservation strategies should also include establishment and maintenance of herbal gardens as well as plant nurseries, ensure availability of quality planting material, and education and awareness programs for large-scale cultivation.

• *In situ* conservation

Natural habitat of the species is the best place for its conservation, and has been widely suggested (Rana *et al.,* 2010; Balkrishan *et al.*, 2012). In addition to providing stringent protection and maintaining number of such target areas within the natural habitat of the species, fresh in-fillings of saplings in depleted areas can speed up attaining a healthy status of the species. *In-vitro* micropropagation, which has been worked out for *P. verticillatum* (Bisht *et al.*, 2012; Saboon *et al.*, 2016) can be resorted to for generating large supplies of quality planting material.

G. CULTIVATION ASPECTS

• Climate

The species are generally adapted to temperate to alpine regions on altitudes ranging from 1500 m to 3700 m with moderate to high rainfall and receiving regular snowfall every year. However, they have been found to be very sensitive to soil moisture and temperature conditions in their natural habitats (Sharma, *et al.*, 2013).

• Soil

The plants grow best in sandy, loamy, clay soils rich in humus, under partially shaded, sloppy areas and moist areas.

• Nursery raising and planting

The seeds of *Polygonatum* species have deep morpho-physiological epicotyl dormancy and hence require chilling temperature for breakingit. Germination, without any treatment, is poor and takes 35-45 days. No standard hormonal seed treatments have so far been developed.The best method, however, is the use of vegetative means. For propagation, rhizomes are cut into 1.25 cm to 2.00 cm pieces in size transplanted immediately into raised beds (5 tons/ha) containing adequate quantity of manure, during March to May. These produce sprouts within 20 to 30 days.

Development of Agro-techniques and cultivation

Polygonatum species prefer temperate climatic conditions. So far, commercial growing is not popular except in China where it is claimed to be grown in Anhui Province of Eastern China. It is the major production area where 11 species of *Polygonatum* are commercially grown with a production of 20 to 100 tons annually (Wujisguleng *et al.*, 2012). Recently, need to develop agro-techniques for cultivation has been emphasized bymany workers (Annual Report NMPB, 2014; Nautiyal and Nautiyal, 2004).

Means of Propagation

The species are propagated by (i) rootstock cuttings, and (ii) seeds. However, propagation by vegetative methods is quick and easier.Larger divisions can be planted out direct into their permanent positions.

Field transplantation

Nearly 40,000 to 45,000 cuttings are required for planting an area of 400 square meters (one acre). For field plantation, the plants are given 20 cm spacing and row to row spacing is kept at 30 cm. Intercropping is the most desirable to fulfill the need of partial shady condition.

Irrigation

To maintain sufficient soil moisture level in the soil, frequent irrigation at weekly intervals is required if rains are scanty.

• Manure

Well decayed forest litter, farmyard manure and vermicompostare suggested to enhance growth and yield.It is suggested to improve physical, chemical and microbial properties of soil for enhanced productivity.

Weed control: Weeding after every 15 to 20 days is sufficient to keep the crop weed free. Karimi *et al.* (2016) reported leaf spot disease caused by *Phoma odoratissimi* and *Colletotrichum dematium* on *P. odoratum* in Italy.

- **Maturity and harvesting:** The crop is ready for harvest after three years when there is yellowing of foliage after flowering and fruits have ripened by the end of September (Annual Report NMPB, 2014). Harvested roots should be cleaned to remove foreign matter and shade dried. Thicker roots can be cut into pieces and dried under continuous air flow. Oven or microwave drying is not recommended. Dried roots should be stored in paper bags.

H. COMMERCIAL ASPECTS

• Demand

There is apparently high demand for crude drug obtained from medicinal plants in the country, which is primarily met by extraction from the wild and import from the neighboring countries. A rough estimate suggested annual import of well over 4 lakhs kg of crude drug material from SAARC countries (Olesen and Hellens, 1997).

• Market trends

About 70 per cent of these plants are used in Ayurveda. An earlier estimate for the year 1998 has put this annual demand at around Rs. 45 billion for the year (Subrat *et al.,* 2002). With the gradual popularity and acceptance, the demand for plant based Ayurvedic drugs is believed to have increased by many folds.

Trade data

In India, medicinal plants trade is believed to be extremely complex, unregulated and badly organized. Amidst variable estimates, economic value of medicinal plant trade has been high (ITCOT, 1999), and estimated to be around Rs.10 billion/year and the world trade over US $ 60 billion (Srivastava*et al.*, 1996).

• Major users

Of about 9,000 flowering plants reported in the country, more than 1,500 species are used for medicinal purposes under various indigenous systems of medicines (ISMs).With the renewed and growing interest in nutraceutical at national and international levels, the users of species of *Polygonatum* are expected to increase by many folds.

I. FUTURE DIRECTIONS FOR RESEARCH / VALUE ADDITION

Future directions for Research /Value addition: The Indian species of *Polygonatum*, as indicated earlier, have been used as important ingredients of *Chayavanprash* for centuries. However, to add scientific credibility for its usage a comprehensive chemical analysis for nutritional constituents of shoots and rhizomes is needed before these are considered as candidate for developing health foods or nutraceutical products. Phytochemicals of medicinal significance also require appropriate clinical evaluation. For conservation of the species and their sustainable use, commercial cultivation is recommended. It should be made mandatory that the species of *Polygonatum* for use in pharmaceutical industries be procured from the cultivated source, following standardized agro-techniques.

J. PATENTS

(https://patents.google.com/patent/US20040161524A1/en?q=Polygonatum &q=granted &q=patents&oq=Polygonatum+granted+patents) Total 257 in number.

1. Sakai, Y., Yokoo, Y. Process for producing a plant extract containing plant powder US20040161524A1 2009-04-21.
2. Yang, J., Foong, Y., Furukawa, F. Hair care composition comprising a polypropylene glycol. US6849252B1. 2005-02-01.
3. Dreher, F. Antioxidant Compositions and Methods of Using the Same. US20140315995A1. 2017-07-25.
4. Snyder, M.A., Yang, J. Hair conditioning composition comprising carboxylic acid/ carboxylate copolymer and moisturizing agent. US6767875B1. 2004-07-27.
5. Xia, Y.C. Herbal composition for treatment of neuronal injuries and neuronal degeneration, methods to prepare the same and uses thereof. US20030211178A1. 2007-04-17.
6. Yang, J. Hair care composition containing a polyalkylene (n) alkylamine which provide hair volume reduction. US6946122B2. 2005-09-20.
7. Kato, C., Sakuma, K. Moisture-retaining plant extract, and external preparation, cosmetic, bathing agent and detergent each containing the extract. JP2005145891A. 2008-11-12.
8. Arakawa, K., Hatakeyama, M., Hiraki, J. Polylysine formulation and cosmetic composition containing the same. JP2009108065A. 2012-06-06.
9. Kato, C., Sakuma, K. Moisture-keeping plant extract, and external preparation, cosmetic, bath agent and detergent all containing the extract. JP2004323424A. 2008-06-18.
10. Poon, T.P.M. Po-man's photosynthesis in mens blood to cure cancer and Poon's production function synthesis. GB2226494A. 1991-10-09.

Literature Cited

1. Ahn M.J., Kim C.Y., Yoon K.D., Ryu M.Y., Cheong J.H., Chin Y.W. and Kim J.W. 2006. Steroidal Saponins from the Rhizomes of *Polygonatum sibiricum*. *J. Nat. Product* 69, 360-364.
2. Alluri, V.K., T.V.N. Rao, D. Sunderaraju, M. vanisree, H.S. Tsay, G.V. Subbaraju. 2006. Biologicl screening of medicinal plants collected from Eastern Ghats of India using Artemia salina (Brine Shrimp test). *Int. J Appl. Sci. Eng.*, 4 (2): 115-125.
3. Annual Reports, 2014 submitted by the HRG to the National Plant Medicinal Board, Govt. of India, Department of AYUSH, Min. of Health and Family Welfare, New Delhi.
4. Artsdatabanken. 2010. Red List Database (Informasjon om rødlistede arter er nå i Artsportalen). Trondheim Available at: http://www.biodiversity.no/Article.aspx?m=39&amid=1864.
5. Baek, S. Lee, J. G., Park S. Y. *et al.* 2012. "Gas chromatographic determination of azetidine-2-carboxylic acid in rhizomes of Polygonatum sibiricum and Polygonatum odoratum," Journal of Food Composition and Analysis, vol. 25, no. 2, pp. 137–141,.
6. Balkrishan, A, A. Srivastava, R.K. Mishra, S.P. Patel, R.K. Vasshistha, A. Singh, V. Jadon, P. Saxena. 2012. Ashtavarga plants – threatened medicinal herbs of northwest Himalaya. *Int. J Med Aroma Plants*, 2 (4): 661-676.

7. Bhatt, D., R. Kumar, L.M. Tiwari, G.C. Joshi. 2014. Polygonatum cirrhifolium Royle and *Polygonatum verticillatum* (L) Allioni: Status aassessment and medicinal uses in Uttarakhand, India. *J Med Plants Res*., 8 (5): 253-259.
8. Bisht S, Bisht N.S., Bhandari S. 2012. In vitro micropropagation in *Polygonatum*
9. Bist, P., Pratti P, Prasad, N.B. 2011. *Polygonatum verticillatum* (Linn.) All. and *Polygonatum cirrhifolium* (Wall.) Royle: two threatened vital healers from Asthaverga nurtured by Garhwal Himalaya, *India. J Plant Dev*: 18: 159-67.
10. Butola J.S. 2011. Post-cultivation evaluation of germplasm in Himalayan threatened medicinal herbs: implication for ex-cultivation and conservation. *Natl Acad Sci Lett*; 34: 49-58.
11. Cheffings, C.M. and Farrell, L. (eds). 2005. The vascular plant Red Data List for Great Britain. Species status No. 7. Joint Nature Conservation Committee, Peterborough.
12. Chen SF. 1989. Karyotype analysis of eight species of *Polygonatum* Mill. *Acta Phytotaxonomica Sinica* 27: 39–48.
13. Chen Y., Yin L., Zhang X., Wang Y., Chen Q., Jin C, Hu Y., Wang J.2014. Optimization of Alkaline Extraction and Bioactivities of Polysaccharides from Rhizome of *Polygonatum odoratum*. *BioMed Res. International.*; Vol. 20, pages 8, Article ID 504896.
14. Chiang, Ni.Na, C.T. horng, S.S. Chang, C.I. Lu, C.Y. Su, H.Y. Wqang, F.A. Chen. 2013. Hypolipidemic activity of *Polygonatum alte-lobatum* Hayata extract in hamsters with hypolipidemia induced by by high fat diet. *Life Sci,J*., 10 (2): 939-942.
15. Chinmay, R., Kumari, S., Dhar, B., Mohanty, R.C., Tiwari, R.K.,Padhi, M., M. 2009. Evaluation and Identification of Meda and Mahameda to Common Substitutes and Adulterants In the Crude Drug Markets of India *Pharmacog. J.* 1(1), 56–63.
16. Colling, G., 2005. Red List of the Vascular Plants of Luxembourg. *Ferrantia* 42.
17. Debnath, T., F.R. Park, D.H. Kim, J.E. Jo, D..O Lim. 2013. Antioxidant and anti-inflammatory activity of Polygonatum sibiricum rhizome extracts. *Asian Pac J Trop Dis*., 3 (4): 308-313.
18. Deng X,Y, Wang Q., He X.,J. 2009. Karyotypes of 16 populations of eight species in the genus *Polygonatum* (Asparagaceae) from China. Botanical Journal of the Linnean Society 159: 245–254.
19. Deng, Y. He, K. Ye X. *et al.* 2012. Saponin rich fractions from *Polygonatum odoratum* (Mill.) Druce with more potential hypoglycemic effects, *J. Ethnopharmacol.*, 141(1) 228–233.
20. Devi, U., Pankaj Sharma, J..C.. Rana, Aman Sharma. 2014. Phytodiversity assessment in Sangla valley, Northwest Himalaya, India. *Check List* 10 (4): 740-760.
21. Devi,U., M. K. Seth, Pankaj Sharma, J.C.. Rana. 2013. Study on ethno medicinal plants of Kibber Wildlife Sanctuary: A cold desert in Trans Himalaya, India. *J. Med. Plants Res.* 7(47):3400-3419.
22. Gu, H.M., YW Meng, Q. Pu. 2003. Polysaccharide from cytonema Hua against herpes simplex virus in vitro. Chinese J Appl. Environ. Biol., 8 : 21-23.
23. Horng, Chi Ting, J.K. Huang, H.Y. Wang, C.C. Huang, F.D. Chen. 2014. Antioxidant and antifatigue activities of *Polygonatum alte-lobatum* Hayata rhizomes in rats. *Nutrients*, 6 (11): 5327-5337. India. *Physiol Mol Biol Plants*; 18: 89-93.
24. ITCOT (Industrial and Technology Consultancy Organisation of Tamil Nadu Limited). 1999. Report on Ayurvedic Medicines. ITCOT, Chennai.
25. Jin J, Zhang Y., Li H.., Yang C. 2004. Cytotoxic steroidal saponins from *Polygonatum zanlanscianense*. Journal of Natural Product 67, 1992-1995.
26. Kala C. P., Dhyani P. P., Sajwan B. S. 2006. Developing the medicinal plant sector in northern India: challenges and opportunities. *J. Ethnobiol. Ethnomed.*, 2: 32.
27. Karimi, K., Khodaei, S., Rota Stabelli, O. , Arzanlou, M., Pertot, I. 2016. Identification and Characterization of two new Fungal Pathogens of Polygonatum odoratum (Angular Solomon's seal) in Italy. *J Phytopathol*, 164: 1075-1084. doi:10.1111/jph.12528

28. Kazmi, Imran, M. Afzal, M. Rahman, G. Gupta. 2012. Aphrodisiac properties of *Polygonatum verticillatum* leaf extract. *J. Tropical Disease*, Supplement: 5841-5845.
29. Khan H, Saeed M, Gilani AH, Muhammad N, Haq I, Ashraf N, et al. 2013. Antipyretic and anticonvulsant activity of *Polygonatum verticillatum*: comparison of rhizomes and aerial parts. *Phytother Res.*;27(3):468–71.
30. Khan H., Saeed M., Muhammad N., Khan F., Ibrar M., Hassan S., Shaw W.A. 2012c. Comprehensive nutrients analysis of rhizomes of *Polygonatum verticillatum. Pak J Pharm Sci,* 25: 871-5.
31. Khan, H., M. Saeed, A.H. Gilani, M.A. Khan, A. Dar, and L. Khan. 2010. The antinoceptive activity of *Polygonatum verticillatum* rhizome in pain models. *J Bhnopharmacology*, 127 (2): 521-527.
32. Khan, H., M. Saeed, N. Muhammad, 2012b. Pharmacological and phytochemical updates of genus *Polygonatum*. *Phytopharmacology* 2012, 3(2) 286-308.
33. Khan, H., M. Saeed, N. Muhammad, R. Ghaffar, S.A. Khan, Hassan,S. 2012a. Antimicrobial activities of *Polygonatum verticillatum* attributed to its total flavonoidal and phenolic contents. *Pak. J Pharm Sci*., 25 (2): 465-467.
34. Khan, Haroon, Abdur Rauf. 2015. Phytochemistry of genus Polygonatum: A review. American J of Biomedical and Life Sciences, 3 (1): 5-20.
35. Khela, S. 2012. *Polygonatum odoratum*. The IUCN Red List of Threatened Species 2012: e.T202966A2758297.
36. Kim, J.H., D.K. Kim, F. Forest, M.F. Fey, Chase. M.W. 2010. Molecular phylogenetics of Ruscaceae sensu lato and related families (Asparagales) based on plastid and nuclear DNA sequences. Annals of Botany doi: 10.1093/aob/meq/67 (on line).
37. Kim, J.S. Lee, S. C. Lee, B., Cho, K. Y. 2006. Biological activity of l-2-azetidinecarboxylic acid, isolated from *Polygonatumodoratum* var. *pluriflorum*, against several algae," *Aquatic Bot.*85 (1) 1–6.
38. Kumari, P., Joshi, G.C., Tewari, L.M. 2014. Taxonomical description, nativity, and endemism with particular reference to threatened ethnomedicinal plants of Almora district of western Himalaya. *J.Phytology*, 6: 14-25
39. Lattoo S. K., Khan S., Dhar A. K. 2005. A new chromosome number in *Polygonatum cirrhifolium* Royle: an endangered liliaceous medicinal herb. is, 89(7): 1080-1081.
40. Lattoo, S.K., A.K. Dhar, A. Jiasrotia. 2001. Epycotyl seed dormancy and phenology of germination in *Polygonatum cirrhifolium* Royle. *Curr. Sci.,* 81 (11): 1414-1417.
41. Lewis, W.H., Elvin-Lewis, M.P.F. 2003. Medical Botany - Plants affecting human health. John Wiley & Sons, Inc., Hoboken, New Jersey.
42. Liu,X., W. Zhen-jiang, S. Lin, D. Kan.2011. Carbohydrate polymers, 83 : 737-742.
43. Lohani N., Kumar R., Tewari L.M., Joshi G.C. 2012. Ex-situ conservation of Polygonatum verticillatum (L.) Allioni under different types of organic treatments. *Int J Biodivers Conserv*; 4: 22-31.
44. Lohani N., Tewari L.M., Kumar R., Joshi G.C., Chandra J., Kshore K., Kumar, S., Upreti, B.M. 2013. Population studies, habitat assessment and threat categorization of *Polygonatum verticillatum* (L.) Allioni in Kumaun Himalaya. *J Ecol Nat Environ*; 5: 74-82
45. Moser, D., Gygax, A., Bäumler, B., Wyler, N., Palese, R. 2002. Red List of the Threatened Ferns and Flowering Plants of Switzerland (Rote Liste der gefährdeten Farn- und Blütenpflanzen der Schweiz). Bundesamt für Umwelt, Wald und Landschaft, Bern; Zentrum des Datenverbundnetzes der Schweizer Flora, Chambésy; Conservatoire et Jardin botaniques de la Ville de Genève, Chambésy.
46. Nautiyal, M.C., B.P. Nautiyal. 2004. Agrotechniques for High Altitude Medicinal and Aromatic Plants. Bishen Singh mahendra Pal Singh, Dehradun.
47. NERI. 2007. The Danish Red Data Book. Roskilde: National Environmental Research

Institute, Danish Ministry of the Environment Available at: http://www2.dmu.dk/1_Om_DMU/2_Tvaerfunk/3_fdc_bio/projekter/redlist/redlist_en.asp.

48. Olesen, C.S., Hellens, F. 1997. Medicinal plants, markets and margins in the Nepal Himalaya-trouble in paradise. Mountain Research and Development: XVIII (4), pp363–374.

49. Pandey M. M., Govindarajan R., Khatoon S, Rawat A. K. S., Mehrotra, S. (2006) *J. Herbs, Spices & Med. Plants*, 12:01-02.

50. Prakash, O., A.K. Jyoti, P. Kumar, N.K. Sharma. 2013. Adulteration and substitution in Indian medicinal plants: An overview. *J. Med. Plant Studies,* 1 (4): 127-132.

51. Qian, Y. Liang, J. Y. Qu, W., Che, Y. Y. 2010. Two new homoisoflavanones from *Polygonatum odoratum* (Mill.) Druce, *Chinese Chemical Letters*, vol. 21, no. 6, pp. 706–708.

52. Rafi M.M., Vastano B.C. 2007. Identification of a structure specific Bcl-2 phosphorylating homoisoflavone molecule from Vietnamese coriander (*Polygonatum odoratum*) that induces apoptosis and G2/M cell cycle arrest in breastcancer cell lines. *Food Chem.*, 104: 332-340.

53. Rana J.C., Archna Singh, Yogender Sharma, K. Pradheep, Nisha Mendiratta. 2010. Dynamics of Plant Bioresources in Western Himalayan Region of India – Watershed Based Case Study. *Curr. Sci.*, 98 (2): 192-203

54. Rana J.C., Dutta M., Rathi R.S. 2012 Plant genetic resources of the Indian Himalayan region – an overview. *Indian J. Genet.,* 72 (2): 115-129

55. Saboon, B., Y., Arshad, M., Sabir, S., Shoaib M., Amjad, Ahmed, E., Chaudhari, S.K. 2016. Pharmacology and biochemistry of *Polygonatum verticillatum*: A review. *J. Coastal Life Med.*. 4(5): 406-415

56. Saeed, M., H. Khan, M.A. khan, S.U. Sinju, N. Muhammad, S.A. Khan.2010. Phytotoxic, insecticidal and leishmanicidal activities of aerial parts of *Polygonatum verticillatum. Africal J Biotech.*, 9 (8): 1241-1244.

57. Sagar, P.K. 2014. Adulteration and substitution in endangered costly herbal medicinal plants of India, investigates their active phytochemical constituents. Int. J Pharmacy and Therapeutics, 5 (4): 243-268.

58. Saini, R., S. Chaturvedi, MS Yar, H.P., Bharitya, P Singh. 2010. Chemical investigation of Polygonatum verticillartum. Asian J Resarch Chem., 3 (1): 20-21.

59. Savita Rani, J.C. Rana. 2014. Ethnobotanical Uses of Some Plants of Bhattiyat Block in District Chamba, Himachal Pradesh (Western Himalaya). *Ethnobot. Res. Applications* 12:407-414 (2014)

60. Sharma Pankaj, J.C. Rana, Devi, U., Randhawa SS, Rajesh Kumar. 2014. Floristic Diversity and Distribution Pattern of Plant Communities along Altitudinal Gradient in Sangla Valley, Northwest Himalaya. *The Scientific World Journal,* ID 264878, 11 pages http://dx.doi.org/10.1155/2014/264878

61. Sharma, B.D., 2014. Himalayan Medicinal Plants: Science and Traditional Wisdom. Bishen singh Mahendra Pal Shingh, Dehradun, Uttarakhand, India.

62. Sharma, B.D., Balkrishan. 2005. Vitality Strengthening Astavarga Plants (Jeevaniya and Vyasthapan Paudhe). Divya Yog Mandir (Trust), Kripalu Bagh Ashram, Kankhal, Haridwar.

63. Sharma, B.D., K Pal. 2007. Antidiabetic activity of *Polygonatum verticillatum* (L.) Allioni in streptozotocine induced non-insulin dependent diabetic rats. Envis Forestry Bulletin, 7 (2): 99-102.

64. Sharma, B.D., Lal Singh, Maninder Jeet Kaur. 2015. Seed biology, dormancy and phenophases of seed germination of three rare Himalayan Herbs. Journal of Non-Timber Forest Products, 22(3) 143-147.

65. Sharma, B.D., MJ Kaur, Lal Singh, Dakesh. 2013. Climate changes threat to Himalayan herbs with specia reference to Astavarga plants. In "Environmental Pollution: Assessment

and Control Techniques (eds. DR Khanna, AK Chopra, Vikash Singh, Gagan Matta, Rakesh Bhuttiani), pages 55-58, Biotech Books, New Delhi.

66. Sharma, B.D., Singh L., Maninder J.K. 2014. Nutritional Composition of Rare Himalayan Herbs Constituting the World's First Health Food, *International Journal of Agriculture and Food Science Technology.*; Vol.5, No. 2 pp 75-80,
67. Singh S.K., Singh S., Verma S.K., Jain P., Dixit V.K., Solanki S. 2013. A review on plants of genus polygonatum, *International J. Res. Develop. Pharmacy Life Sci.*; Vol. 2, No.3, pp 387-397.
68. Singh, A.P., 2006. Ashtavarga- Rrare medicinal plants. Ethnobot. Leaflets, 10: 104-108.
69. Srivastava, J., Lambert, J., Vietmeyer, N. 1996. Medicinal Plants: An Expanding Role in Development. World Bank Technical Paper Number 320.
70. Subrat, N., Iyer, M.Prasad, R. 2002. The Ayurvedic Medicine Industry: Current Status and Sustainability, Ecotech Services (India) Pvt Ltd, and International Institute for Environment and Development, New Delhi.
71. Sugiyama, M. Nakano, K. Tomimatsu, T., Nohara, T. 1984. Five steroidal components from the rhizomes of Polygonatum odoratum var Pluriflorum, Chemical and Pharmaceutical Bulletin, vol. 32, no. 4, pp. 1365–1372.
72. Tamura MN. 1993. Biosystematic studies on the genus *Polygonatum* (Liliaceae) III. Morphology of staminal filaments and karyology of eleven Eurasian species. *Botanische Jahrbücher für Systematik* 115: 1–26.
73. The Ayurvedic Pharmacopoeia of India, 2016, Government of India, Ministry of AYUSH, Pharmacopoeia Commission For Indian Medicine & Homoeopathy, Ghaziabad.

75. Tiwari, Lalit. 2001. Traditional Himalayan Medicine System and its Materia Medica. www.infinityfoundation.com/mandal/...htm.
76. Tomoda, M. Yoshida, Y. Tanaka, H., Uno, M. 1971. Plant mucilages. II. Isolation and characterization of a mucous polysaccharide, "Odoratan," from *Polygonatum odoratum* var. *japonicum* rhizomes, *Chem. Pharmaceutical Bull.*, vol. 19, no. 10, pp. 2173–2177.
77. Ved, DK, Kinhal, GA Ravikumar, K Prabhakaran, V Ghate, U Sankar RV., Indresh, JH. 2003. CAMP report: Conservation assessment and management prioritisation for the medicinal plants of Jammu and Kashmir, Himachal Pradesh and Uttaranchal. FRLHT, Bangalore. *verticillatum* (L.) All. an important threatened medicinal herb of Northern
78. Wang, Dongmei, ZHU WeilI, Juaneli. 2007. Study on chemical constituents of *Polygonatum cirrhifolium* rhizome and their fungicidal activities. Journal of Sichuan University (Natural Science Edition) DOI CNKI SUNSCDNO.2007-04-04 (on line).
79. Wang, Dongmei. 2008. Studies on the chemical constituents and bioactivity of two *Polygonatum* plants of Qinling Mountain. Doctrate Thesis (Botany) submitted to Northwest A & F University.
80. World Checklist of Selected Plant Families (Internet) Kew Royal Botanic Gardens, 2014.
81. Wujisguleng, W., Y Liu, C long. 2012. Ethnobotanical review of food uses of *Polygonatum* (Convallariaceae) in China. Acta Societatis Botanicorum Poloniae, 81 (4): 239-244.
82. Yang J, Wang JW, Li MX. 1988. Cytotaxonomic studies on the genus of Polygonatum II. Karyotypes of 4 species from Sichuan Jin-Fo-Shan. *J. Wuhan Bot. Res.* 6: 311–314.
83. Zhao, P., Zhao, C., Li, X, Gao, Q, Huang, L, Xiao, P, Gao, W. 2018. The genus Polygonatum: A review of ethnopharmacology, phytochemistry and pharmacology. *J. Ethnopharmacol.*, 214: 274-291.

4

Picrorhiza kurroa

Hindi: Karu, Kutki

Sanjay Kumar and Rakesh Kumar

***Picrorhiza kurroa* Royle *ex* Benth**

(A) Herbarium sheet, **(B)** Cluster of plants, **(C)** Flowering stage, **(D)** Nursery plants ready for transplantation, **(E)** Cultivation in field, **(F)** Crude drug material (rhizomes and roots)

Introduction

Picrorhiza kurroa Royle *ex* Benth is an important alpine medicinal plant of the Himalayan region. The root and rhizome of *P. kurroa* are source of the crude drug, which is characteristically bitter in taste and is linked to the medicinal properties. In Indian system of medicine, the species popularly known as *Kutki,* is an important constituent of several drugs items. This plant also finds a place in Chinese system of medicine where it is known as *Hun-hung lien*. Chemically, compounds of interest in Picrorhiza are iridoid glycosides: picroside-I and picroside-II. Material for crude drug has traditionally been extracted from the wild. Natural stands have sharply dwindled due to over extraction and habitat degradation. The species was listed as "endangered" in 1997 by International Union for Conservation of Nature (IUCN). Commercial cultivation is not popular, but is urgently recommended.

A. BIOLOGICAL ASPECTS

• Systematic classification

Kingdom : Plantae

Division : Dicotyledonae

Class : Angiospermae

Order : Scrophulariales

Family : Scrophulariaceae

Genus : *Picrorhiza*

Species : *kurroa*

• Synonyms

According to the plant list, which is a working list of all plants, information for synonyms of *P. kurroa* remains unresolved (http://www.theplantlist.org/).

• Common names (in different Indian languages)

In India, it is identified by different commercial and regional names but "kutki" is the most prevalent. Linguistic variants are: Sanskrit- Katuka, Katurohini; Bengali- Katki; Gujrati- Kadu; Hindi- Karu, Kutki: Marathi- Kutaki; Malayalam- Kadugurohani, Katukarogani; Tamil- Katukurogani; Telgu- Katukarogani; Kashmiri- Kour; Himachali- Kadu; Uttrakhandi- Kadvi.

• Diversity within genus

The genus *Picrorhiza* was earlier considered monotypic, with only one species *P. kurroa.* Another species, *P. scrophulariiflora,* was distinguished by Pennell in 1943. Hong (1984) brought this new species under a separate genus and named it *Neopicrorhiza scrophulariiflora.* Both *Picrorhiza* and *Neopicrorhiza* belong to the tribe Veroniceae of the family Scrophulariaceae. *P. kurroa* and

N. scrophulariiflora differ in length of stamens: the former possess long stamens and the latter short stamens. Both the species are prevalent in the Indian Himalayan region and in China. *P. kurroa* is found mainly in the western Himalaya at altitude ranging from 3000-4300 meters amsl. *N. scrophulariiflora* occurs in the Eastern Himalaya at 4300-5200 meters amsl (Pennell, 1943).

• Geographical distribution

P. kurroa is known to occur wild in alpine Himalayas from Kashmir to Kumaon (Pennell, 1943). In Himachal Pradesh, it has been reported from around 3800 m amsl at Rohtang Pass (Uniyal *et al.,* 1989; Aswal and Mehrotra, 1994), and above 3500 m amsl at Pangi-Bharmour (Chauhan, 1988). In Uttarakhand, *P. kurroa* is distributed mainly around Agora-Dodital, Uttarkashi district (Chandra and Pandey, 1983), Bhillangna Valley, Sahasru Tal, Tehri-Garhwal up to 5,300 m amsl (Gupta, 1957; 1989), and between 3000-4500 m amsl in Buhna, Nar Parbat, Garhwal, (Rau, 1961; 1975). In Jammu & Kashmir, the species is reported from localities such as Apharwat (Rau, 1975), Burzil Pass, Kamri Pass, Sonamarg (Pennell, 1943), Gumri (Jee *et al.,* 1987), Lipper Valley, northwest of Kashmir Valley (Singh and Kachroo, 1976), Pir Panjal range, Krishan Ganga Valley, upper Lidder Valley (Kapahi *et al.,* 1993; Kapil, 1995), and Simthan, Jammu (Sharma and Kachroo, 1981).

• Habitat and ecology

P. kurroa is restricted in distribution in sub-alpine regions/ alpine region in Himalaya. The species grows in open habitat conditions and is well adapted to fluctuating conditions of temperature and light, high radiation load, and other parameters of high altitude environment. It prefers moist, sandy soil and grows well in rock crevices. Fine patches of populations are seen due to rhizomatic nature of its propagation.

• Taxonomic enumeration

P. kurroa belongs to the family Scrophulariaceae. Plants are perennial, herbaceous with creeping rootstock and aerial parts represented by basal leaves and flowering scape (a leafless flower stalk growing directly from the ground). The simple leaves arise from the tip of the upturned stolons in rosettes or whorls. From tip of the rhizome, a whorl of radical leaves arises (Raina *et al.,* 2010). A mean height of 16.0 to 17.5 cm is attained by flowering scape. Cauline leaves (leaves arising from the upper part of a stem or when the leaves are found on node of stem) are present only during the flowering phase. The inflorescence is indeterminate terminal spike that forms a more or less triangular head (Nautiyal *et al.,* 2001; Arya *et al.,* 2013).

• Phenological period

P. kurroa (2n=34) is an alpine herb and reaches a height of upto 20 cm at flowering stage. The plant possesses stoloniferous rootstock which perennates during most part of the year. Leaves are basal, elliptic, coarsely saw-toothed, and arise in rosettes or whorls from the tip of the upturned rhizome. The rhizomes are up to 30-45 cm long and 0.3-1.0 cm in diameter, branched, run on the top layers of soil, and roots at nodes. The apparently separate looking plants above ground are actually joined together by rhizomes beneath. The inflorescence is indeterminate, and terminal spike forms a triangular head. The flowers are sessile, zygomorphic, bilipped, bisexual, purple in colour and are borne in the axils of small green coloured bracts. The flower architecture mostly favours cross-pollination. Fruit is an ovoid capsule with an average dimension of 9 x 5 mm. It dehisces by means of lateral splits. The seeds are extremely small in size, each measuring 1.3 x 1.0 mm in dimension and weighs about 50-60 mg. A large bladdery loose hyaline reticulate testa surrounds the embryo.

Depending upon the altitude of the growing site, flowering occurs in one or two stages. At 3500 m amsl of elevation, the first stage flowering starts in 1st week of May and lasts upto 3rd week of June. The second flowering starts in August and extends upto September end. In alpine regions (>4000 m, amsl), the species flowers only once in July-August and seeds are develop by September (Kaul and Kaul, 1996; Raina *et al.,* 2010).

• Cytotypes

Cytotype is characteristic of a cellular organism of same species which differ in chromosome structure, ploidy, mitochondrial genome and karyotype. It was reported by Federov (1969) that the primary base numbers in the family Scrophulariaceae form a series from x=6 up to 12 and in taxa with 2n=12 are up to 168. Literature is scanty regarding chromosome number: 2n=34, appears to be the first report by Raina *et al.* (2010) for *P. kurroa.*

B. REPRODUCTIVE BIOLOGY

• Reproductive phenology

The inflorescence is indeterminate terminal spike forming a more or less triangular head. The flowers are sessile, zygomorphic, bilipped, bisexual and purple in colour. These are borne in the axils of small green bracts, which open during June to September (Nayar and Shastry, 1990; Kaul and Kaul, 1996; NIIR, 2006; Pandit *et al.,* 2012). Sepals are unequal in size. Filaments are colorless at bud stage and turn purplish after anthesis in order to attract insects for pollination. The four free stamens are epipetalous and alternate with the petal lobes. Stamens are didynamous with 2 long (15.50 mm) and 2 short (13.25 mm) stamens. Although the stamens are exerted, anthers are borne on incurved filaments, which make their position introrse. The growth of style keeps pace with that of the filaments and at the stage of attaining maximum length both are of almost equal length. The process of shedding pollen starts once the anthers comes out of the corolla cavity, and completes till the maximum growth of the filament is attained (Nautiyal *et al.,* 2001; Raina *et al.,* 2010). Fruit is an ovoid capsule with an average dimension of 9 x 5 mm. The lateral splits are responsible for its dehiscence.

• Life Cycle

P. kurroa has a life cycle of three years. During the first year, the plant grows vigorously. Flowering occurs in one or two stages, depending upon elevation of occurrence. After the completion of flowering, fruiting starts in August and continues up to September, leading to seed development. Plant needs yet another year for complete maturity of the seeds. The roots and rhizomes are manually harvested in September when the shoots or the aerial parts begin to wither and dry. Plants raised through stem cuttings mature almost a year earlier than those raised from the seeds. However, to get higher active contents, plants must be collected before flowering occurs.

• Pollination and pollinators

P. kurroa is a cross pollinated plant but self-pollination has also been observed. The anthers start shedding pollen once they come out of the corolla cavity and dehiscence last still the filaments attain maximum growth (Raina *et al.,* 2010). However, introrse anthers ensure that some pollen may fall on the flowering spike itself so as to affect self-pollination. *Bombus* spp. and *Apis florae* are commonly seen as pollinators. In case of *P. kurroa*, exerted stamens as well as introrse anthers ensure both allogamy and autogamy (Hagerup, 1951). Autogamy is often the only possible means of sexual reproduction in high alpine or arctic region because of paucity of pollinators due to unfavorable weather conditions (Hagerup, 1932; 1951).

C. MEDICINAL ASPECTS

• Plant material used as crude drug

• General appearance

Crude drug is mainly composed of small pieces of cylindrical rhizomes, though whole plant (buds, aerial stems along with the roots) also constitutes the drug. Rhizome pieces, grey to brown in colour, are cylindrical in shape and may be straight or slightly curved. The surface may bear a few small root scars and numerous scale leaves.

• Organoleptic properties

The crude drug displays slightly unpleasant odour and tastes very bitter.

• Microscopic characteristics

The periderm of the rhizomes constitutes 8-10 layers of thin walled cork. The phelloderm cells are full of brownish contents. The cortical cells are rounded with occasional inter-cellular spaces. The secondary phloem is many layered and xylem is made up of vessels, fibres and tracheids. Pith consists of thick parenchymatous cells.

• Powdered plant material

Grayish-brown colored powdered plant material of *P. kurroa* is extremely bitter in taste (Raina *et al.,* 2010). Microscopic examination reveals numerous pits in pith cells and fragments of cork cell, which are associated with cortex. Simple or compound scattered starch grains are with distinct central hilum and fragments of vessels (Keshari *et al.,* 2015). For use, 400-1500 mg of powdered picrorhiza (encapsulated) is recommended, or 1-2 ml of fluid extract, two times a day. To improve its palatability and to avoid its bitter taste, it can be consumed in tea or in combination with ginger root powder, or in capsules (Bone, 1995).

• Substitutes and adulterants

The drug samples sold in Indian markets are mostly genuine. However, different herbs, such as *P. scrophularia, Actaea spicata, Cimcifuga foetida, Coptis teeta, Coscinium fanestratum, Swertia chirata and Latotis cashmiriana* etc. are used as adulterants (Kapahi *et al.,* 1993; Vaidya and Adarsa, 1994; Warrier, 1996; Gogte, 2000; Handa, 2000; Banwarilal, 2004). The roots of *Acacia spicata, Cimicifuga foetida* and *Coptis teeta* are sold under the name of 'kutki' or 'karu' in Mumbai. At Bengal drug shops, the roots of *Gentiana kurroo, Coscinium fenestratum* and *Swertia chirata* are available as 'karu'. *Lagotis cashmiriana* is an adulterant of commercial samples of *P. kurroa* in Kullu markets of H.P. Quality wise, there is a lot of variance, which can be attributed to collection time, age of the plant and region of collection.

• General identity tests

Rhizomes of *P. kurroa* are sub-cylindrical and straight, surrounded by a tufted crown of leaves. The roots are thin, cylindrical in pieces, 5 to 10 cm long, 0.5 to 1.0 mm in diameter, moderately curved with a few elongated grooves, spotted blotches. These are dusky grey in color with pleasant odor and bitter taste (Kumar *et al.,* 2012; Keshari *et al.,* 2015). Demand for Picrorhiza has increased in recent years due to its high medicinal importance and declining availability. The species has been included under negative listed drugs, so that its export as plant parts, other derivatives or extracts as such, is prohibited. The prepared formulation however can be exported. General identification tests for purity of powdered *P. kurroa* are as follows (Keshari *et al.,* 2015).

- Powder gives brownish fluorescence under UV light when treated with NaOH in methanol.
- Powder gives yellowish green florescence when treated with NaOH in methanol and planted in nitrocellulose.
- Powder gives dirty green fluorescence when mounted in nitrocellulose.
- Powder as such gives dirty green fluorescence.

• Purity tests

The common approach to identify impurities in medicinal plant samples is thin layer chromatography (TLC) and high performance thin layer chromatography (HPTLC). In order to distinguish genuine plant from adulterants, HPTLC and HPLC fingerprints are commonly used (Ganeshan and Bhatt, 2008; Wang *et al.,* 2009). Different analytical methods such as TLC, HPLC, Gas Chromatography (GC), quantitative TLC (Q-TLC) and HPTLC can establish the constant composition of herbal preparations and for determining the homogeneity of different plant extracts. Over-pressured layer chromatography (OPLC), infrared and UV-visible spectrometry, GC-MS (Gas chromatography-mass spectrometry), LC-MS (Liquid chromatography-mass spectrometry), volumetric analysis, gravimetric determination, gas chromatography and column chromatography, HPLC and spectroscopic methods are convenient methods for standardization and to regulate the quality of product (Wang *et al.,* 2006; Zhao *et al.,* 2009; Sagar, 2014; Keshari *et al.,* 2015; Patel *et al.,* 2015).

• Medicinal usage

• General uses

In ayurvedic as well as other indigenous systems of medicine, herbal preparations of *P. kurroa* are used as stomachic, purgative, anti-allergic, dyspepsia and brain tonic (Hussain, 1984). Iridoid component is used to treat viral hepatitis and it possesses anti HBs Ag (anti-hepatitis B virus surface antigen) like activity (Mehrotra *et al.,* 1990). The plant is also found beneficial in disorders associated with liver, fever, asthma and jaundice (Puri *et al.,* 1992; Sharma *et al.,* 1994).

P. kurroa has been used extensively for skin lesions, lung diseases, urinary disorders, rheumatic diseases, worm infections, heart ailments and against snake/scorpion bites (Halliwel and Gutteridge, 1999). The crude extracts of *P. kurroa* has been reported to protect liver against poisoning from *Amanita* sp. (Floersheim *et al.,* 1990; Dwivedi *et al.,* 1992), and a wide range of chemicals such as, carbon tetrachloride (Dwivedi *et al.,* 1990; Saraswat *et al.,* 1993; Santra *et al.,* 1998), galactosamine (Dwivedi *et al.,* 1992; Dwivedi *et al.,* 1993), ethanol (Rastogi *et al.,* 1996), aflatoxin B1 (Dwivedi *et al.,* 1991), acetaminophen (Singh *et al.,* 1992), thioacetamide (Dwivedi *et al.,* 1993), oxytetracyline (Saraswat *et al.,*1997) and monocrotaline (Dwivedi *et al.,* 1991) in both *in vitro* and *in vivo* experiments. Iridoid glycosides mixture compound, Picroliv was reported to restrict hepato-carcinogenesis induced through drugs (Stupner and Wagner, 1989; Kaul and Kaul, 1996; Visen *et al.,* 1998; Jia *et al.,* 1999). The extracts were suggested to lower the serum lipids (total VLDL and LDL cholesterol), phospolipids and triglycerides, in both normal and hyperlipidemic animals (Khanna *et al.,* 1994). Anticancer activity was also reported for *P. kurroa* (Joy *et al.,* 2000), and its extracts to selectively strengthen the neuron growth (Li *et al.,* 2000; 2002). The plant extract was cytotoxic to carcinoma cells and was able to target the cells towards apoptosis (Rajkumar *et al.,* 2011). Significant anti-cholestatic, anti-asthematic, anti-inflamatory, anti-ulcerogenic, anti-diabetic, anti-neoplastic, anti-oxidant activities and immuno-regulatory functions have been attributed to iridoid glycosides (Ram, 2001; Singh *et al.,* 1993). Anti-fungal activity of ethanolic extract of *P. kurroa* was also reported against *Candida albicans* (Mandloi *et al.,* 2010).

• Ethnobotanical uses

P. kurroa has been found by native peoples to cure diseases like high fever and stomachache. Mildly boiled 10 g root decoction savoured with honey is given to cure stomachache for adults. 10 g root powder in combination with 1 g black pepper and honey is given to adult patients for curing fever. 1/4 g Kutki powder with mother's milk is recommended for curing stomachache of infants (Arya *et al.,* 2013).

D. CHEMISTRY ASPECT

• Chemical composition

The active constituent of *P. kurroa* is picrorhizin, a glucoside which yields picrorhizetin and dextrose on hydrolysis. It also contains kutkin, a glucosidal bitter principle, picroside-I, picroside-II, picroside-III, D-mannitol, vanilic acid, kurrin, kutkiol, kutkisterol, apocynin, 6-feruloylcatalpol, vernicoside, kutkoside, minecoside, picein, androsin, β-D-6-cinnamoylglucose, arvenin III, phenolic glycosides picein and androsin and seven cucurbitacin glycosides.

Major chemical constituent

The therapeutic property of *P. kurroa* is mainly associated with an active constituent called Kutkin (a terpenoid moiety) which is predominantly accumulated in the roots/rhizomes. Kutkin is a crystalline compound, reported to be a blend of two major C-9-iridoid glycosides, picroside-I (6-O-trans cinnamoyl catapol) and picroside-II (6-vanilloyl catapol) (Jia *et al.,* 1999). The chemical structure of picroside-I and picroside-II is given below. Other identified active chemical compounds are apocynin, drosin and cucurbitacin (Weinges *et al.,* 1972).

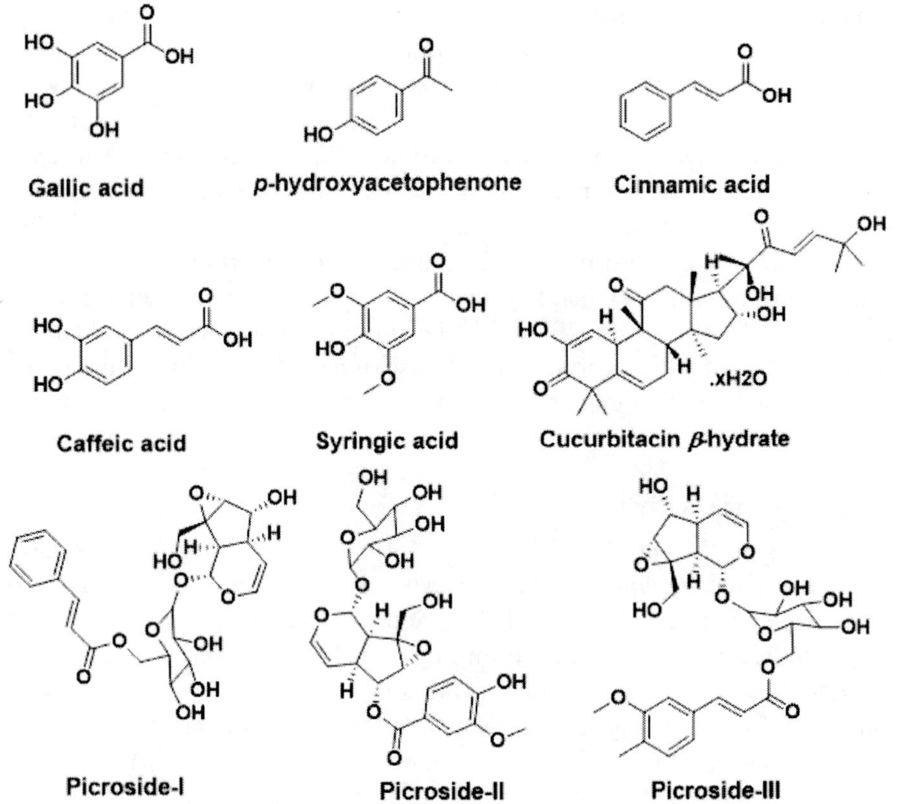

Gallic acid **p-hydroxyacetophenone** **Cinnamic acid**

Caffeic acid **Syringic acid** **Cucurbitacin β-hydrate**

Picroside-I **Picroside-II** **Picroside-III**

Chemical constituents of *Picrorhiza kurroa*
(Chemical structure is contributed by Dinesh Kumar)

• Marker compound

Picrosides I and II are specifically present in *P. kurroa* and hence are used as chemical markers for identification purpose.

E. PHARMACOLOGICAL ASPECTS

• Anti-cancer activity

The rhizome of *P. kurroa* has been reported to contain iridoid glycoside like picroside I, picroside II, terpene like cucurbitacins and flavonoids like apocynin which are responsible for the anticancer activity of the plant (Stupner and Wagner, 1989; Duangmano *et al.,* 2012; Zhu *et al.,* 2012; Alghasham, 2013; Aribi *et al.,* 2013; Kong *et al.,* 2014). Ghisalberti (1998) reported that *P. kurroa* contains iridoid glycoside which showed promising anti-cancer activity in different cell lines. Gaidhani *et al.,* (2013) conducted an experiment to examine the anticancerous activity of standardized hydro alcoholic extracts against 14 different cell lines. *P. kurroa* was found active against only one cell line i.e. colon (Colo205). Some other workers reported that the extract enhanced the life spell of ascites tumor bearing mice by reducing the proportion of sub cultured solid tumors caused by Dalton's lymphoma ascites (DLA) tumor cell lines (Joy *et al.,* 2000; Mallick *et al.,* 2015). The Dichloromethane fraction of *P. kurroa* displayed potent anticancer activity (Rajeshkumar, 2000; 2001; Uddin *et al.,* 2015). It can be further explored for the development of a potential candidate for cancer therapy.

• Anti-bacterial activity

To investigate the anti-bacterial activity of *P. kurroa,* various studies have been conducted. *P. kurroa* ethanolic extracts have been evaluated against different strains of Gram-positive bacteria like *Bacillus subtilis* and *Staphylococcus aureus,* and Gram-negative bacteria like *Pseudomonas aeruginosa* and *Escherichia coli*. The minimal inhibitory concentration (MIC) values of 65 to 260 mg mL^{-1} proved its effectiveness against all assayed organisms (Kumar *et al.,* 2010; Usman *et al.,* 2012). Phytochemical screening by HPTLC studies showed the presence of high amounts of iridoid glycosides in the rhizomes of this plant that conferred antimicrobial activity (Hari venkatesh and Chethana, 2013). Methanolic extract was found effective against *P. aeruginosa* and *S. aureus*; while aqueous extract showed moderate activity against *E. coli*, *B. subtilis* and *M. luteus* (Kumar *et al.* 2010; Sharma and Kumar, 2012; Kaigongi *et al.,* 2014). Gahlaut and Chhillar (2013) reported that different picrorhiza extracts showed good inhibitory effects against different bacterial strains *viz. B.subtilis, S. aureus, Salmonalla typhi, E. coli, Klebsiella pneumonia* and *P.aeruginosa.* Similar antibacterial effect was also reported by Ghansar *et al.* (2012).

• Anti-parasitic activity

Studies have been conducted to assess the anti-malarial activity of *P. kurroa* using extracts prepared by maceration and percolation of different plant parts. Aqueous, cold alcoholic and hot alcoholic extracts of *P. kurroa* restricted growth of *Plasmodium falciparum* (Irshad, 2011; Singh and Banyal, 2011). Rathee *et al.,* (2012) also reported that *P. kurroa* extract restores declined glutathione level in *Plasmodium berghei* infected rats, by boosting detoxification and antioxidation process. Crude extracts or various iridoid glycosides such as, *Picroliv*, kutkin, kutkosides (picrosides I, II, III & IV) are reported to be effective against a wide range of helminthes i.e. *Leishmania donovanipromastigote* (Puri *et al.,* 1992; Mittal *et al.,* 1998). Methanolic extract is effective against earthworms (Singh *et al.,*1993; Rajani *et al.,* 2013).

• Anti-inflammatory/ anti-arthritic activity

P. kurroa extracts restricted proliferation of agents such as macrophages, neutrophils and mast cells which cause inflammation. Potent anti-inflammatory actions on different types of inflammatory cells have been seen by apocynin, a catechol fraction obtained from *P. kurroa*. Apocynin restricts the neutrophil oxidative burst *in vitro* conditions without influencing beneficial activities such as phagocytosis, chemotaxis and intracellular elimination of bacteria (Bruynzeel *et al.,* 2007).

• Gastro-protective/ anti-ulcer effect

P. kurroa is an important component of promising anti-ulcerogenic formulations. *P. kurroa* (20 mgkg^{-1}, PO × 3 days) could effectively cure stomach ulcer induced by indomethacin in mice, by lowering oxidative stress, and promoting secretion of mucin, prostaglandin synthesis and increasing assertion of cyclooxygenase enzymes and growth elements (Banerjee *et al.,* 2008). Nadkarni (1954) reported that about 10 g of its root powder with sugar and warm water was useful as a mild purgative and about 1.2 g powder mixed with pepper, asafoetida, *triphala* and salts helped overcome constipation caused by scanty intestinal secretions.

• Antiviral activity

Iridoids isolated and purified from *P. kurroa* exhibit many bioactivities. One of its important compounds that forms major constituent of commercial formulation known by the name *Picroliv* showed antiviral activity against hepatitis B virus. This active compound was reported to have anti-HBsAg activity that inhibited purified HBV antigens prepared from healthy HBsAg carriers (Bhattacharjee *et al.,* 2013).

• Antioxidant activity

The antioxidant and stabilizing actions of *P. kurroa* has been attributed its hepato-protective activity. *Picroliv* (drug prepared from the *P. kurroa extract*)was affective as oxygen-free radical scavenger that restricts lipid peroxidation involved in membrane injury induced by hepatotoxins such as aflatoxin B1. *Picroliv* is known to enhance action of hepatic cytochrome P-450 enzyme system and protects the liver from hepatotoxicity induced through CCl_4. It is also reported to help normalize elevated lipid peroxide levels. It also hinders the depletion of reduced glutathione, which is required for the GST detoxification and increasing the amount of lipid peroxides in the liver.

Chander *et al.* (1992) found that iridoid glycosides of *P. kurroa* restricted production of O^{2-} anions in a phenazin methosulphate NADH system, thus restraining oxidative malonaldehyde production by both ascorbate-Fe^{2+} and NADPH-ADP-Fe^{2+} systems, and scavenged superoxide anions originated in a xanthine-xanthine oxidase system. The antioxidant activity *P. kurroa* was found to be similar to that of metal-ion chelators, superoxide dismutase and xanthine oxidase inhibitors (Kant *et al.,* 2013). *P. kurroa* was reported to restore depleted glutathione levels, as well as the enzymes associated with glutathione function, including glutathione-S-transferase, glutathione peroxidase and glutathione reductase in African desert rats infected with *Plasmodium berghei* (malaria) (Kidd, 1997). Glutathione maintains a variety of intracellular functions, such as detoxification, configuration of tertiary proteins; antioxidation and redox balance (Bhandari *et al.,* 2010).

• Anti-obesity activity

Diabetes is frequently associated with obesity. *P. kurroa* improves gallbladder secretions, thus aiding in the digestion and metabolism of fats. It is very effective in regulating fat metabolism in the liver (Verma and Paraidathathu, 2014). In a study on mice raised on a high-fat diet, daily doses of water extract of *P. kurroa* significantly reduced total cholesterols, triglycerides, and LDL levels in 12 weeks (Thakur *et al.,* 2010).

• Hepato-protective

The extracts from *P. kurroa* produce hepatoprotective activity against rifampicin, galactosamine, thiocetamide and cadmium-induced liver toxicity seen in experimental animal and cell cultures (Dwivedi *et al.,* 1991; Saraswat *et al.,* 1993; Yadav *et al.,* 2006; Kawoosa *et al.,* 2010). *Picroliv*, a formulation containing *P. kurroa,*was reported to show hepatoprotective activity at doses of 6 and 12 $mgkg^{-1}$ against CCl_4 induced variation in biochemical parameters, *viz.* aspartate transaminase (AST), alanine transaminase (ALT), bilirubin, protein, lipoprotein X and cholesterol triglycerides (Dwivedi *et al.,* 1990; Girish *et al.,* 2009).

P. kurroa showed anticholestatic effect against paracetamol and ethinylestradiol-induced cholestasis. It alienates the changes in bile volume along with bile salts and bile acids. *P. kurroa* extract was found to be more potent choleretic and anticholestatic agent than silymarin and flavonolignan (Shukla *et al.,* 1991). It lowers LDL receptors cell surface expression induced by paracetamol and increases the conjugated dienes in hepatocytes. Bile salt-dependent fractions are influenced by *P. kurroa* extract, which further increases the synthesis of bile salts and bile acids, and increase protein conjugation (Saraswat *et al.,* 1997). Singh et al. (1992) reported that liver regenerative activity in rats resulted from stimulating nucleic acid and protein synthesis. A combination of membrane stabilizing, antioxidant and hypolipidemic properties may contribute towards hepatoprotective effect. These properties are also attributed to be responsible for the effects on the immune system (Negi *et al.,* 2007).

Hepato-protective effects of *P. kurroa* manifested in chronic alcohol-intoxication by raising the levels of alcohol-metabolizing enzymes in hepatocytes, suggesting its restricting role on the cumulated acetaldehyde. It is seen that antioxidant activity of the extract might reduce lipid peroxidation and free radical damage. Its extract was found to have restricting effect on pro-inflammatory cells including macrophages, mast cells and neutrophils (Pandey and Das, 1989). It was reported that extract of *P. kurroa* restrained the membrane mediated activation of these cells (inhibited 8-adrenergic receptors) (Pandey and Das, 1988a; 1988b). However, the exact mechanism of hepatoprotection is yet to be fully understood.

• Beneficial role in respiratory /Cardio problems

Traditionally, picrorhiza rhizome was used for the cure of different allergic disorders *viz*., eczema, bronchial asthma, psoriasis etc. (Dorsch and Wagner, 1991). *In vivo* studies with *P. kurroa* showed that one of its chemical constituents, drosin, prevented allergies and acted as a potent phospholipid activator (platelet activating factor) that induce bronchial congestion in guinea pigs. Plant extract also inhibits the release of histamine under *in vitro* conditions (Muller *et al.,* 1997; Anonymous 2001). Orally administered picrorhiza extract, when given to rats and mice, showed a concentration-dependent lowering in mast cell degranulation and cardioprotective effect against adriamycin induced cardiomyopathy (Kapahi *et al.,* 1993; Baruah *et al.,* 1998). Membrane stabilizing action, free radical enzymes and lowered level of glutathione are responsible for its distinctive protective action, which inturn protects the myocardial membrane against peroxidative impairment by reducing lipid peroxidation (Kumar *et al.,* 2001; Rajaprabhu *et al.,* 2007).

Clinical Studies

• Role in management of specific ailments

P. kurroa is commonly used in conventional as well in modern medicine (Ansari *et al.,* 1988; Chaturvedi and Singh, 1996). It is also beneficial for curing gastrointestinal and renal disfunctions, inflammatory infection, and poisonous bites (Weinges *et al.,* 1972; Hussain, 1984; Kirtikar and Basu, 1984; Visen *et al.,* 1998; Halliwel and Gutteridge, 1999; Rajaprabhu *et al.,* 2007; Verma *et al.,* 2009). It possess hepatoprotective (Dhawan, 1995; Vaidya *et al.,* 1996), anti-inflammatory (Del Carmen Recio *et al.,* 1994), immunomodulatory (Puri *et al.,* 1992), free radical scavenging (Chander at al., 1992), gastric ulcer (Ray *et al.,* 2002; Banerjee *et al.,* 2008), anti allergic and anti-anaphylactic activities (Baruah *et al.,* 1998). Anti-hepatitis B surface antigen activity was also found in *P. kurroa* (Mehrotra *et al.,* 1990). Picrorhiza is very effective for relieving constipation, flatulence, colic, anorexia, inflammation, burning sensation, skin diseases, fever, diabetes, spleen and liver problems including anemia, jaundice, hemorrhoids and general debility etc. (Kumar *et al.,* 2010; Thapliyal *et al.,* 2012; Kaigongi *et al.,* 2014; Sharma *et al.*, 2014; Keshari *et al.,* 2015).

• Toxicity related information

P. kurroa extract were shown non-toxic in experiment conducted for histo-pathological parameters in rats. Similar studies conducted on adult rhesus monkeys reported no irregularity in food intake, body weight, daily activities, and haematology. Its use is also found safe for human beings though bitterness is intolerable to some (Girish *et al.,* 2008).

• Formulations

A decoction of *P. kurroa* is generally not preferred as its active compounds are poorly soluble in water (even though these are soluble in ethanol). The crude drug is extremely bitter and unpalatable (Anonymous, 2001). It may, however, be taken as an encapsulated standard extract (4% kutkin). Dose of 400–1500 mg day^{-1} is usually given to adults, although 3.5 g day^{-1} has been proposed for fever. In rats, 2.5 g kg^{-1} PO (orally) showed no mortality. Comparison of different studies showed that the maximum PO dose of *P. kurroa* root is about 3–6 mg kg^{-1} body weight (Luper, 1998; Negi *et al.,* 2007).

F. THREAT AND CONSERVATION ASPECT

Population status

• Threats

Since time immemorial, the Himalayas have been a treasure place for important medicinal plants, and for local medicinal practitioners, *Vaidyas* and plant explorers (Samant *et al.,* 2001). This curative significance of several medicinal plant species realized by natives and others has resulted in high extraction pressure from the wild. For sustainable utilization, it is important that a critical mass of species population and its habitat remains intact. *P. kurroa* regenerates both vegetatively (through rhizomes) and through seeds. Seed setting and survival of seedling under alpine conditions is generally poor (Pandey *et al.,* 2000), leading to poor regeneration response.The indiscriminate collection from the wild and lack of organized cultivation has therefore adversely hit the natural status of the species (Nayar and Sastry, 1990; Rai *et al.,* 2000; Samant *et al.,* 2001). The species was listed as endangered by the International Union for Conservation of Nature (IUCN) and was also included in Appendix II of the Convention on International Trade in Endangered Species of Wild Fauna and Flora (CITES) in 1997 (Purohit, 1997). Consequently, *P. kurroa* export in any form is not permitted now by the Export Authority of India, and the species figures in the negative list of Indian medicinal plants [by Ministry of Commerce, GOI, Vide Notification No. 03 (RE-2003)/2002–2007 (Appendix II) dated 31st March 2003 issued by Director General of Foreign Trade, Govt. of India].

• Habitat loss/degradation/fragmentation

Various environmental upheavals like landslides, forest fire, and human interference *viz*., Jhum cultivation, human settlement, and various other development activities has led to habitat loss of many such important plant species. Grazing and trampling are the other common factors that affect most herbaceous species (Pandey *et al*., 2000; Nautiyal *et al*., 2001; Singh and Chowdhery, 2002; Baig *et al*., 2013).Due to these factors many species have entered the threatened categories and *P. kurroa* is one of them (Ved *et al*., 2003; CUTS, 2004; Gajurel *et al*., 2015).

• Over-exploitation

Over the past century, because of the increasing trade demand by different pharmaceutical industries, *P. kurroa* has been regularly harvested from its natural habitat, at times well before seed setting and often entire plants are uprooted by unskilled labourers (Rai *et al.,* 2000). The untimely extraction eliminates possibility of proper seed development and seed maturation thus reducing future generation of the species (Sheldon *et al.,* 1997). Also, no proper cultivation of the plant is practiced.

Therefore, overexploitation and illegal collection has shrunken the population of *P. kurroa*, and has become a great challenge to restore its status in the wild (Jacobson *et al.,* 1991; Sheldon *et al.,* 1998; Dhar *et al.,* 2000). Other factors responsible for its declining population in the wild are the restricted distribution of the species, increasing anthropogenic disturbances and ill-harvesting of complete population patches (Semwal *et al.,* 2007). Generally, the medicinal plant collectors are unaware about proper collection and storage practices (Maikhuri *et al.,* 1998; Nautiyal *et al.,* 2001), which results in destructive harvesting and finally leads to poor natural regeneration. Consequently, this leaves very little scope for the natural regeneration of the species (Rao *et al.,* 2000).

• Climate change

Over time, global mean temperatures have risen by 0.6°C, and are forecasted to increase further by 1.5–4.5°C around 2100 AD (IPCC, 2007). The high altitude mountain sites are even more susceptible. Oerlemans (2005) reported that the Himalaya and surrounding areas have warmed by approximately 0.68°C since the middle of the 19th century in response to global environmental changes. Visible effects in the Himalayan region can be seen even now in the form of melting glaciers; rising sea levels and increasing temperatures.

Different researchers have reported that the influence of climate change on natural ecosystems is extending and have attributed it to changes in weather events, seasonal patterns, temperature ranges and the related phenomenon (Parmesan and Yohe, 2003; Root *et al.,* 2003; IPCC, 2007; Thomas *et al.,* 2004). Range shifts of individual species due to local extinction and dispersal/migration has resulted in change in community composition (Hansen *et al.,* 2001; Benning *et al.,* 2002). Studies have revealed that along with shifting phenology, plants have evolved to adapt to recent climate changes through altered species ranges (Parmesan, 2006). Climatic changes could alter the major chemical constituents and survival of different medicinal plants in higher altitudes such as *P. kurroa*, which are mainly responsible for their medicinal activity (Salick *et al.,* 2009).

Conservation strategies

Use of both conventional (vegetative propagation and seed germination) as well as *in vitro* methods are needed for rapid and mass propagation of *P. kurroa* which offer possibilities of recovering this endangered species, thus reducing the risk of its extinction (Nadeem *et al.,* 2000; Chandra *et al.,* 2006).

• *In situ conservation*

Poor relative density of *P. kurroa* in almost all the observed population of the species at different geographical locations calls for immediate action for conservation of the species. Owing to its narrow ecological range, it cannot be assumed that the species can perform well under *ex situ* conditions. Therefore, *in situ* conservation is a better option but the area needs to be well defined. Furthermore, to preserve the important medicinal plants, it is necessary to enhance awareness among local people and general masses to minimize over exploitation of species from its natural habitat (Arya *et al.,* 2013).

• *Ex situ* conservation

Efforts for *ex situ* conservation include maintaining propagules in plant tissue culture archives for long duration preservation (Rands *et al.,* 2010). Overall, conservation of endangered species needs adequate protection by managing wild populations in their natural habitats, augmented by *ex situ* propagation (Fay, 1992; Negash *et al.,* 2001 Sarasan *et al.,* 2006; Kumar *et al.,* 2008).

In vitro methods for conservation can offer some distinct advantage over alternative strategies. Some of these are: (1) independent of flowering period, collection can be made at anytime (this assumes that seed material is not required), (2) for virus infected populations, elimination through meristem culture, (3) clonal material for the maintenance of elite genotypes, (4) rapid multiplication may occur at any time where stocks are required through micropropagation procedures, and (5) for breeding programmes, germination of immature seed or embryo may be encouraged.

Other advantage of *in vitro* conservation is the reduced requirement of storage space compared with field storage. Storage facilities may be established at any geographical location, and are not subject to environmental disturbances such as temperature fluctuation, cyclones, insect, pests, and pathogen (Bhojwani and Dennis, 1999; Shibli *et al.,* 2006; Verma *et al.* 2012; Arya *et al.,* 2013).

G. CULTIVATION ASPECTS

• Climate

Cool, shaded and moist climate are most suitable for growth of *P. kurroa.* Prolonged and intense rains cause high mortality. Shaded areas (canopy of small shrubs) are excellent for optimal growth and productivity. It has been demonstrated that the crop can be raised across an altitude range of 1800 m to 2800 m, but at lower elevation the yield is generally poor.

• Soil

Sandy loamy soil is ideal for cultivation of *P. kurroa*. Organic carbon rich site with thick layer of humus or litter, and high moisture content supports good growth and high yield. Good water drainage is important for a healthy and productive crop because it cannot withstand water logging.

• Nursery raising and planting

Propagation techniques

P. kurroa can be propagated sexually or asexually through seeds or through rhizomes and stolon segments. Each plant of *P. kurroa* bears five to fourteen capsules and in each capsule there are 32-65 seeds. Seeds are viable for one month only under dry storage conditions, and for five to ten days under moist conditions in nature. The viability of seeds can be retained for a longer period (4-6 months), when stored at low temperature (4-6°C) and absence of moisture. Incubation of seeds at 20°C was found to improve the germination of Picrorhiza.

Cultivation through rhizomes is considered faster and better method of propagation than through seed. Underground propagules, left during harvesting can easily regenerate. For better performance, cultivation of *P. kurroa* is recommended in high altitude zones.

Propagation through seed

The crop raised through seed takes more time for establishment since seed germination takes time and plant growth is poor. Styrofoam seedling trays are used for sowing seeds on soil surface. For retaining moisture the tray is covered with thin layer of moss. This arrangement increases seed germination up to 58% at lower altitude in polyhouses. Germination was observed to be better in polyhouses at 15-20°C temperature in sandy soil supplemented with litter and high moisture content (Nautiyal *et al.,* 2001). Seeds collected from the wild around September generally perform well when sown immediately after harvest at places having comparatively high temperature (15–20°C). At lower altitude, seeds are sown during November-December in polyhouses and the raised seedlings are transplanted in nursery beds and tended for up to 6 months during winter. The plantlets can later be transplanted at higher altitude during spring. For raising the crop through seed at middle altitude (around 2000 m), the sowing is done in beds around March-April and in alpine areas, seed can be sown directly in May.

Three kilograms freshly harvested seed are required for raising a one hectare plantation. One acre of land can accommodate about 4000 plants (NIIR, 2006; Patial *et al.,* 2012). Plants so raised in the nursery are transplanted in the fields at spacing 15 x 15 cm, and can accommodate 2-3 lakh plants in one hectare.

Propagation through rhizomes or stolons

Vegetative propagation is a more appropriate and successful method for propagation through rhizomes and it is considered faster and better than cultivation through seed. For propagation through rhizomes or stolons, segments of 5-7 cm length are taken and are dipped in water or in a solution of plant growth regulators like IBA and then planted in the field. The time for planting these segments is April-May or September-October. The later dates give better results. Rooting and sprouting in stolen will appear within 10 to 12 days of planting. Planting can be done either directly in the field or in nursery beds. Plants from nursery beds can be transplanted later in the field. In areas where snow fall occurs, it is recommended to raise the nursery in poly house. Maximum sprouting and survival is given by 6 cm long stolon cuttings obtained from top portion of the plant. The maximum rootstock yield was obtained from 10 cm long stolon cuttings taken from mid portion of the plant. Maximum rootstock yield ha^{-1} is given by stolon cuttings planted horizontally at 7.5 cm depth with 30 x 30 cm spacing (Thakur *et al.,* 2010; Sharma, 2013).

In vitro propagation

In vitro propagation is carried out successfully by using MS medium with added agar and sucrose or Hoagland nutrients. Lal *et al.* (1988) reported that MS medium with kinetin (3.0 to 5.0 mg L^{-1}) was good for rapid multiplication of shoots from the explants and supplementing indole-3-acetic acid (1.0 mg L^{-1}) to the kinetin containing medium showed marked enhancement in the growth of regenerated shoots. Shoot multiplication rate moderately improves when the number of subculture increases (Upadhyay *et al.,* 1989). Chandra *et al.* (2006) observed that *in vitro* shoot multiplication from nodal segments was attained through sprouting of axillary buds on MS medium containing 1.0 μM BAP in three weeks. Three plant growth promoting rhizobacteria *viz*., *Bacillus subtilis, B. megaterium* and *Pseudomonas corrugate* were used for biological hardening of micro propagated plantlets.Rhizobacteria positively influenced survival and growth rate of *Picrorhiza* plants under greenhouse conditions (Trivedi and Pandey, 2006). Shoot buds showed plantlet development under both *in vitro* and *in vivo* conditions. *In vitro* shoot multiplication was attained through sprouting of axillary buds using nodal segments and leaf tissue (Sharma *et al*., 2010). A hormone combination containing kinetin (2.0 mg L^{-1}) and kinetin + Indole-3-butyric acid (IBA) (2.0 mg L^{-1} + 0.50 mg L^{-1}) for leaf ex-plant was found beneficial for shoot regeneration. Interestingly, a higher percentage of direct proliferation (99.94%) was achieved by using basal MS medium with nodal explants.

• Varieties/Accessions

The growing demand for the kutki drug has prompted researchers to search for new genotypes of *P. kurroa* which are richer in picroside content (Katoch *et al.,* 2011). While comparing 18 accessions of *P. kurroa* for growth and marker compound accumulation, Kumar *et al.* (2012) found that under western Himalayan conditions, Picroside-I (P-I) content varied from 0.20 to 4.14% and P-II varied from 0.83 to 7.29% in different accessions. Higher P-I in rhizome was found in IHBT- PK-16 (4.14%). Rhizomes were found to contain higher amount of P- II in comparison to P-I.

• Time of planting

Crop raised through seed takes more time than that raised through stolon propagation. The seed raised nursery will be ready in 3-4 months for transplantation. Rooted seedling can be transplanted in the month of February - March in well prepare field. In case of vegetative propagated crop, transplanting can be done in the month of September or October either directly or through nursery raised plants, which have developed 5 to 6 leaves. In areas experiencing snow fall, planting can be delayed till the snow melts.

• Planting density

Planting, for both type of propagation, is carried out at a distance of 30x15 cm. During rainy season, a planting distance of 5-10 cm apart is found better for plant survival since dense planting avoids soil erosion as well as weed competition.

• Intercropping

P. kurroa plants get better microclimatic conditions when intercropped with *Foeniculum vulgare*, which can enhance yield by moisture augmentation for extended duration under its canopy. Economically beneficial plants *viz.*, *F. vulgare*and *Solanum tuberosum* (potato) were found most suitable for intercropping (Nautiyal *et al.,* 2001). Further, intercropping with *Digitalis purpurea* has also been practiced and found to be beneficial.

• Nutrient management

The speciesresponds well to nutrient application. Soil enriched with high amount of litter is found preferable. Generally, 15 t manure $ha^{-1}/acre^{-1}$ is recommended at lower altitudes and 10 t ha^{-1} at mid-altitudes to attain best productivity for three years. Organically rich sites require only 18 q $acre^{-1}$ manure for three years (Nautiyal and Nautiyal, 2004). Application of manure is suggested during the winter months or before transplanting (in the rainy season) by digging 6 inch deep trenches (Patial *et al.,* 2012). It is recommended to add about 20 t ha^{-1} of well decomposed FYM before transplanting. Since it is a perennial crop, supplemental doses of FYM may also be added during winter months (Kumar *et al.,* 2012). Livestock manure, used in large quantities however, invites powdery mildew and needs excessive irrigation during the initial phase of manure application. Biofertilizers showed better response when applied in combination with FYM or vermicompost. Use of FYM or vermicompost, either in combination with Azotobacter +PSB (B)+VAM or in combination with Azotobacter +PSB (F)+VAM was best for the optimum root yield of *P. kurroa* (Thakur *et al.,* 2013). Root yield of 400 kg ha^{-1} and 947 kg ha^{-1} was obtained with former while 392 and 939 kg ha^{-1} was observed for latter, for two or three year old plants, respectively.

- **Water management**

P. kurroa requires adequate moisture for growth and needs regular irrigation to prevent mortality. Seedlings, as well as stolon cuttings, during the early developmental stages in beds, need irrigation after every 24 h at low altitude (say 1800 m). Once crop is established, it should be irrigated once every fortnight or as and when required. During winter months, watering should be carried out at interval of two days (Punetha *et al.,* 2006).

- **Weeding**

P. kurroa is very sensitive to weed competition and requires weed-free condition during early stage of crop growth. It is recommended that weeding should be done at monthly interval during the initial stages of crop growth while during the second and third year weeding should be done as and when required.

- **Pest and diseases**

P. kurroa is commonly found infected with powdery mildew disease at early stages of growth. Disease incidence is higher under high humidity and moderate temperature conditions (Sinha, 2005; Sharma *et al.,* 2014). Powdery mildew starts appearing in the months of May-June and infection index increases with increase in wetness. At the initiation of symptoms, 2-3 sprays of fungicide are applied in order to control powdery mildew disease. Semilooper of *Thysanoplusia orichalcea* (moth) and *Bemisia tabaci* (white fly) are common pests, which infect Picrorhiza plant (Sharma, 2004; Ramdas and Pokhriyal, 2012).

- **Harvesting**

Economic yield of crop can be achieved after 3-4 years when sown by seed, while crop planted through nodal cutting or rhizome gives yield after 2nd year. The crop should only be harvested after completion of reproductive phase because active contents are highest at this stage. At different elevations, the duration for completion of reproductive phase differ, as stated earlier. For harvesting, the plants are uprooted and rhizomes are separated and washed (Punetha *et al.,* 2006).

- **Commercial viability**

At lower altitudes (1800 m), after three years, maximum yield of 450 kg ha^{-1} is obtained. At mid altitude (2200 m), production of 600-700 kg ha^{-1} rhizome yield can be obtained by using high doses of forest litter. The broad leaved varieties grow rapidly and yield more than narrow leaved ones. Further, content of active ingredients is also higher in the broad leaved varieties. Cost benefit analysis of Picrorhiza cultivation at 1800-2200 m is found to be favourable (Nautiyal *et al.,* 2001).

- **Post-harvest management**

Harvested stolons and roots are washed to remove adhering soil and then dried. Drying at room temperature (15-20°C) retains high picrotin and picrotoxin contents of upto three and two percent, respectively. Drying in direct sun or in an oven rapidly decreases the active constituents. Dried stolons and roots are stored in air tight containers or cloth bags for marketing.

H. COMMERCIAL ASPECTS

- **Demand**

In terms of economic value of traded materials, *P. kurroa* is one of the top 15 plant species traded in India (Malaisamy and Ravindran, 2003; Ved and Goraya, 2008). In recent years, 20% annual growth rate of herbal medicines has increased the demand of *P. kurroa* by 11.1% (Subrat, 2002). More than 10,000 kg annual transaction of *P. kurroa* takes place in Delhi market alone (Uniyal *et al.,* 2011). Annual demand of the species is more than 500 t. However, global supply is about 375 t including 70 t met from India (Shitiz *et al.,* 2013).

Market trends

Market price of *P. kurroa* rhizomes during 2006-07 was Rs. 230 to 265 per kg (AYUSH 2008). It was Rs 220 to 340 per kg during 2007-10 (Uniyal and Uniyal, 2008; Shitiz *et al.,* 2013). Several national and international firms, such as Life Extension Foundation, Gero-Vita International and Nature's Sunshine, use picrorhiza in as a constituent to cure various disorders in humans. Demand for picrorhiza has shown a consistent upwards trend, i.e., 220 t in 2001–02, 317 t during 2004–05 and 500 t during 2006-07 (Singh *et al.,* 2011; AYUSH, 2008).

• Trade data

Global trade estimates suggest that the price of *P. kurroa* is more than most other medicinal herbs grown or collected at lower altitudes (Olsen and Helles, 1997; Olsen, 1998; Larsen, 2002). The annual supply of kutki from Nepal, India and Bhutan has been estimated as 375 MT (Olsen, 2001). The legal annual collection of kutki from Nepal has been reported as 25 MT (Malla *et al.,* 1995) and from India as 50– 300 MT (Olsen, 2005; Sarin, 2008). The deployment of kutki in different Indian sectors has been estimated as 415 MT year^{-1} (Rai *et al.,* 2000; Virdi, 2004; Uniyal *et al.,* 2006; Ved and Goraya, 2008). The declared quantum of material collected from the wild, however, gives only an indication of demand, because illegal collection and secretive trade is rampant. The ever-increasing market demand, together with overexploitation of the species, has raised concern about its declining status in the wild (Dobriyal *et al.,* 1997; Samant *et al.,* 2001; Rai and Sharma, 2002; Olsen and Larsen, 2003). Consequently, *P. kurroa* is included in the Negative List of Export (GOI), Appendix II of CITIES, Red Data Book of Indian Plants and Prioritized List of Medicinal Plants for Conservation (Nayar and Sastry, 1990; Kala, 2004; Kala *et al.,* 2006) and EXIM (Export and Import Policy) 1998 and Foreign Trade (Development and Regulation) Act, 1992 banned the export in raw form , excluding manufactured parts or derivatives (Sharma *et al*., 2008; Raina *et al.,* 2010; Uniyal *et al.,* 2011).

• Major users

The following are the major industries in India using *P. kurroa* in different herbal preparations (Bhandari *et al.,* 2008, Biren and Seth 2010).

- Katuki, Zandu Pharma Works Ltd., Mumbai, India
- Arogya, Zandu Pharma Works Ltd., Vapi, GJ, India
- Kutaki, Tansukh Herbal Pvt. Ltd., Lucknow, UP, India
- Livocare, Dindayal Aushdi Pvt. Ltd., Gwalior, MP, India
- Livomap, Maharishi Ayurveda Pvt. Ltd., New Delhi, India
- Livomyn, Charka Pharma Pvt. Ltd., Mumbai, India
- Livplus, BACFO Pharmaceuticals Ltd., Noida, UP, India
- Pravekliv, Pravek Kalp Herbal products Pvt. Ltd., Noida, UP, India
- Vimliv, OM Pharma Ltd., Bangalore, India
- Ansar drug laboratories, Surat, India
- Acis laboratories, Kanpur, India

- Amil pharmaceutical, New Delhi, India
- Allen laboratories, Kolkata, India
- Bharti rasanagar, Kolkata, India
- Dabur india Ltd., Ghaziabad, India
- Dattataraya krishan sandhu Bros., Mumbai, India
- ALRASIN Marketing, Mumbai, India
- Herbals Pvt. Ltd., Patna, India
- Herbo-med (P) Ltd., Kolkatta, India
- The Himalaya Drug Co., Banglore , India
- Indian Herb & Research Supply co., Shharanpur, India
- J & J Dechane Laboratories pvt. Ltd., Hyderabad, India
- Madona Pharmaceutical Reaearch Pvt. Ltd., Kolkatta, India
- Kruzer Herbals, New Delhi, India
- Shilpachem, Indore, India
- Hamdard (Wakf) Laboratories, Delhi, India
- Zandu Pharmaceutical Works Ltd., Mumbai, India
- Baidyanath Ayurveda Bhavan, Jhansi, India
- Charak Pharmaceuticals, Mumbai, India

I. FUTURE DIRECTIONS FOR RESEARCH / VALUE ADDITION

In order to promote cultivation of *P. kurroa*, and thereby production of the crop, identifying new areas could be of paramount importance. While efforts have been made to cultivate the species at elevations lower than its natural distribution, the plants thus produced have not been well evaluated for secondary metabolites *vis-à-vis* growth locations differing in elevation or other environmental factors. What is required further is to develop at least some idea of production of biomass and harvest of bioactive ingredients, keeping in view the growing significance of biological extracts in pharmaceutical industry. Further, there is a need to maintain focus on a kind of balance between *in situ* and *ex situ* conservation, particularly on the participation of local communities in the later case. While shift from collection from the wild to cultivation has become imperative, a healthy model of crop production is required to be developed, keeping in view the practical realities of available space and production of crop in term of bio-molecules.

J. PATENTS

- Title - A simplified process for purification of kutkin from *Picrorhiza kurroa* Royle ex Benth

  Inventor - Bikram Singh

  Grant Date - 2011-04-25

  Patent No. - 247574

- Title -*Picrorhiza kurroa* extract for prevention, elimination and treatment of DNA based viruses *in humans and in* biotech industry

  Inventor - Satyasayee Babu Divi, Munisekhar. Medasani

  Grant Date - Aug 8, 2012

  Patent No. - EP2482832 A2

- Title -Process for the preparation and composition of a fraction containing picroside I and kutkoside

  Inventor - Bacchan S. Aswal, Ramesh Chander, Sunil K. Chatterji, Bhola N. Dhawan, Yogesh Dwivedi, Narendra K. Garg, Poonam Jain,Narinder K. Kapoor, Dinesh K. Kulshreshtha,Bishan N. Mehrotra, Gyanendra K. Patnaik,Ravi Rastogi, Jagat P. S. Sarin, Krishna C. Saxena, Shekhar C. Sharma, Shri K. Sharma, Binduja Shukla, Pradeep K. S. Visen

  Grant Date - Sep 8, 1992

  Patent No. - US5145955 A

- Title -*Picrorhiza kurroa* extract for prevention, elimination and treatment of dna based viruses in humans and in biotech industry

  Inventor - Munisekhar Medasani, Satyasayee Babu Divi

  Grant Date - Mar 10, 2011

  Patent No. - WO2011027364 A2'1d in biotech industry

- Title -A process for preparation of synergistic herbal preparation of extracts *Picrorhiza kurroa* and *Glycyrrhiza glabra* having medicinal value

  Inventor - Ramana Rao

  Grant Date - Mar 28, 2013

  Patent No. - WO2013042131 A1

- Title -Iridoid glycoside composition

  Inventor - Khajuria Anamika, Gupta Amit, Singh Surjeet,Malik Fayaz, Singh Jaswant, Lal Bedi Kasturi, Suri Krishan Avtar, Satti Naresh Kumar, Prakash Suri Om, Qazi Ghulam Nabi,Srinivas Vellimedu Kannappa, Ella Krishna

  Grant Date - Sep 26, 2007

  Patent No. - EP1837027 A1

- Title -Extracting glycosides from *Picrorrhiza kurroa*-useful - for treating liver disorders

  Inventor - Peter Dr Kloss, Willmar Dr Med Schwabe

  Grant Date - Aug 2, 1973

  Patent No. - DE2203884 A1

- Title -Preparation containing a mixture of picrorrhiza and ginkgo for the relief of disease Inventor - Nicholas John Larkins

  Grant Date - Nov 6, 2003

  Patent No. - WO2003090769 A1

LITERATURES CITED

1. Alghasham A.A. 2013. Cucurbitacins—a promising target for cancer therapy. Int. *J. Health Sci.*, 7: 77-89.
2. Anonymous, 2001. *Picrorhiza kurroa*. *Altern. Med. Rev.*, 6: 319-321.
3. Ansari R.A. A swal B.S., Chander R., Dhawan B.N., Garg N.K., Kapoor N., Kulshrestha D.K., Mehdi H.S., Mehrotra B.N., Patnaik G.K. 1988. Hepatoprotective activity of kutkin – the iridoid glycoside mixture of *Picrorhiza kurroa*. *Indian J. Med. Res.*, 87: 401-404.
4. Aribi A., Gery S., Lee D.H., Thoennissen N.H., Thoennissen G.B., Alvarez R., Ho Q., Lee K., Doan N.B., Chan K.T., Toh M., Said J.W., Koeffler H.P. 2013. The triterpenoid cucurbitacin B augments the antiproliferative activity of chemotherapy in human breast cancer. *Int. J. Cancer,* 132: 2730-2737.
5. Arya D., Bhatt D., Kumar R., Tewari L.M., Kishor K., Josh G.C. 2013. Studies on natural resources, trade and conservation of Kutki (*Picrorhiza kurroa* Royle ex Benth., Scrophulariaceae) from Kumaun Himalaya. *Sci. Res. Essays*, 8: 575-580.
6. Aswal B.S., Mehrotra B.N. 1994. Flora of Lahaul-Spiti: a cold desert in North West Himalaya. Bishen Singh Mahendra Pal Singh, Dehra Dun. 22: 484-485.
7. AYUSH. 2008. Agrotechnique of selected medicinal plants. National medicinal plant board. Department of AYUSH. Vol. 1 pp. 139-144.
8. Baig B.A., Ramamoorthy D., Bhat T.A. 2013. Threatened medicinal plants of Menwarsar Pahalgam, Kashmir Himalayas: Distribution pattern and current conservation status. *Proc.Int. Acad. Ecol. Environ. Sci.*, 3: 25-35.

9. Banerjee D., Maity B., Nag S.K., Bandyopadhyay S.K., Chattopadhyay S. 2008. Healing potential of *Picrorhiza kurroa* (Scrofulariaceae) rhizomes against indomethacin-induced gastric ulceration: a mechanistic exploration. *BMC Complem.Altern. Med.*, 88: 3-17.
10. Banwarilal M. 2004. Dravyaguna Hastâmlaka. 4th revised edition, Publication scheme, Jaipur. 207: 383-384.
11. Baruah C.C., Gupta P.P., Nath A., Patnaik L.G., Dhawan B.N. 1998. Antiallergic and anti-anaphylactic activity of picroliv—a standardised iridoid glycoside fraction of *Picrorhiza kurroa*. *Pharmacol. Res.*, 38: 487-492.
12. Benning T.L., LaPointe D., Atkinson C.T., Vitousek P.M. 2002. Interactions of climate change with biological invasions and land use in the Hawaiian Islands: modeling the fate of endemic birds using a geographic information system. *Proc. Natl. Acad. Sci.,* 99: 14246-14249.
13. Bhandari P., Kumar N., Singh B., Ahuja P.S. 2010. Online HPLC-DPPH method for antioxidant activity of *Picrorhiza kurroa* Royle ex Benth and characterization of kutkoside by Ultra - Performance LC-electrospray ionization quadrupole time-of-flight mass spectrometery. *Indian J. Exp. Biol.,* 48: 323-328.
14. Bhandari P., Kumar N., Singh B., Kaul V.K. 2008. Simultaneous determination of sugars and picrosides in *Picrorhiza* species using ultrasonic extraction and high-performance liquid chromatography with evaporative light scattering detection. *J. Chromatogr. A.,* 1194: 257-261.
15. Bhattacharjee S., Bhattacharya S., Jana S., Baghel D.S. 2013. A review on medicinally important species of picrorhiza. *Int. J. Pharmaceut. Res. Bio. Sci.,* 2: 1-16.
16. Bhojwani S.S., Dennis T.T. 1999. *In vitro* conservation of plant genetic resources. Botanica., 49: 47-52.
17. Biren S., Seth A.K. 2010. Test book of pharmacognosy and phytochemistry. Drug Adulteration. pp. 107-109.
18. Bone K. 1995. Picrorhiza [sic]: Important modulator of immune function. Townsend letter for doctors. pp. 88-94.
19. Bruynzeel A.M., Abou-El-Hassan M.A., Schalkwijk C., Berkhof J., Bast A., Niessen H.W., Van der Vijgh W.J. 2007. Anti-inflammatory agents and monoHER protect against DOX-induced cardiotoxicity and accumulation of CML in mice. *Br. J. Cancer.*, 96: 937-943.
20. Chander R., Kapoor N.K., Dhawan B.N. 1992. *Picroliv*, picroside-I and kutkoside from *Picrorhiza kurroa* are scavengers of superoxide anions. *Biochem. Pharmacol.*, 44: 180-183.
21. Chandra B., Palni L.M.S., Nandi S.K. 2006. Propagation and conservation of *Picrorhiza kurroa* Royle ex Berth: an endangered Himalayan medicinal herb oh high commercial value. *Biodiver. Conserv.*, 15: 2325-2338.
22. Chandra K., Pandey H.C. 1983. Collection of plants around Agora-Dodital in Uttarkashi district of Uttar Pradesh, with medicinal values and folklore claims. *Int.J. Crude Drug Res.*, 21: 21-28.
23. Chaturvedi G.N., Singh R.H. 1996. Jaundice of infectious hepatitis and its treatment with an indigenous drug *Picrorhiza kurroa. J. Res. Ind. Med.*, 1: 1-14.
24. Chauhan N.S. 1988. Endangered Ayurvedic pharmacopoeial plant resources of Himachal Pradesh. In: Indigenous medicinal plants – including microbes and fungi. Today and Tomorrow's Printers and Publishers, New Delhi. pp.199-205.
25. CUTS. 2004. Data Base on Medicinal Plants, Centre for International Trade, Economics & Environment, http://www.cuts-international.org/pdf/Database-fullreport.pdf.
26. Del Carmen R.M., Maria G.R., Manez S., Rios J.L. 1994. Leaf extracts of some Cordia species. *Planta Med.*, 60: 232-234.

27. Dhar U., Rawal R.S., Upreti J. 2000. Setting priorities for conservation of medicinal plants: a case study in the Indian Himalaya. *Biol. Conserv.*, 95: 57-65.
28. Dhawan B.N. 1995. A new hepatoprotective agent from an Indian medicinal plant, *Picrorhiza kurroa*. *Med. Chem. Res.*, 5: 595-605.
29. Dobriyal R.M., Singh G.S., Rao K.S., Saxena K.G. 1997. Medicinal plant resources in Chhakinal watershed in the Northwestern Himalaya. *J. Herbs Spices Med. Plants*, 5: 15-27.
30. Dorsch W., Wagner H. 1991. New antiasthmatic drugs from traditional medicine? *Int. Arch. Allergy. Immunol.*, 94: 262-265.
31. Duangmano S., Sae-Lim P., Suksamrarn A., Domann F.E., Patmasiriwat P. 2012. Cucurbitacin B inhibits human breast cancer cell proliferation through disruption of microtubule polymerization and nucleophosmin/B23 translocation. *BMC Complem. Altern. Med.,* 12: 185-197.
32. Dwivedi Y., Rastogi R., Mehrotra R., Garg N.K., Dhawan B.N. 1993. *Picroliv* protects against aflatoxin B1 acute hepatotoxicity in rats. *Pharmacol. Res.*, 27: 189-199.
33. Dwivedi Y., Rastogi R., Chander R., Sharma S.K., Kapoor N.K., Garg N.K., Dhawan B.N. 1990. Hepatoprotective activity of picroliv against carbon tetrachloride-induced liver damage in rats. *Indian J. Med. Res.*, 92: 195-200.
34. Dwivedi Y., Rastogi R., Garg N.K., Dhawan B.N. 1992. *Picroliv* and its components kutkoside and picroside I protect liver against galactosamine-induced damage in rats. *Pharmacol. Toxicol.*, 71: 383-387.
35. Dwivedi Y., Rastogi R., Sharma S.K., Garg N.K., Dhawan B.N. 1991. *Picroliv* affords protection against thioacetamide-induced hepatic damage in rats. *Planta Med.*, 57: 25-28.
36. Fay M.F. 1992. Conservation of rare and endangered plants using *in vitro* methods. *In Vitro Cell Dev. Biol. Plants*, 28: 1-4.
37. Federov 1969. Chromosome number of flowering plants. Ottokaeltz Science Publishers West Germany. pp. 670-684.
38. Floersheim G.L., Bieri A., Koenig R., Pletscher A. 1990. Protection against Amanita phalloides by the iridoid glycoside mixture of *Picrorhiza kurroa* (kutkin). *Agents Actions*, 29: 386-387.
39. Gahlaut A., Chhillar A.K. 2013. Evaluation of antibacterial potential of plant extracts using resazurin based microtiter dilution assay. *Int. J. Pharma. Pharmcuet. Sci.,* 5: 372-376.
40. Ghansar N.H., Bhopale S.U., Prabhune V.S. 2012. *In vitro* studies of antimicrobial properties of extracts from unani medicinal plants. *Int.J. Pharma Bio Sci.*, 3: 240-249.
41. Ghisalberti E.L. 1998. Biological and pharmacological activity of naturally occurring iridoids and secoiridoids. *Phytomed.*, 5: 147-163.
42. Gaidhani S.N., Singh A., Kumari S., Lavekar G.S., Juvekar A.S., Sen S., Padhi M.M. 2013. Evaluation of some plant extracts for standardization and anticancer activity. *Indian J. Tradit. Know.*, 12: 682-687.
43. Gajurel P.R., Ronald K., Buragohain R., Rethy P., Singh B., Potsangbam S. 2015. On the present status of distribution and threat of high value medicinal plants in the higher altitude forest of Indian eastern Himalayas. *J. Threat. Taxa*, 7: 7243-7252.
44. Ganeshan S., Bhatt R.Y. 2008. Qualitative nature of some traditional crude drugs available in commercial markets of Mumbai, Maharashtra, India. *Ethnobot. Leaflets,* 12: 348-360.
45. Girish C., Koner B.C., Jayanthi S., Ramachandra Rao K., Rajesh B., Pradhan S.C. 2009. Hepatoprotective activity of picroliv, curcumin and ellagic acid compared to silymarin on paracetamol induced liver toxicity in mice. *Fundam. Clin. Pharma.*, 23: 735-745.
46. Girish C., Pradhan S.C. 2008. Drug development for liver diseases; focus on picroliv, ellagic acid and curcumin. *Fundam. Clin. Pharmacol.*, 22: 623-632.

47. Gogte V.M., 2000. Ayurvedic pharmacology and therapeutic uses of medicinal plants. 1st English edition. Bharatiya vidya bhavan Mumbai. pp. 325-327.
48. Gupta R.K. 1957. Botanical explorations in the Bhillangna Valley of the erstwhile Tehri Garhwal State – II. *J. Bombay Nat. Hist. Soc.,* 54: 878-886.
49. Gupta R.K. 1989. The living Himalayas. Today and Tomorrow's Printers and Publishers New Delhi. Vol. 2, pp. 296.
50. Hagerup O. 1951. Pollination in the Faeroes – in spite of rain and poverty of insects. *Danske Biologiske Meddelelser*, 18: 1-48.
51. Hagerup O. 1932. On pollination in the extremely hot air at Timbuctu. *Dansk Bot. Ark.*, 8: 1-20.
52. Halliwel B., Gutteridge M.C. 1999. Free radicals in biology and medicine. In study of generalized light emission. 3rd edition. University Press Oxford. pp. 387-388.
53. Handa S.S., 2000. Pharmacognosy. 2nd edition. 11th reprint. Vallabh Prakashan Delhi. pp. 110-111.
54. Hansen A.J., Neilson R.R., Dale V.H., Flather C.H., Iverson L.R., Currie D.J., Shafer, S., Cook, R., Bartlein, P.J. 2001. Global change in forests: Responses of species, communities, and biomes. *Biosci.*, 51: 765-779.
55. Hong D.Y. 1984. Taxonomy and evolution of the Veroniceae (Scrophulariaceae) with special reference to palynology. *Opera Bot.*, 75: 5-60.
56. Hari Venkatesh K.R., Chethana G.S. 2013. Screening of *Plumbago zeylanica* and *Picrorhiza kurroa* for their antibacterial activity against multi-drug resistant organisms (MDROs). *J. Biotech. Biosafety*, 1: 34-37.
57. Hussain A., 1984. Conservation of genetic resources of medicinal plants in India. In: Jain S.K. (editor). Conservation of Tropical Plant Resources. Botanical Survey of India Howrah. pp. 110-117.
58. IPCC. 2007. Intergovernmental Panel on Climate Change. Working Group I Report "The Physical Science Basis", Working Group II Report "Impacts, Adaptation and Vulnerability". Working Group III Report "Mitigation of Climate Change". http://www.ipcc.ch.
59. Irshad S., Mannan A., Mirza B. 2011. Antimalarial activity of three Pakistani medicinal plants. *Pak J. Pharm. Sci.*, 24: 589-591.
60. Jacobson G.L., Jacobson H.A., Winne J.C., 1991. Conservation of rare plant habitat: insights from the recent history of vegetation and fire at Crystal Fen, northern Maine, USA. *Biol. Conserv.,* 57: 287-314.
61. Jee V., Dhar U., Kachroo P. 1987. Addition to the cytological conspectus of alpine subalpine flora of Kashmir Himalaya. *Chromosome Information Service*, 43: 7-9.
62. Jia Q., Hong M.F., Minter D. 1999. Pikuroside: a novel iridoid from *Picrorhiza kurroa*. *J. Nat. Prod.*, 62: 901-903.
63. Joy K.L., Rajeshkumar N.V., Kuttan G., Kuttan R. 2000. Effect of *Picrorrhiza kurroa* extract on transplanted tumours and chemical carcinogenesis in mice. *J. Ethnopharmacol.*, 71: 261-266.
64. Kaigongi M.M., Dossaji S.F., Nguta J.M., Lukhoda C.W., Musila F.M. 2014. Antimicrobial activity, toxicity and phytochemical screening of four medicinal plants traditionally used in msambweni district, Kenya. *J. Biol. Agric. Healthcare*, 4: 6-12.
65. Kala C.P. 2004. Assessment of species rarity. *Curr. Sci.,* 86: 1058-1059.
66. Kala C.P., Dhyani P.P., Sajwan B.S. 2006. Developing of medicinal plant sector in northern India- challenges and opportunities. *J. Ethnobiol. Ethnomed.*, 2: 1-15.
67. Kant K., Walia M., Agnihotri V.K., Pathania V., Singh B. 2013. Evaluation of antioxidant activity of *Picrorhiza kurroa* (leaves) extracts. Indian *J. Pharm. Sci.*, 75: 324-329.

68. Kapahi B.K., Srivastava T.N., Sarin Y.K. 1993. Description of *Picrorhiza kurroa*, a source of the Ayurvedic drug Kutaki. *Int. J. Pharmacogn.*, 31: 217-222.
69. Kapil R.S. 1995. Studies on Ayurvedic plants. *J. Non-Timber Forest Prod.*, 2: 32-36.
70. Katoch M., Fazli I.S., Suri K.A., Ahuja A., Qazi G.N. 2011. Effect of altitude on picroside content in core collections of *Picrorhiza kurroa* from the north western Himalayas. *J. Nat. Med.*, 65: 578-582.
71. Kaul M.K., Kaul K. 1996. Studies on medicoethnobotany, diversity, domestication and utilization of *Picrorhiza kurroa*. In: Supplement to cultivation and utilization of medicinal plants (Eds. Handa, S. S. and Kaul, M. K.). RRL and CSIR Jammu. pp. 333-348.
72. Keshari D.S., Bishnupriya M., Panda P.K., Arun J., Sandeep B. 2015. A study of various market samples of picrorhiza/ hellebore (*Picrorhiza kurroa* Royle Ex Benth) with special reference to its pharmacognostic & phytochemical aspects. *Int. J. Ayur. Altern. Med.*, 3: 33-38.
73. Khanna A.K., Chander R., Kapoor N.K., Dhawan B.N. 1994. Hypolipidaemic activity of *Picroliv* in albino rats. *Phytother. Res.*, 8: 403-407.
74. Kawoosa T., Singh H., Kumar A. Sharma S.K., Devi K., Dutt S., Vats S.K., Sharma M., Ahuja P.S., Kumar S. 2010. Light and temperature regulated terpene biosynthesis: hepatoprotective monoterpene picroside accumulation in *Picrorhiza kurroa*. *Func. Integ. Genom.*, 10: 393-404.
75. Kidd P.M. 1997. Glutathione: systemic protectant against oxidative and free radical damage. *Altern. Med. Rev.*, 1: 155-176.
76. Kirtikar K.R., Basu B.D. 1984. Indian Medicinal Plants. Bishen Singh Mahendra Pal Singh, Dehra Dun. 3: 1824-1826.
77. Kong Y., Chen J., Zhou Z., Xia H., Qiu M.H., Chen C. 2014. Cucurbitacin E induces cell cycle G2/M phase arrest and apoptosis in triple negative breast cancer. *PLoS ONE*, 9: 1-8.
78. Kumar P.V., Sivaraj A., Madhumitha G., Saral A.M., Kumar B.S. 2010. In-vitro anti bacterial activities of *Picrorhiza kurroa* rhizome extract using agar well diffusion method. *Int.J. Curr. Pharmaceut. Res.*, 2: 30-33.
79. Kumar R., Bhandari P., Singh B., Ahuja P.S. 2012. Evaluation of *Picrorhiza kurroa* accession for growth and quality in north western Himalayas. *J. Med. Plants Res.*, 6: 2660-2665.
80. Kumar R, Singh B, Ahuja P.S. 2008. *Ex situ* conservation of *Picrorhiza kurroa*: an endangered medicinal plant species- A pilot study in Indian western Himalayas". In: National Conference on Increasing Production and Productivity of Medicinal and Aromatic Plants through Traditional Practices, GBPUAT, Pantnagar, India, September 18-20. pp. 59.
81. Kumar S.H.S., Anandan R., Devaki T., Kumar M.S. 2001. Cardio protective effects of *Picrorhiza kurroa* against isoproterenol-induced myocardial stress in rats. *Fitoterapia*, 72: 402-405.
82. Lal N., Ahuja P.S., Kukreja A.K., Pandey B. 1988. Clonal propagation of *Picrorhiza kurroa* royle ex benth. by shoot tip culture. *Plant Cell Report*, 7: 202-205.
83. Larsen H.O. 2002. Commercial medicinal plant extraction in the hills of Nepal: Local management systems and ecological sustainability. *Environ. Manage.*, 29: 88-101.
84. Li J., Wang J., Wang J., Nawaz Z., Liu J.M., Qin J., Wong J. 2000. Both corepressor proteins SMRT and N CoR exist in large protein complexes containing HDAC3. *Embo. J.*, 19: 4342-4350.
85. Li P., Yu X., Ge K., Melamed J., Roeder R.G., Wang Z. 2002. Heterogeneous expression and functions of androgen receptor co factors in primary prostate cancer. *Am. J. Pathol.*, 161: 1467-74.

86. Luper N.D.S. 1998. A review of plants used in the treatment of liver disease: part one. *Alt. Med. Rev.*, 3: 410-421.
87. Maikhuri R.K., Nautiyal S., Rao K.S., Saxena K.G. 1998. Medicinal plants cultivation and biosphere reserve management: a case study from Nanda Devi Biosphere Reserve. *Curr. Sci.*, 73: 777-782.
88. Malaisamy A., Ravindran C. 2003. Medicinal plants: Where do we stand globally? *Science Tech Entrepreneur Magazine* 2: 42-49.
89. Malla S.B., Shakya B.P.R., Rajbhandari K.R., Bhattarai N.K., Subedi N.N. 1995. Minor Forest Products of Nepal. General Status and Trade. Forest Resource Information System Project, Kathmandu, Nepal.
90. Mallick M.N., Singh M., Parveen R., Khan W., Ahmad S., Najim M.Z., Husain S.A. 2015. HPTLC analysis of bioactivity guided anticancer enriched fraction of hydroalcoholic extract of *Picrorhiza kurroa*. *Biomed Res. Int.*, pp. 1-18. http://dx.doi.org/10.1155/2015/513875
91. Mandloi D., Agarwal A., Goyal S., Wadhwa S., Patel R., Rawal H. 2010. Anti-fungal Potential of alcoholic Extract of *Picrorhiza kurroa*. *Pharmacol. online,* 3: 882-885.
92. Mehrotra R., Rawat S., Kulshreshltha D.K., Patnaik G.K., Dhawan B.N., 1990. In vitro studies on the effect of certain natural products against hepatitis B virus. *Indian J. Med. Res.*, 92: 133-138.
93. Mittal N., Gupta N., Saksena S., Goyal N., Roy U., Rastogi A.K. 1998. Protective effect of *Picroliv* from *Picrorhiza kurroa* against *Leishmania donovani* infections in Mesocricetus auratus. *Life Sci.*, 63: 1823-1834.
94. Muller A., Hake K., Endres A., Schwarthner C., Dunstan A.C., Dorsch W., Wagner H. 1997. Pharmacological investigations on antiasthmatic compounds of plant origin. *Phytomed.*, 3: 145.
95. Nadeem M., Palni L.M.S., Purohit A.N., Pandey H., Nandi S.K. 2000. Propagation and conservation of *Podophyllum hexandrum* Royle: An important medicinal herb. *Biol. Conserv.*, 92: 121129.
96. Nadkarni A.K. 1954. Indian *Materia Medica.* Popular Prakashan Bombay. Vol. 1, pp. 1331.
97. Nautiyal B.P., Prakash V., Chauhan R.S., Purohit H., Nautiyal M.C. 2001. Assessment of germinability, productivity and cost benefit analysis of *Picrorhiza kurroa* cultivated at lower altitudes. *Curr. Sci.*, 81: 579-585.
98. Nautiyal M.C., Nautiyal B.P. 2004. Agrotechniques for high altitude medicinal and aromatic plants. *Picrorhiza kurroa* Royle ex Benth. High altitude plant physiology research centre, Srinagar. pp. 121-133.
99. Nayar M.P., Shastry A.R.K. 1990. Red data book of Indian plants. Botanical survey of India, Kolkatta. Vol. 3.
100. Negash A., Krens F., Schaart J., Visser B. 2001. In vitro conservation of enset under slow-growth conditions. *Pl. Cell Tiss. Organ Cult.,* 66: 107-111.
101. Negi A.S., Kumar J.K., Luqman S., Shanker K., Gupta M.M., Khanuja S.P.S. 2007. Recent advances in plant hepatoprotectives: a chemical and biological profile of some important leads. *Med. Res. Rev.,* 28: 746-772.
102. NIIR. 2006. Kutki. Cultivation and Processing of Selected Medicinal Plants. NIIR Board of Consultants and Engineers. pp. 389-491.
103. Oerlemans J. 2005. Extracting a climate signal from 169 glacier records. *Sci.*, 308: 675-677.
104. Olsen C.S. 1998. The trade in medicinal and aromatic plants from Central Nepal to Northern India. *Econ. Bot.*, 52: 279-292.
105. Olsen C.S. 2001. Trade in Himalayan medicinal plant product Picrorhiza—New data. *Med. Plant Conserv.*, 7: 11-13.

106. Olsen C.S. 2005. Trade and conservation of Himalayan medicinal plants: *Nardostachys grandiflora* DC. and *Neopicrorhiza scrophulariiflora* (Pennell) Hong. *Biol. Conserv.*, 125: 505-514.
107. Olsen C.S., Helles F. 1997. Medicinal plants, markets and margins in Nepal Himalaya: Trouble in paradise. *Mount. Res. Develop.*, 17: 363-374.
108. Olsen C.S., Larsen H.O. 2003. Alpine medicinal plant trade and Himalayan mountain livelihood strategies. *Geographic J.*, 169: 243-254.
109. Pandey B.L., Das P.K. 1989. Immunopharmacological studies on *Picrorhiza kurroa* Royle–Ex Benth. Part IV: cellular mechanisms of anti inflammatory action. *Indian J. Physiol. Pharmacol.*, 33: 28-30.
110. Pandey B.L., Das P.K. 1988a. Immunopharmacological studies on *Picrorhiza kurroa* Royle-Ex-Benth. Part V: Anti-inflammatory action: relation with cell types involved in inflammation. *Indian J. Physiol. Pharmacol.*, 32: 289-292.
111. Pandey B.L., Das P.K. 1988b. Immunopharmacological studies on *Picrorhiza kurroa* Royle-ex-Benth. Part III: Adrenergic mechanisms of anti-inflammatory action. *Indian J. Physiol. Pharmacol.*, 32: 120-125.
112. Pandey N., Nautiyal B.P., Bhatt A.B. 2000. Studies on vegetation analysis, plant form and biological spectrum of an alpine zone of north-west Himalaya. *Trop. Ecol.*, 40: 163-166.
113. Pandit S., Shitiz K., Sood H., Chauhan R.S. 2012. Differential biosynthesis and accumulation of picrosides in an endangered medicinal herb *Picrorhiza kurroa*. *J. Plant Biochem. Biotechnol.*, 22: 335-342.
114. Parmesan C. 2006. Ecological and evolutionary responses to recent climate change. *Annu. Rev. Ecol. Evolution Syst.*, 37: 637-639.
115. Parmesan C., Yohe G. 2003. A globally coherent fingerprint of climate change impacts across natural systems. *Nature*, 421: 37-42.
116. Patel D.N., Qiao L., Yuk J., Yu K. 2015. Screening Synthetic Adulterants from Herbal and Dietary Supplements (HDS) Using the Natural Products Application Solution with UNIFI. Waters the science of what's Possible, Waters Corporation, USA. pp. 1-8.
117. Patial V., Devi K., Sharma M., Bhattacharya A., Ahuja P.S. 2012. Propagation of *Picrorhiza kurroa* Royle ex Benth: an important medicinal plant of western Himalaya. *J. Med. Plants Res.*, 6: 4848-4860.
118. Pennell F.W. 1943. The Scrophulariaceae of the Western Himalayas. Reprint 1997, Bishen Singh Mahendra Pal Singh, Dehra Dun. 6: 63-66.
119. Punetha H., Bist R., Gaur A.K., Bist L.D. 2006. Indian farmer digest. Kutki (*Picrorhiza kurroa)* An important endangered plant of the Himalayan region. Medicinal and Agrotechnology perspective(s). pp. 27-31.
120. Puri A., Saxena R.P., Sumati Guru P.Y., Kulshreshtha D.K., Saxena K.C., Dhawan B.N. 1992. Immunostimulant activity of *Picroliv*, the iridoid glycoside fraction of *Picrorhiza kurroa*, and its protective action against Leishmania donovani infection in hamsters. *Planta Med.*, 58: 528-532.
121. Purohit A.N., 1997. Medicinal Plants—need for upgrading technology for trading the tradition. In: Harvesting herbs-2000: medicinal and aromatic plants and action plan for Uttarakhand, Dehradun. pp. 49-75.
122. Rai L.K., Prasad P., Sharma E. 2000. Conservation threats to some important medicinal plants of the Sikkim Himalaya. *Biol. Conser.*, 93: 27-33.
123. Rai L.K., Sharma E. 2002. Diversity and indigenous uses of medicinal plants of Sikkim. In: Himalayan Medicinal Plants. Potential and Prospects. Almora, India. Gyanodaya Prakashan Nainital. pp. 157-163.

124. Raina R., Mehra T.S., Chand R., Sharma Y.P. 2010. Reproductive biology of *Picrorhiza kurroa* a critically endangered high value temperate medicinal plant. *J. Med. Arom. Plants*, 1: 40-43.
125. Rajani A., Hemamalini K., Satyavati D., Begum S.K.A., Saradhi N.D.V.R. 2013. "Anthelmintic activity of methanolic extract of rhizomes of *Picrorrhiza kurroa* royal ex. benth." *Int. J. Res. Dev. Pharm. Life Sci.*, 2: 705-707.
126. Rajaprabhu D., Rajesh R., Jeyakumar R., Buddha S., Ganeshan B., Anandan R. 2007. Protective effect of *Picrorhiza kurroa* an antioxidant defence status in andriamycin induced cardiomyopathy in rats. *J. Med. Plants Res.*, 1: 80-85.
127. Rajeshkumar N.V., Kuttan R. 2000. Inhibition of N nitrosodiethylamine- induced hepatocarcinogenesis by *Picroliv*. *J. Exp. Clin. Cancer Res.*, 19: 459-465.
128. Rajeshkumar N.V., Kuttan R. 2001. Protective effect of *Picroliv*, the active constituent of *Picrorhiza kurroa*, against chemical carcinogenesis in mice. *Teratog Carcinog. Mutagen*, 21: 303-313.
129. Rajkumar V., Guha G., Kumar R.A. 2011. Antioxidant and antineoplastic activities of *Picrorhiza kurroa* extracts. Food Chem. *Toxicol.*, 49: 363-369.
130. Ram V.J. 2001. Herbal preparations as a source of hepatoprotective agents. *Drug News Perspect.*, 14: 353-363.
131. Ramdas G.K., Pokhriyal P. 2012. Seeds germination and seedlings analysis of *Picrorhiza kurroa* royle ex benth. in Genwala and Bagori (Harsil) of district Uttarkashi (Uttarakhand). *ARPN J. Sci. Tech.,* 1: 11-17.
132. Rands M.R.W., Adams W.M., Bennun L. 2010. Biodiversity conservation: challenges beyond 2010. *Sci.*, 329: 1298-1303.
133. Rao K.S., Nautiyal S., Maikhuri R.K., Saxena K.G. 2000. Management Conflicts in the Nanda Devi Biosphere Reserve, India. *Mountain Res. Dev.*, 20: 320-323.
134. Rastogi R., Saksena S., Garg N.K., Dhawan B.N. 1996. Effect of picroliv on antioxidant system in liver of rats, after partial hepatectomy. *Phytotherapy Res.*, 9: 364-367.
135. Rathee D., Rathee P., Rathee S., Rathee D. 2012. Phytochemical screening and antimicrobial activity of *Picrorhiza kurroa* an Indian traditional plant used to treat chronic diarrhea. *Arabian J. Chem.*, 1-7.
138. Rau M.A. 1961. Flowering plants and ferns of North Garhwal, Uttar Pradesh, India. *Bull. Bot. Survey India*, 3: 215-251.
139. Rau M.A. 1975. High altitude flowering plants of West Himalaya. *Bot. Survey India*, 159: 31.
140. Ray A., Chaudhuri S.R., Majumdar B., Bandyopadhyay S.K. 2002. Antioxidant activity of ethanol extract of rhizome of *Picrorhiza kurroa* on indomethacin induced gastric ulcer during healing. *Indian J. Clin. Biochem.*, 17: 44-51.
141. Root T.L., Price J.T., Hall K.R., Schneider S.H., Rosenzweig C., Pounds A. 2003. Fingerprints of global warming on wild animals and plants. *Nature*, 421: 57-60.
142. Sagar P.K. 2014. Adulteration and substitution in endangered, ASU herbal medicinal plants of India, their legal status, scientific screening of active phytochemical constituents. *Int. J. Pharmaceut. Sci. Res.*, 5: 4023-4039.
143. Salick J., Fang Z., Byg A. 2009. Eastern Himalayan alpine plant ecology, Tibetan ethnobotany, and climate change. *Global Environ. Change*, 19: 147-155.
144. Samant S.S., Dhar U., Palni L.M.S. 2001. Himalayan Medicinal Plants: Potential and Prospects. Gyanodaya Prakashan Nainital, India.
145. Santra A., Das S., Maity A., Rao S.B., Mazumder D.N. 1998. Prevention of carbon tetrachloride-induced hepatic injury in mice by *Picrorhiza kurroa*. *Indian J. Gastroenterol.*, 17: 6-9.
146. Sarasan V., Cripps R., Ramsay M.M. 2006. Conservation in vitro of threatened plants-progress in the past decade. *In vitro Cellular Dev. Biol. Pl.*, 42: 206-214.

147. Saraswat B., Visen P.K., Patnaik G.K., Dhawan B.N. 1993. Anticholestatic effect of picroliv, active hepatoprotective principle of *Picrorhiza kurroa*, against carbon tetrachloride induced cholestasis. *Indian J. Exp. Biol.*, 31: 316-318.
148. Saraswat B., Visen P.K., Patnaik G.K., Dhawan B.N. 1997. Hepatoprotective effect of picroliv against rifampicin induced toxicity. *Drug Develop. Res.*, 40: 299-303.
149. Sarin Y.K., 2008. Principal crude herbal drugs of India. An illustrated guide to important largely used and traded medicinal raw materials of plant origin. Dehradun, India. pp. 222.
150. Semwal D.P., Saradhi P.P., Nautiyal B.P., Bhatt A.B. 2007. Current status, distribution and conservation of rare and endangered medicinal plants of Kedarnath Wildlife Sanctuary, Central Himalayas. *Curr. Sci.*, 92: 1733-1738.
151. Sharma A., Shanker C., Tyagi L.K., Singh M., Rao C.V. 2008. Herbal medicine for market potential in India: An overview. *Academic J. Plant Sci.*, 1: 26-36.
152. Sharma B.M., Kachroo P. 1981. Flora of Jammu and plants of neighbourhood.Bishen Singh Mahendra Pal Singh, Dehra Dun. 1: 243.
153. Sharma M.L., Rao C.S., Duda P.L. 1994. Immunostimulatory activity of Picrorhiza kurroa leaf extract. *J. Ethnopharmacol.*, 41: 185-192.
154. Sharma R. 2004. Agro-techniques of medicinal plants. Daya Publishing House.
155. Sharma R., Shivling V.D., Kumar D., Sharma A.K. 2014. A beta regression model for Himalayan medicinal plant diseases prediction. *Indian J. Sci. Technol.*, 7: 776-780.
156. Sharma S. 2013. Multiple Nursery Planting Bar: a tool for maintaining optimum spacing in the field. *Int. J. Farm Sci.*, 3: 77-80.
157. Sharma S., Katoch V., Rathour R., Sharma, T.R. 2010. *In Vitro* Propagation of Endangered Temperate Himalayan Medicinal Herb *Picrorhiza kurroa* Royle ex Benth using Leaf Explants and Nodal Segments *J. Plant Biochem. Biotechnol.*, 19: 111-114.
158. Sharma S.K., Kumar N. 2012. Antimicrobial Screening of *Picrorhiza kurroa* Royle Ex Benth Rhizome. *Int. J. Curr. Pharmaceut. Rev. Res.*, 3: 60-65.
159. Sheldon J.W., Balick M.J., Laird S.A. 1997. Medicinal plants: can utilization and conservation coexist? Advances in economic botany. *Econ. Bot.*, 12: 101-104.
160. Sheldon J.W., Balick M., Laird S.A. 1998. Is using medicinal plants compatible with conservation? *Plant Talk*, 13: 29-31.
161. Shibli R.A., Shatnawi M.A., Subaih W.S., Ajlouni M.M. 2006. In vitro conservation and cryopreservation of plant genetic resources: a review. *World J. Agri. Sci.*, 2: 372-382.
162. Shitiz K., Pandit S., Chauhan P.S., Sood H. 2013. Picrosides content in the rhizomes of *Picrorhiza kurroa* Royle ex Benth traded for herbal drugs in the markets of North India. *Int. J. Med. Arom. Plants*, 3: 226-233.
163. Shukla B., Visen P.K., Patnaik G.K., Dhawan B.N. 1991. Choleretic effect of picroliv, the hepatoprotective principle of *Picrorhiza kurroa. Planta Med.*, 57: 29-33.
164. Singh G.B., Sarang B., Singh S., Khajuria A., Sharma M.L., Gupta B.D., Banerjee S.K. 1993. Anti inflammatory activity of the Iridoids kutkin, picroside-I and Kutkoside from *Picrorrhiza kurroa. Phytotherapy Res.*, 7: 402-407.
165. Singh H., Gahlan P., Dutt S., Ahuja P.S., Kumar S. 2011. Why uproot *Picrorhiza kurroa* an endangered medicinal herb? *Curr. Sci.,* 100: 1055-1059.
166. Singh V., Banyal H.S. 2011. Antimalarial effects of *Picrorhiza kurroa* Royle Ex Benth Extracts on *Plasmodium berghei. Asian J. Exp. Biol. Sci.*, 2: 529-532.
167. Singh G., Kachroo P. 1976. Forest flora of Srinagar and plants of neighbourhood. Reprint 1994, Bishen Singh Mahendra Pal Singh, Dehra Dun. pp. 21.
168. Singh N.P., Chowdhery H.J. 2002. Biodiversity conservation in India. In: Perspectives of Plant Biodiversity, Dehradun, India. pp. 501-527.
169. Singh V., Kapoor N.K., Dhawan B.N. 1992. Effect of picroliv on protein and nucleic acid synthesis. *Indian J. Exp. Biol.*, 30: 68-69.

170. Sinha P. 2005. Epidemiology and development of forecasting model for powdery mildew (*Oidium erysiphoides*) in jujube (*Ziziphus mauritiana*) for Delhi. *Indian J. Agric. Sci.*, 75: 272-276.
171. Stupner H., Wagner H. 1989. Minor iridoid and phenol glycosides of *Picrorhiza kurroa. Planta Med.,* 55: 467-469.
172. Subrat N. 2002. Ayurvedic and herbal products industry: An overview. Conservation and Management of Himalayan Medicinal Plants Workshop, Kathmandu, Nepal. pp. 15-20.
173. Thakur M.K., Chauhan R., Dutt B. 2010. Effect of different propagation and planting techniques on the performance of *Picrorhiza kurroa*. *J. Hill Agric.*, 1: 43-46.
174. Thakur M.K., Chauhan R., Pant K.S. 2013. Bioresources for productivity enhancement in kutki, *Picrorhiza kurroa* Royle ex Benth. *Int. J. Bio-Resource Stress Manag.*, 4: 482-486.
175. Thapliyal S., Mahadevan N., Nanjan M.J. 2012. Analysis of Picroside I and Kutkoside in *Picrorhiza kurroa* and its Formulation by HPTLC. *Int. J. Res. Pharmaceut. Biomed. Sci.*, 3: 25-30.
176. Thomas, C.D., Cameron, A., Green, R.E. 2004. Extinction risk from climate change. *Nature*.
177. Trivedi P., Pandey A. 2006. Biological hardening of micropropagated *Picrorhiza kurroa*Royel ex Benth., an endangered species of medical importance. *World J. Microbiol. Biotechnol.*, 23: 877-878.
178. Uddin Q., Amanullah, Siddiqui K.M., Rahman R. 2015. Unani Medicine for Cancer Care: An Evidence-Based Review. *Int. J. Ayur. Herbal Med.*, 5: 1811-1825.
179. Uniyal A., Uniyal S.K. 2008. ENVIS Bulletin (Wildlife and Protected Areas). 11: 55-62.
180. Uniyal A., Uniyal S.K., Rawat G.S. 2011. Commercial Extraction of *Picrorhiza kurroa* Royle ex Benth. in the Western Himalaya Patterns of Collection, Processing, and Conservation Threats. *Mount.Res. Develop.*, 31: 201-208.
181. Uniyal S.K., Singh K.N., Jamwal P., Lal B. 2006. Traditional use of medicinal plants among the tribal communities of Chhota Bhangal, Western Himalaya. *J. Ethnobiol. Ethnomed.*, 2: 14-22.
182. Uniyal M.R. 1989. Medicinal flora of Garhwal Himalayas. Baidyanath, Nagpur. pp. 79.
183. Upadhyay R., Arumugam N., Bhojwani S.S. 1989. *In vitro* propagation of *Picrorhiza kurroa* Royle Ex Benth: an endangered species of medicinal importance. *Phytomorphology,* 39: 235-242.
184. Usman M.R.M., Surekha Y., Chhaya G., Devendra S. 2012. Preliminary screening and antimicrobiak activity of *Picrorhiza kurroa* Royle ethanolic extracts. *Int. J. Pharm. Sci. Rev. R*es., 14: 73-76.
185. Vaidya A.B., Antarkar D.S., Doshi J.C., Bhatt A.D., Ramesh V.V., Vora P.V., Perissond D.D., Baxi A.J., Kale P.M. 1996. *Picrorhiza kurroa* (Kutki) Royle ex Benth as a hepatoprotective agent-experimental & clinical studies. *J. Post Grad. Med.*, 42: 105-108.
186. Vaidya B., Adarsa N. 1994. Chaukhambha Bharati Academy, 1st edition, Varanasi. 1: 172-176.
187. Ved D.K., Goraya G.S. 2008. Demand and supply of medicinal plants in India. Bangalore, India: BSMPS, Dehradun, and FRLHT. http://nmpb.nic.in/FRLHT/Contents.pdf
188. Ved D.K., Kinhal G.A., Haridasan K., Ravikumar K., Ghate R.U., Vijayasankar R., Indresha J.H. 2003. CAMP for the Medicinal Plants of Arunachal Pradesh, Assam, Meghalaya and Sikkim - Proceedings of the Workshop. Foundation for Revitalisation of Local Health Traditions (FRLHT).
189. Verma P.C., Basu V., Gupta V., Saxena G., Rahman L.U., 2009. Pharmacology and chemistry of a potent hepatoprotective compound *Picroliv* isolated from the roots and rhizomes of *Picrorhiza kurroa* Royle ex Benth. (kutki). *Curr. Pharm. Biotechnol.*, 10: 641-649.

190. Verma R.K., Paraidathathu T. 2014. Herbal medicines used in the traditional Indian medicinal system as a therapeutic treatment option for overweight and obesity management: A review .*Int. J. Pharm. Pharmacaeut. Sci.*, 6: 40-47.

191. Verma P., Mathur A.K., Jain S.P., Mathur A. 2012. *In Vitro* conservation of twenty-three overexploited medicinal plants belonging to the Indian sub continent. *Sci. World. J.*, 12: 1-10.

192. Virdi M. 2004. Wild plants as resource: New opportunities or last resort? Some dimensions of the collection, cultivation and trade of medicinal plants in the Gori Basin. In: Searching Synergy Stakeholder Views on Developing a Sustainable Medicinal Plant Chain in Uttaranchal, India. Royal Tropical Institute, Amsterdam. pp. 41-54.

193. Visen P.K., Saraswat B., Dhawan B.N. 1998. Curative effect of picroliv on primary cultured rat hepatocytes against different hepatotoxins: an in vitro study. *J. Pharmacol. Toxicol. Methods*, 40: 173-179.

194. Wang J., Van der Heijden R., Spruit S., Hankermeier T., Chan K., Van der Greef J., Xu G., Wang M. 2009. Quality and safety of Chinese herbal medicines guided by a systems biology perspective. *J. Ethnopharma.*, 126: 31-41.

195. Wang Z., Wen J., Xing J., He Y. 2006. Quantitative determination diterpenoid alkaloids in four species of Aconitum by HPLC. *J. Pharm. Biomed. Anal.*, 40: 1031-1034.

196. Warrier P.K. 1996. Indian medicinal plants. Orient and Longman Ltd., Madras. 4: 269-272.

197. Weinges K., Kloss P., Henkels W.D. 1972. Natural products from medicinal plants. XVII. picroside-II, a new 6-vanilloyl-catapol from *Picrorhiza kurroa* Royle ex Benth. Justus Liebigs. *Ann. Chem.*, 759: 173-182.

198. Yadav N., Khandelwal S. 2006. Effect of picroliv on cadmium-induced hepatic and renal damage in the rat. *Human Exp. Toxicol.*, 25: 581-591.

199. Zhao Y.Y., Zhang Y., Lin R.C., Sun W.J. 2009. An expeditious HPLC method to distinguish *Aconitum kusnezoffii* from related species. *Fitoterapia*, 80: 333-338.

200. Zhu J.S., Ouyang D.Y., Shi Z.J., Xu L.H., Zhang Y.T., He X.H. 2012. Cucurbitacin B induces cell cycle arrest, apoptosis and autophagy associated with G actin reduction and persistent activation of cofilin in Jurkat cells. *Pharmacol.*, 89: 348-356.

5

Dactylorhiza hatagirea

English: Himalayan marsh orchid

Hindi: *Salam panja*

Brij Lal and Dinesh Kumar

Dactylorhiza hatagirea **(D. Don) Soo**

(A) Herbarium sheet *Dactylorhiza hatagirea*, (D. Don) Soo **(B)** Whole plant **(C)** Roots (source of crude drug), **(D)** Flower (early stage), **(E)** Flower (mature stage) **(F)** Flower (dehiscence stage)

Abstract

Dactylorhiza hatagirea (D.Don) Soo, a plant species native to Indian Himalayas (Badola and Aitken, 2003; Ved *et al.*, 2003), is one of 110 orchids known for their medicinal properties (Jain 1991; Lal *et al.,* 2004). Known earlier as *Orchis latifolia* auct. non Linn., *D. hatagirea* has traditionally been used in indigenous system of medicine in India as '*Hatajari*' or '*Salam Panja*'. The plant has therapeutic, ornamental and ecological significance (Gupta *et al.,* 1998; Vij *et al.,* 1992; Kala, 2000a). Due to over extraction from nature for medicinal use, the species has become threatened and categorized as critically endangered species (Ved and Tandon, 1998; Kala, 2000b; Ved *et al.,* 2003; Bhattarai *et al.*, 2014). This contribution provides insight into the botanical background, distribution, medicinal importance, threat status, cultivation and commercial aspects, and strategies for its conservation and sustainable utilization.

A. Biological Aspects

Systematic classification

Kingdom	:	Plantae
Division	:	Tracheophyta
Class	:	Liliopsida
Order	:	Asparagales
Family	:	Orchidaceae
Genus	:	*Dactylorhiza*
Species	:	*hatagirea* (D. Don) Soo

• Synonyms: *Orchis hatagirea* D. Don

Orchis latifolia auct. non Linn.

• Common names (in different languages)

Sanskrit	*Salam pamisri, Mumjatak*
Hindi	*Salam panja*
English	Himalayan marsh orchid
Unani	*Sa'lab, Khusyat-us salab, Buzidan, Salab misri*
Nepali	*Panch-aonle*

Bhoti/ Tibetan	*Wangpo-lakpa*
Garhwali/ Kumaoni	*Hatajari or hata jadee*
Kashmiri	*Salam panja*
Tamil	*Salam misri*
Trade name	*Salampanja, Salep*

• Diversity within genus

The generic name *Dactylorhiza* is derived from the Greek words *Dactylos* (means finger) and *rhiza* (means root) in reference to the digitate tubers of the plant species (Vij *et al.*, 2013). The specific epithet *'hatagirea'* is derived from a local name 'Hatajari' or 'Hattha jadee', commonly used in Garhwal and Kumaon regions of Uttarakhand in India, which means hand like roots. *Dactylorhiza* belongs to the family Orchidaceae, which is one of the top ten dominant families of flowering plants, represented by around 1000 species in India (Jain and Mehrotra, 1984). The genus is represented by about 50 species worldwide, distributed widely in north hemisphere, with four species reported in India and only two species from Himachal Pradesh (Vij *et al.,* 2013). About 1520 names, including 937 scientific names of infra-specific rank and 583 names of species rank are recorded under this genus. Of the total 583 scientific names of species rank, only 115 names are accepted as synonyms for this genus (http://www.theplantlist.org.).

• Geographical distribution

Globally, *D. hatagirea* is widely distributed across Europe, North Africa and temperate Asia, including Afghanistan, Bhutan, China, India, Nepal, Pakistan and Tibet (Kaul, 1997; Kumar *et al.,* 1997; Vij *et al.,* 2013). In India, it has a habitat specific distribution between the elevation range of 2500 to 5000 m amsl in open grassy slopes and marshy alpine meadows from Jammu & Kashmir to Arunachal Pradesh (Dhar and Kachroo, 1983; Aswal and Mehrotra, 1994; Hajra and Balodi, 1995, Samant *et al.,* 2001; Bhatt *et al.,* 2005; Khadka *et al.,* 2016). Different workers have reported its distribution in Indian Himalayan states (Fig. 1) of Jammu & Kashmir (Dhar and Kachroo, 1983; Kaul, 1986; 1997; Kumar *et al.*, 1997; Sharma and Jamwal, 1998; Kaul and Handa, 2002), Himachal Pradesh (Shabnam, 1964; Gupta, 1966; Aswal and Mehrotra, 1994; Kumar *et al.,* 1997; Kala, 2000b; Samant *et al.,* 2001; Uniyal *et al.,* 2006) and Uttarakhand (Rawat *et al.,* 1985; Kala *et al.,* 1998; Uniyal *et al.,* 2001; Hajra and Balodi, 1995; Kumar *et al.*, 1997). In eastern Himalaya, the species occurrence has been reported from Sikkim and Arunachal Pradesh (Vij *et al.,* 1992; Srivastava and Mainera, 1994; Rai and Sharma, 2002).

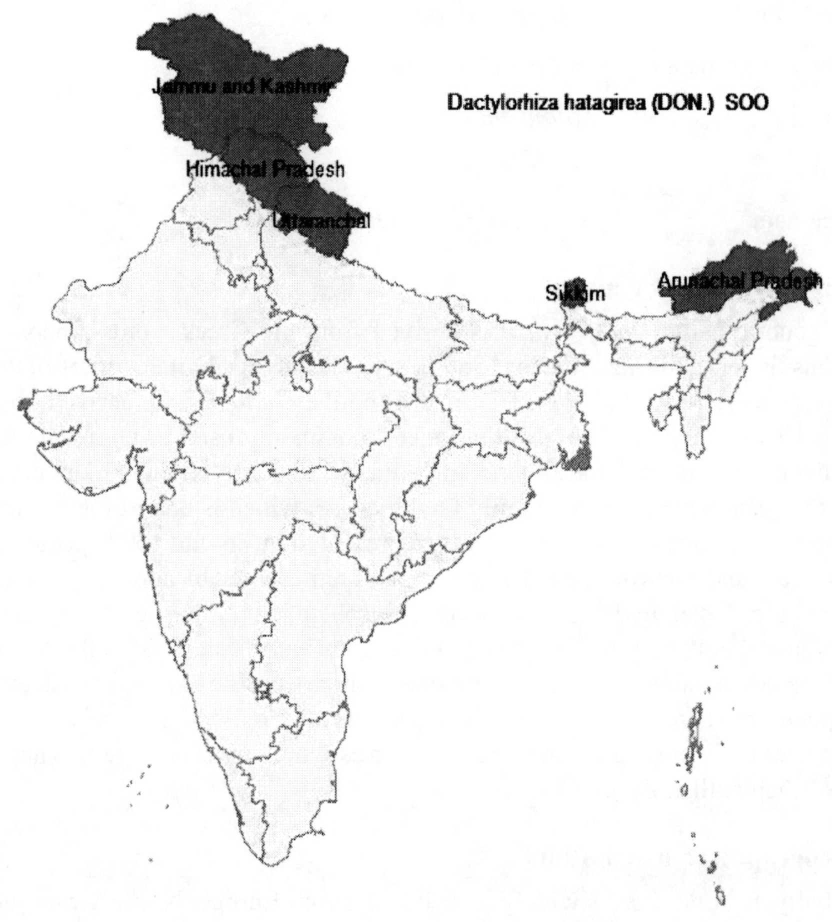

Fig. 1. Map showing distribution of *Dactylorhiza hatagirea* in India

• Habitat and ecology

D. hatagirea is a terrestrial and perennial orchid that grows in open watered slopes in alpine meadows, or under subarctic climatic conditions. It prefers shrubberies, moist grassy slopes and alpine meadows in high altitude valleys or along river-banks in the Himalayas, between 2500 to 5000 m above msl (Bhatt *et al.*, 2005). Perhaps, this is the only orchid recorded from such high elevation in the western Himalayan region (Rau, 1975). It generally grows in association with *Aconitum heterophyllum, Caltha palustris, Equisetum arvense, Fragaria vesca, Hippophae rhamnoides, Juncus* spp., *Sinopodophyllum hexandrum, Potentilla atrosanguinea*, species of *Rumex, Salix* and some grasses (Vij *et al.,* 2013; Aswal and Mehrotra 1994).

• Taxonomic enumeration

D. hatagirea exhibits great variability in plant size, floral arrangement and flower colour at intra-and inter-population levels (Vij *et al.,* 2013). Irrespective of the altitude and habitat, flower colour exhibits high level of variability, ranging from pink to rosy purple to maroon. The species is an erect, glabrous, terrestrial, leafy perennial herb, 20-60 cm tall with fleshy, flattened, palmately divided into 2 to 3 (rarely 5) lobs or finger like, cream coloured, tuberous roots, 1-4 mm - thick at the base; roots are clustered above tuber near the stem base, long, 1-1.5 mm thick. Leaves are 3-6, membranous, arranged along with the whole length of the stem, linear-oblong, lanceolate to elliptic with sheathed base, apex concave or acute, 5-17 (-20) x 2-5 cm in dimension. Inflorescence is apical raceme, densely flowered, 5-15 cm long; bracts linear-lanceolate, acuminate, 10-40 x 2-7 mm, longer than the flowers; lips oblong, obtuse, 3 lobed, lowest one longest and then decreasing in length towards apex. Flowers pinkish or purplish-Lilac or spotted rosy purple coloured, rarely white, 8-10 mm across arranged in dense, terminal cylindrical spikes borne on the robust leafy stem; sepals sub-equal; the dorsal 6-8 x 4-5 mm, ovate, oblong; the laterals slightly longer than the dorsal, 7-9 x 4-5 mm, obliquely ovate-lanceolate. Petals are slightly shorter than sepals, 6-7 x 3-4 mm, broadly lanceolate. Lip broader than long, almost flat, oval, 10-12 mm long, usually spotted with dark purple, slightly 3-lobed, mid-lobe acute, broadly triangular, spur cylindrical, more or less straight, equal or shorter than the ovary. Column is short and pollinia 2 in number. Fruit is capsule, fusiform, erect, ribbed, with marcescent tips; seeds are minute, many, black or dark-brown in colour (Kaul, 1986; Aswal and Mehrotra, 1994; Lal *et al.,* 2004; Selvam, 2012; Vij *et al.,* 2013) (http://www.flowersofindia.net/catalog/slides/Himalayan/ Marsh/Orchid.html).

• Phenological period

D. hatagirea takes around four monthsfrom flowering to fruiting or seed setting. Flowering usually starts in the mid-June and endsin the month of August. Fruiting starts in the end of August and seed setting takes place in the month of October (Kaul, 1986; Kumar *et al.,* 1997; Selvam, 2012).

• Cytotypes

Chromosome number 2n = 40, 80 (Vij *et al.* 2013)**.**

B. REPRODUCTIVE BIOLOGY

Reproductive phenology

D. hatagirea, a perennial herb with long flowering stalks, is an inherently poorly regenerating and slow-growing species owing toits minute seeds, pollinator specificity and mycorrhizal association. The species has been reported to have less reproductive capacity and low fruit setting (28.14±15.65%) much due to its minute non-endospermic seeds (0.098±0.018mm^2) besides having smaller area of occurrence (Thakur *et al.,* 2018). The tubers are lobed and each lobe can regenerate into a fresh plant (Kaul, 1986).

Life Cycle

The above ground growth starts in the month of May when snow start to melt at high altitudes. The senescence starts at the end of September, flowering takes place from beginning of July to end of August, and fruit-setting starts just afterthe flowering is over. The peak flowering period of this orchid is around one month in high altitude areas of Indian Himalaya (Kumar *et al.,* 1997; Vij *et al.*, 2013).

Pollination and Pollinators

D.hatagirea is a self-incompatible,insect pollinated and food deceptive orchid (Kaul, 1997; Thakur *et al.,* 2018). The reproductive success of this plant in nature is generally low (28%), but it has high percentage of fruit setting (95%) in hand-pollinated flowers indicating its pollination limitation (Thakur *et al.,* 2018). The reason behind the pollinator limitation is that the flowers are without nectar (deceptive feature) that often fails to attract pollinators. The pollinator insects for *D. hatagirea* are unknown (Thakur *et al.*, 2018).

C. MEDICINALASPECTS

• Plant material used as crude drug

The dried tuberous rhizomes of *D. hatagirea* generally constitute the crude-drug. A few studies have also reported use of roots for treatingsome ailments. *D. hatagirea* does not find a mention in the Ayurvedic Pharmacopoeia of India (API), yet some information onpharmacognostic characteristic features of this drug is available in literature (Singh, 1983; Selvam, 2012).

• General appearance

The dried rhizomesare slightly flattened, palmately divided into 2-3 or 2-5 finger-like lobes, each lobe 2-8 cm long, 2-5 cm wide and 1-4 mm thick at the base. The powder of the rhizomatous tubers is pale yellow to yellowish in colour (Kumar *et al.,* 1997; Vij *et al.*, 2013).

• Organoleptic Properties

The surface of the fresh tuberous rhizomes is fleshy, smooth and purple-white in colour, whereas dried rhizomes are hard and rough, slightly wrinkled or nearly smooth, translucent pale-yellow or creamy white in colour and has powdery or starchytexture. The taste of the drug is slightly sweet or salty or sometimes tasteless, odourless, at times with no characteristic odour (Kumar *et al.,* 1997; Shukla and Sinha, 2000; Selvam, 2012; Vij *et al.,* 2013).

• Microscopic Characteristics

Microscopic study of the cross section of rhizomes of *D. hatagirea* reveals that outer line of the rhizome is rough, circular and elliptic followed by a thin and continuous layer of epidermis. Parenchymatous cells, vessel elements and crystals are visible under microscopic observation.The epidermis hasa thin layer of spindle shaped sub-epidermis. The aerenchymatous region represents major part of ground tissue having wide air chambers. Parenchymatous tissue having a thin wall is loaded with starch grains ranges from 5-10 μm, which forms thin partition of air chambers. A limited number of vascular strands are present within the ground tissue. The xylem/ vessel elements (10-35 μm) are narrow, thin walled and scattered within the vascular bundles. Phloem elements occur in small nests in between the xylem strands (Selvam, 2012).

• Powdered Plant Material

The microscopic study of the powdered rhizomes exhibits the presence of vessel elements, parenchyma cells, crystals etc. The broken vessel elements, with scalariform and spiral lateral wall thickening, are conspicuous in the powdered material. Particles of orange-red colour, of different shapes and sizes with polyhedral or circular thin walled compact parenchyma cells are visible in the rhizomatous powder. Microscopic examination reveals different types of crystals in the powdered drugsamples, which may be sphaero-crystals, rod shaped or triangular shaped prismatic crystals. The last two shapes are relatively rarely seen (Selvam, 2012).

• Major Chemical constituents

D. hatagirea is rich source of bitter substances, mucilage, volatile oil, starch, sugar, proteins, lipids etc. The dactylorhins A-E and dactyloses A & B are the major chemical markers present in *D. hatagirea* (Kizu *et al.,* 1999; Dutta and Karn, 2007).

• Substitutes and Adulterants

Mixing of adulterantsor substitute in any herbal drug/ crude drug sold in market is a common practice. Due to high demand of the drug in the market and non-availability of required raw materials, a number of closely allied species are the cheap substitute sold under the name '*Salep*' or '*Salam Panja*'. These are dried tubers of various species (*O. slatifolia* Linn., *O. laxiflora* Lam., *O. maculate* Linn., *O. mascula* Linn., *Eulophia* sp.) belonging to the family Orchidaceae (Aswal and Mehrotra, 1994; Shukla and Sinha, 2000; Selvam, 2012). Roots of species of *Asparagus* and *Polygonatum* are also other cheap substitutes and adulterants of *D.hatagirea* crude drug. A substantial quantity of dried material of *D. hatagirea*is imported from Afghanistan, Iran and some European countries under the name of '*Salep*' to meet its demand in the domestic market (Shukla and Sinha, 2000; Selvam, 2012).

• Medicinal usage

D. hatagirea has been widely used for ages in different indigenous systems of medicine, including folk medicine. Known as *Mujjat* in Ayurveda, it is most effective in cough, bile and as a tonic for general weakness (Shukla and Sinha, 2000; Giri and Tamta, 2010). The tuberous roots of the plant are widely used to cure fever, stomachache, dysentery, diarrhea, wounds, cuts, burns and general weakness, particularly in debilitated women after delivery (Anonymous, 1966; Kumar *et al.,* 1997;). The tubers of *D. hatagirea* are also used in some local preparations as nervine tonic for astringent and aphrodisiacs and to increase immunity (immunomodulator) (Vij *et al.,* 1992; Baral and Kurmi, 2006).The decoction taken with milk and honey serves as an aphrodisiac, as well as a tonic (Rawat *et al.,* 1985). There is a mention of its use to treat seminal debility, paralytic affections, and as demulcent (Devi and Thakur, 2011).The tubers can cure bone fractures, relieve hoarseness (Singh, 1983). Its mucilaginous roots can cure diarrhoea, dysentery and chronic fevers (Rawat *et al.,* 1985; Vij *et al.,* 2013).

• General uses

The tubers are important ingredient of many Ayurvedic and Unani preparations and therefore are collected by the local people for sale. The large scale exploitation from nature may lead to its extinction from the natural habitat (Aswal and Mehrotra, 1994).The Juice extracted from tuber is used as tonic and also used for the treatment of pyorrhea (inflammation of the gum). Root paste is generally applied externally as poultice on cuts and wounds and the extract is given in intestinal disorders. The term *Hatta Haddi* was likely coined because of it's ability for treating bone fractures. The tubers of *D.hatagirea* are considered to be nutritious and used as strengthening and soothing agent. The tubers of *D. hatagirea* contain glucoside (a bitter substance), starch, mucilage, albumen, a trace of volatile oil, and ash (Dutta and Karn, 2007). Tubers of *D. hatagirea* are rich in starch, mucilage, sugar, phosphate, chloride and glucoside-loroglossin (Rawat *et al.,* 1985). Young leaves and shoots are consumed as vegetables. Paste of root is used in promoting growth and darkening of hair.

• Ethnobotanical uses

Natives dwelling in high altitude areas of Indian Himalaya use *D. hatagirea* for different purposes (Gupta, 1966; Aswal and Mehrotra, 1994; Singh, 1999; Chauhan, 1999; Kala, 2000a; Sharma *et al.,* 2003; Lal *et al.,* 2004; Singh and Lal, 2008; Guleria and Vasishtha, 2009; Singh *et al.,* 2009; Devi and Thakur, 2011; Sharma *et al.,* 2011; Dutt *et al.*, 2011; Vij *et al.,* 2013). Local inhabitants in Lahaul-Spiti, Chamba, Mandi and Kullu districts of Himachal Pradesh use paste of tubers, mixed with other ingredients like milk, honey and other herbs, as a tonic to increase vigour-vitality, and to treat ailments like fractured bones, fever, burns, bleeding, colic pain, headache, etc. Different ethnic groups inhabiting high altitude areas of Jammu and Kashmir use the species for treating different gynaecological disorders, kidney problems, diarrhoea, dysentery, fever, or use as aphrodisiac, as material to rejuvenate, as a tonic and as expectorant (Kaul, 1997; Sharma and Jamwal, 1998; Dar *et al.,* 2017). In Uttarakhand, *D. hatagirea* is used to treat bone fracture, bilious fever, cuts and wound, diabetes etc. (Gupta, 1981; Gaur *et al.,* 1983; Rawat *et al.,* 1985; Shukla and Sinha 2000; Samal *et al.*, 2010; Negi *et al.* 2011; Rawat and Jalal 2011). The people of north-eastern region of India particularly from the state of Sikkim use tubers in healthcare (Bennet, 1985). In Nepal, tubers of *D. hatagirea* are used to treat fever, cuts, wounds, intestinal pain and as nervine tonic. Rhizome is eaten raw as tonic (Kunwar and Nirmal, 2005; Ranpal, 2009).

D.CHEMISTRY ASPECT

Chemical composition

- **Leaves:** The leaves contain glucoside and loroglossin (Bhutya, 2011), albumin, militarrin, pyranoside, pyrocatechol and volatile oil.
- **Tubers and roots**: Both contain mucilage, starch, glucoside, a bitter substance, starch, albumen, a trace of volatile oil and ash (Anonymous, 1966; Dutta and Karn, 2007). The tubers of *D. hatagirea* contain dactylorhins A-E (dactylorhin A, dactylorhin B, dactylorhin C, dactylorhin D, dactylorhin E), dactyloses A and B and lipids, etc., (Dutta and Karn, 2007). The tubers also contain (2A)-2-*β*-D-glucopyranosyloxy-2-(2-methoxypropyl) butandipoic acid, bis(4-*β*-D-glucopyranosyloxybenzyl) ester, (2R,3S)-2-*β*-D-glucopyranosyloxy-3-hydroxy-2-(2-methylpropyl) butanedioic acid, bis(4-*β*-D- glucopyranosyloxybenzyl) ester, (2R)-2-*β*-D- glucopyranosyloxy-3-hydroxy-(2-methylpropyl)butanedioic acid, (2R.3S)-2-*β*-D- glucopyranosyloxy-3-hydroxy-(2-methylpropyl)butanedioic acid 1-(4-*β*-D- glucopyranosyloxybenzyl) ester, (2R)-2-*β*-D-glucopyranosyloxy-2-(2-methylpropyl), butane-dioic acid 1- (4-*β*-D-

glucopyranosyloxybenzyl) ester,1-deoxy-1-(4-hydroxyphenyl)-L-sorbose,1-deoxy-1-(4-hydroxyphenyl)-L-tagatose (Kizu *et al.*, 1999; Dutta and Karn, 2007).

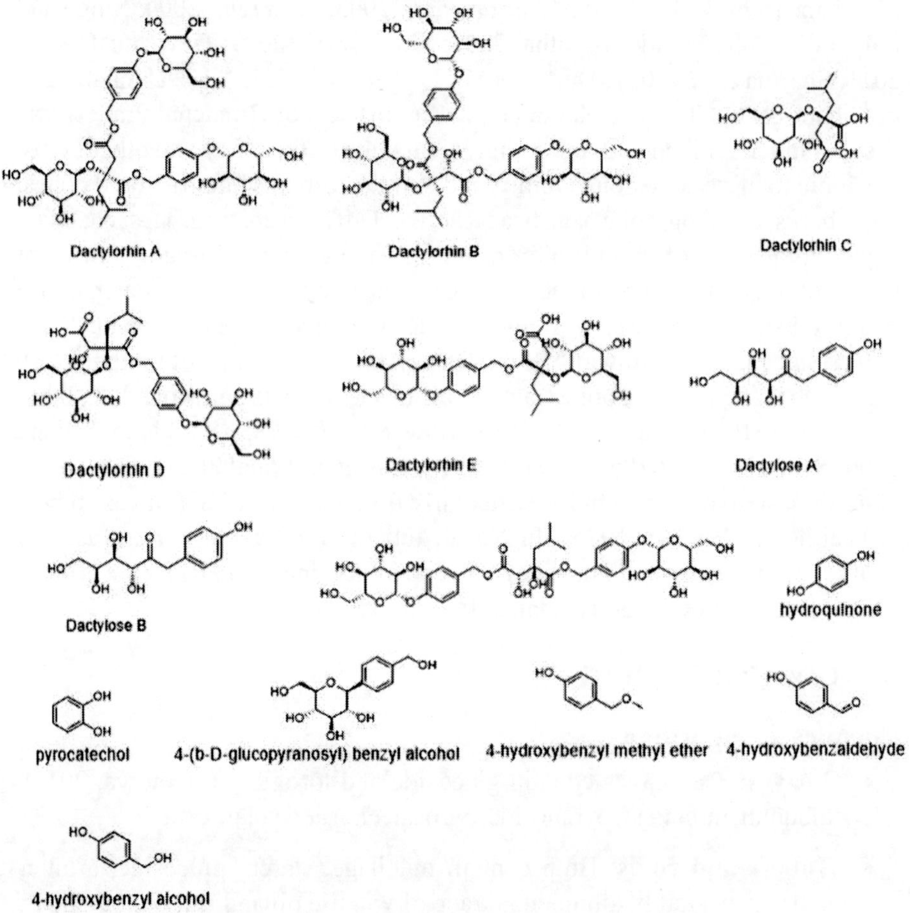

Dactylorhin A, Dactylorhin B, Dactylorhin C, Dactylorhin D, Dactylorhin E, Dactylose A, Dactylose B, hydroquinone, pyrocatechol, 4-(b-D-glucopyranosyl) benzyl alcohol, 4-hydroxybenzyl methyl ether, 4-hydroxybenzaldehyde, 4-hydroxybenzyl alcohol

Chemical constituents of *Dactylorhiza hatagirea*
(Chemical structure is contributed by Dinesh Kumar)

- **Chemical markers**
- Dactylorhins A-E and dactyloses A & B

E. PHARMACOLOGICAL ASPECTS

The traditional knowledge pertaining to *D. hatagirea* has contributed significantly for its pharmacological evaluation in modern times. The tubers, traditionally suggested as a metabolic booster, serve as an aphrodisiac agent (Singh, 1983). The belief that it can increase sexual ability and maintain the level of testosterone has prompted itsuse to treat male impotency (Thakur and Dixit, 2007).

Bioactivity

Anti-bacterial activity

Ranpal (2009) evaluated the antibacterial activity of extracts of *D. hatagirea* against *Bacillus subtilis, Escherichia coli, Pseudomonas aureginosa, Staphlococcus aureus,* and *Shigella flexineri.* Comparing zonation of inhibitions (ZOIs) between rhizome and aerial parts of *D. hatagirea*, the former was more effective than the aerial part against all tested organisms, except *E. coli. E. coli*, one of the very resistant bacteria to synthetic drugs, was reported to be very susceptible to the extract of this plant. Hence, this species can be a potential source for evolving newer antimicrobial compounds for treating dysentery caused by *E. coli.*

Anti-oxidant activities

In a recent study, methanolic extracts of 25 commonly used herbs were screened by using 1,1–diphenyl–2–picrylhydrazyl (DPPH) radical and Ascorbic acid (EC50: 2.6 ìg/mL) as a positive control for their antioxidant activities. Among these 25 drugs tested for DPPH antioxidant activity, *D. hatagirea* has not shown significant results. Extracts at concentrations between 5µg/ml and 50µg/ml showed very less potent antioxidant activities (Baral and Basnet, 2013). Anti-oxidant activity has also been reported in ethanolic extracts using DPPH radicals (Phani *et al.,* 2010).

Sexual dysfunctions

Male rats fed with 200 mg/kg of alyophilized aqueous extract of *D. hatagirea* for 28 days caused significant anabolic effect and influenced the behavior of the treated animals. The extract enhanced male rat attraction towards females (2.5-fold) and was able to modify mount latency (36% reduction) and post-ejaculatory latency (36%) in a manner suggesting libido enhancement (Thakur and Dixit, 2007). The study showed that treated animals indulged in copulation and number of bouts increased remarkably. In another study, male rats when treated with 100 mg/kg of defatted water extract of *D. hatagirea* and fed daily for two weeks, showed an increase in the penile erection index and increase in sperm count by 28.22%. A similar response was observed earlier with 200mg/kg of the root water extract given over 28 days (Thakur and Dixit, 2007). There is also some evidence that the plant extract increased testosterone level in adult male rats, and even slightly increased level of testosterone resulted in an increased sexual desire and arousal (Bancroft, 2005). The plant apparently harbors effectiveness in improving the functionality of sexual organ and sexual behavior/ performance. *D. hatagirea* therefore is a competent candidate drug, for sexual dysfunction or impotency

(Thakur and Dixit, 2007). Further studies however are required for detailed chemical characterization of this plant to better understanding its effect on sexual behaviour and functions (Chauhan *et al.*, 2014a).

Clinical Studies

Crude powder of the tubers of *D. hatagirea* in 3 g doses, two times in a day with water is beneficial for chronic diarrhoea and dysentery while with *Piper longum* 500 mg with she the got's milk instead of water is beneficial for asthma. The tuber powder mix of *D. hatagirea, Eulophia dabia, Asparagus adscendens, Cucurligo orchioides* along with milk is useful in leucorrhea and spermatorrhea. 10 g mixture of *D. hatagirea, A. adscendens, A. racemosus, C. orchioides* tubers along with seeds of *Mucan pruriens, Tamarindus indica* and *Hygrophila auriculata*, flower buds of *Butea monosperma* and gum of *Acacia nilotica, Lepedium iberis and Centaurea behen* when taken twice with milk, helps in general debility, lumbago, penile dysfunction (Bhutya, 2011).

• Formulations

The dried and processed tuberous rhizomes are used extensively as a single drug or as an ingredient in various Ayurvedic and Unani formulations.The major Ayurvedic formulations containing this drug are *Napunsaktawari* (Singh, 1999), *Chyavanprash* (Shukla and Sinha, 2000), *Vajikaran* drugs or aphrodisiac preparations, given as a tonic to ladies after child delivery (Singh, 1983; Baral and Kurmi, 2006) and as a health drink for sick people (Kaul, 1997). The crude drug is also used as one of the ingredients in some Unani formulations like *Majun Salab* and *Safufflasab* (Singh, 1983). *Saalam paka, kaamoddipaka churna, siddhamaasaadi yoga, navajivan yoga* and *phalaaspaavaleha* are some of formulation of plant. Herbal ice cream containing *D. hatagirea* as one of the ingredients was prepared and assessed for antioxidant activity. Significant activity was found at 4% of herbal powder of selected ice cream (Ali *et al.*, 2014)

(d) General usage

The species is recommended as nervine tonic and aphrodisiac, to treat diseases characterized by weakness and general debility (Singh, 1999).

(e) Dosage

The crude drug isrecommended with a normal dose of 1-3 g (Singh, 1983).

F. THREAT AND CONSERVATION ASPECT

Population status

D. hatagirea is native and near endemic to Himalayan region (Samant *et al.*, 1998; Badola and Aitken, 2003, Ved *et al.*, 2003), spread across Afghanistan to Bhutan. Based on being one of the highly traded medicinal plant, its restricted distribution, fragmented range and 80%decline in its population in last 10 years studied in 1997 (Molur and Walker, 1998; Ved and Tandon, 1998; Ved *et al.*, 2003), the species, was ranked as critically endangered by the Conservation Assessment and Management Plan (CAMP) subgroups.The species now figures in the Appendix II, by the Convention on International Trade in Endangered Species of Wild Fauna and Flora (CITES).

Threats

Habitat destruction, fragmentation, over exploitation and the effects of global climate change are the main threats to the survival of orchids (Bubb *et al.*, 2004). Unlike other orchids, population of *D. hatagirea* is rapidly declining because of uninterrupted and careless collection from nature, much to meet the increasing demand in pharmaceutical industries. All these factors, generally applicable to medicinal plants, have led to severe genetic erosion of many species (Machaka-Houri *et al.*, 2012).

The reason for decline in population of this plant is not only its poor regeneration capacity, but the predation pressure for *D. hatagirea* is also linked to its rootstocks eaten by pheasants (particularly monal), rodents and worms during winters when there is a general scarcity of foods in nature (Dhyani and Kala, 2005).

• Habitat loss/degradation/fragmentation

D. hatagirea is one of the 21 listed plant species obtained from temperate forests (Ved and Goraya, 2008) having an annual demand of about 5000 tons (Mishra, 1998; Kala, 2015). In face of the increasing demand of this species in medicinal and pharmaceutical industries, habitat loss is the primary cause of species declining status at local, regional and global scales. Several factors like developmental and construction activities, animal grazing, human interference and landslide, etc., have reduced the natural habitat of the species especially in its niche area.

• Over-exploitation

D. hatagirea attracts heavy demand by herbal industry on account of its medicinal properties and is therefore invariably over exploited. One study reported that to get 1 kg dried roots 90 to 100 mature plants are uprooted (Kala, 2015).

The natives, including professional collectors, generally over extract from nature since they get meager payments, such as @ Rs. 100 to 200 per kg of dried tubers of *D. hatagirea.* The species is still not cultivated but harvested from nature for trade. Demands for some of prioritized medicinal plants have increased by about 50%, whereas availability has declined by about one-fourth, and if this trend continues, many valuable species, including *D. hatagirea*, may face extinction (Vij *et al.,* 2013). Thus, large-scale exploitation of *D. hatagirea* may lead to its acute shortage and possibly extinction from the natural habitat (Aswal and Mehrotra, 1994).

- **Introduction and spread of exotic/alien species**

D. hatagirea is generally found growing in association with other plants like *Aconitum heterophyllum, Caltha palustris, Equisetum arvense, Fragaria vesca, Hippophe rhamnoides, Juncus* spp., *Podophyllum hexandrum, Potentilla atrosanguinea* (Vij *et al.,*2013; Aswal and Mehrotra 1994). Due to various levels of disturbance, changing land-use practices and climate change, species re-distribution is seen as a phenomenon whereby many alien species like *Argemone, Artemisia, Eupatorium, Trifolium, Rumex,* etc. Would compete better in the absence of predators or pests, and may suppress the growth of *D. hatagirea.*

- **Pollution and diseases**

Currently, no studies are available to suggest the kind of effect of pollution on *D. hatagirea.* Similarly, systemic studies are lacking to account for disease(s) associated with the species in its natural habitat, though planting the species at lower elevation has often been seen to promote fungal/ bacterial attacks, apparently due to either lack of species inherent capacity to resist infestations or getting exposed to pathogens which usually are absent at higher elevations. The misuse of insecticides is believed to negatively impact pollinators,which are important for pollination and reproduction of the species (Dahal and Sharma, 2014).

- **Climate change**

Increase in pollution and upward shifting of alien flora due to warming is a matter of great concern for the survival of high altitude medicinal plants. The threat caused by such factors, are likely to compound the adverse effects of climate change, species over-exploitation, habitat loss, pathogen attack, etc. (Badola and Aitken, 2003).

• Conservation strategies

The drastically declining status of *D. hatagirea* within its restricted natural habitat makes it extremely important to develop as well as implement an effective conservation plan. Nearly two decades ago, the species was termed as critically endangered for all the Himalayan states from Jammu & Kashmir to Arunachal Pradesh (Molur and Walker, 1998). Subsequently, it figured among prioritized top 20 medicinal plants based on CAMP assessment report (for conservation and protection for selected threatened plants), in CITES, Red Data Book and IUCN list of threatened plants (IUCN, 2004), and suggested for conservation and sustainable utilization (Dhar *et al.,* 2000). National Medicinal Plant Board (NMPB), India has also prioritized *D. hatagirea* for conservation and cultivation under its various schemes (Anonymous 2008; Anonymous, 2016). While collection of the species from wild, which still remains the major source of raw material for commercial market, remain unlikely to be checked, rotational harvesting (of 4-5 years rotation) has been recommended (Kunwar, 2006).

• *Ex-situ* Conservation

Ex-situ conservation essentially involves collection, maintenance and conservation of samples of organisms, usually as live whole plants, seeds, pollen, spores, vegetative propagules, tissue or cell cultures in other than the original habitat of the species or any other living entity. Among the options available for conservation of *D.hatagirea* under *ex-situ* conditions are botanical gardens, seed banks/ gene and DNA banks, captive cultivation, tissue culture and cryopreservation. Agro-techniques for captive cultivation are known for several species of *Dactylorhiza* (Xancó *et al*., 2012). In case of *D. hatagirea*, one of the limitations is its low rate (0.2%–0.3%) of seed germination (Vij, 2002). The poor germination response may be due to impediment caused by inner integument for hydration of embryo or associated factors that inhibit its development (Stoutamire, 1974; Veyret, 1969). After germination, it may take as much as one year's time for plantlets development (Agarwal *et al.,* 2008).

• *In-situ* Conservation

The *in-situ* conservation of threatened taxa in its natural habitat, preferably in protected areas, such as at high elevation areas in Himalayan region, by involving natives could be a good strategy (Kala, 2000b). Raising awareness on the ownership of natural resources by local people and encouraging profitable uses of these resources may provide strong incentive for conservation (Child, 2002). To minimize pressure on natural population of threatened species, participatory research on setting up of harvest limits for medicinal plants extracted from protected areas, such as buffer zones around National Parks, may help in conservation of over extracted species. While creation of protected areas for conservation of specific medicinal plant resource in nature may not be a viable solution (Kala, 2015), a better habitat management, population enrichment by planting saplings of the target species, through seeds or vegetative propagules, may be a useful.

Germplasm enhancement through *in-vitro* conservation

Tissue culture provides an important measure for mass propagation for most terrestrial orchids (Jakobsone *et al.*, 2007). As evident from different studies, raising plants through *in vitro* technique is quite slow, and this holds true for many species of *Dactylorhiza* as well (Vij *et al.* 1995; Vaasa and Rosenberg, 2004). In MS medium supplemented with peptone (1.0 g/L), morpholino ethanesulfonic acid (1.0 g/L), and activated charcoal (0.1%), better germination response was reported in *D. hatagirea* (Giri and Tamta, 2012). For developing protocorms of *D. hatagirea*, immature seed embryos were cultured for shoot regeneration and mass multiplication (Warghat *et al.,* 2014). In yet another study, seeds of *D. hatagirea* using mycorrhizal fungi showed 100% germination within 10 days after sowing and plants developed seedlings after 3 months (Aggarwal and Zettler, 2010). It is possible to provide a germplasm storage protocols with the help of *in-vitro* techniques, which have strong possibilities of disease elimination and rapid clonal propagation. Subsequently, virus free (virus-tested) cultures are ideal for exchange and distribution of germplasm at national and international level.

There are two basic methods followed to maintain germplasm collections for *in-vitro*conservation; first is minimal growth and the second one is cryopreservation.The minimal growth conditions are applicable for short to medium term storage. These can be achieved by reducing temperature and light, or by incorporation of sub-lethal levels of growth retardants, or induction of osmotic stress with sucrose or mannitol, or by maintenance of cultures at a reduced nutritional status (particularly reduced carbon), or by reduction of gas pressure over the cultures, or by desiccation and mineral oil overlay. Cryopreservation in liquid nitrogen (-196 °C) offers the possibility for long-term storage with maximal phenotypic and genotypic stability. This method is relatively convenient and economical.

G. CULTIVATION ASPECTS

Agrotechnology has been known for about 100 plant species but agronomical practices have been worked out merely for about 40 medicinal plants only (Hussain, 1994). Also, cultivation of many species of medicinal relevance has remained restricted to a few research sites developed by R & D organizations and to a few interior remote areas maintained by the native people (Silori and Badola, 2000, Haridasan *et al.,* 2002; Kaul and Handa, 2002; Rai and Sharma, 2002; Uniyal *et al.,* 2002; Anonymous, 2018). Various workers have reported efforts made to cultivate and propagate *D. hatagirea* in the Himalayan region (Nautiyal, 1995; Sundriyal and Sharma, 1995; Nautiyal and Nautiyal, 2004; Anonymous, 2016, 2018), including through tissue culture techniques (Aggarwal and Zettler, 2010; Warghat *et al.,* 2014; Anonymous, 2018).

Climate

Climatic conditions of alpine or sub-alpine regions in the Himalaya favour growth of *D. hatagirea*. The plant prefers moist, stony slopes and marshy land along riverbanks where average temperature in summer ranges between 15- 18°C but can go down to below the freezing point in winters. It is cultivated in certain pockets at high altitude areas in Garhwal and Sikkim Himalaya (Rawat *et al.,*1985; Shukla and Sinha, 2000; Anonymous, 2016). However, orchids can be cultivated outdoor at places having perfect agro-climatic environment.

Soil

Generally, the plant prefers to grow in well-drained, moisture retentive and marshy meadow soils. These soils are usually dark grey in colour, acidic in nature, granular in texture and sandy loam, or micaceous sandy at greater depth. It also grows well in soil rich in organic matter and cool partial shades. Such soils are best for the germination of seeds, better growth and yield of the plant (Rawat *et al.,* 1985, Anonymous, 2016).

Nursery Raising

D. hatagirea, propagated through both seeds and vegetative methods (Rawat *et al*., 1985; Vij *et al.,* 1995; Anonymous 2016), is a poorly regenerating and inherently slow growing orchid, likely because of its specificity to pollinators and symbiotic mycorrhizal association (Anonymous, 2016). Propagation of nearly all orchids through seeds is a challenging and arduous task as it takes long time for seed to germinate (Arditti *et al.,* 1982; Arditti, 1992; Singh, 1999). A study on *D. hatagirea* reported that freshly collected seeds from wild may not show any germination even under controlled growth conditions (temperature 10-15°C and humidity (90+5%), likely due to lack of cold treatment and lack of mycorrhizal association (Anonymous, 2016). In nature, seeds remain buried under snow for about 4-5 months, before their germination during spring/summer. The percentage germination varies between 20 to 30%, whereas seeds sown in the soil collected from its natural habitat at lower altitudes have shown better results (Anonymous, 2016). In a study carried out in Nepal Himalaya, seeds collected from nature and sown in nursery beds during the month of April-May, with 40-60 cm spacing showed 85-90% germination (Shrestha and Shrestha, 2004; Kunwar, 2006). For cultivation of *D. hatagirea* over one hectare of land, nearly 5 kg seeds are required.

Vegetative propagation through tuber cuttings or splitting of the sprouting tubers is quite a successful technique to raise plants (Rawat *et al.,* 1985, Vij *et al.,* 1995; Anonymous, 2016). Segments of tuber tops with buds are the most suitable planting material. The tubers are cut into lobes and each lobe can regenerate into a new plant (Kaul, 1986). Vegetative propagation through tubers is beneficial for rapid multiplication and higher biomass yield. The small pieces of tubers of 4.0 mm size containing meristematic tissues can develop plantlets when transplanted at depth of 5-7cm and a spacing of 15cm x 15cm. The plant raised from tuber cuttings develops tubers and roots after one year and grows up to 4-8 cm tall (Anonymous, 2016). The green pod culture technique is more effective *in-vitro* propagation of this medicinal orchid whereby immature seeds collected from nature showed seed germination up to 100% (Vij *et al.,* 1995; Ahuja, 2002). Multiplication of protocorm like bodies generates plants under *in vitro* conditions that showed a survival rate of about 85% (Anonymous, 2018).

Land Preparation and Manuring

Selection of land for cultivation should ensure that it receives enough sunshine, has little possibility of water stagnation, and contains enough organic matter/humus. Like for any crop, the field should be well prepared and made free from weeds. It is advisable that chemical fertilizers and pesticides are applied sparingly. Instead, bio-pesticides prepared from leaves, seeds and kernels of Neem *(Azadiracta indica)*, are used singly or in mixture. Farm Yard Manure, forest litter, vermi-compost, or green manure can be applied as per the need to improve growth,performance and yield. The requirement of manure for cultivation differs at high (5 t/ha) and low (10-20 t/ha) altitudes depending upon native soil fertility (Shrestha and Shrestha, 2004; Kunwar, 2006; Anonymous, 2016). Weeding at every 7-10 days intervals, especially during the rainy season is essentially required to favour optimal growth (Anonymous, 2016).

Irrigation

During the initial stage of germination/growth of *D. hatagirea*, it should be ensured that enough soil moisture exists so as to promote root/shoot initiation and growth. The beds should be irrigated at twelve hourly gap during summer, especially in field locations at lower elevation and once a week during winter months to maintain adequate moisture (Anonymous, 2016).

Harvesting

The plants propagated through tuber cuttings complete their vegetative and reproductive phase in 3-4 years. To fetch maximum yield and harvest of active constituents, five years old tubers are uprooted, generally after seed maturity during late September. Sometimes tubers are also harvested after 2 or 3 years of transplantation.The tubers are treated in hot water for one hour to remove its outer layer. The tubers are sun dried and preserved for future use.The crop yield in India under natural conditions of growth is around 1.8 t/ha. Under captive conditions (greenhouse conditions), productivity may increase up to 2.0 t/ha (Anonymous, 2016). Crop yield of 250-300 kg/ha has also been reported from the wild in Nepal (Shrestha and Shrestha, 2004).

Crop protection

High altitude plants generally complete their reproductive phase during the months of September and October, therefore, harvesting should be done during these months only when plants are rich in active constituents (Nautiyal and Nautiyal, 2004).

H. COMMERCIAL ASPECTS

• Demand

With the increase in size of herbal industry globally (Gruenwald, 2000), the demand of medicinal plants is growing fast across the world and it is increasing approximately US $14 billion per year (Sharma, 2004). The demand for herbal raw material is increasing at the rate of 15 to 25% per annum. As per estimates of World Health Organization, demand for herbal raw material from medicinal plants is likely to be more than US $5 trillion in the year 2050. Trade of medicinal plants in India is estimated to be around US $1 billion per year (Joshi *et al.*, 2004). India and China are the main exporters of medicinal plants and herbal products. Total global market of herbal products is about $ 85 billion of which $45 billion alone is the share of herbal pharmaceuticals. The Asian continent holds about 30 % share in the world herbal market. China is the leading one, followed by India as the second largest exporter of herbal medicines.

The crude drug obtained from *D. hatagirea* has a great demand in national and international markets (Olsen and Helles 1997; Badola and Pal, 2002). An earlier report (Badola and Pal, 2002) estimated its annual consumption in India to be about 7.38 tons, valued at 5,000,000 IR or US $83,333. This demand for the tubers remains very high contrary to its meager production. This shortage in its supply, due to its ever-depleting status in nature,was voiced by herbal-traders in Himachal Pradesh (Anonymous 2008). According to arecent report, the annual demand of this herb is about 5000 tons (Kala, 2015), which is met nearly exclusively from wild population in nature. Therefore, the species is liable to be over-extracted from nature since currently it is not under cultivation in the country (Badola and Pal, 2002; Olsen and Helles, 1997).

• Market trends

Market related information for most medicinal plants remains a broad speculation, made in a few scattered reports, showing wide variability with respect to demand, production and market rate of the produce. There is a lack of reliable information on market trends for *D. hatagirea* as well. The information available from market surveys is just a tip of the iceberg as most of the medicinal plants are sold illegally, and actual data remain far from being reliable. Though agricultural practices for several medicinal herbs including *D. hatagirea* have been developed (Anonymous, 2016), about 90% of the plant material is still collected from nature in a destructive manner.

• Trade Data

There is enormous variation in the selling price of plant based crude drugs, made in different reports spanning different time-period and regions within Himalayas. Studies around the beginning of this century, suggested the sale price of roots of *D. hatagirea* to vary from Rs.80-1200 per kg (Gupta *et al.,* 1998; Silori and Badola, 2000; Chetri *et al.,* 2005). Nearly a decade later, another study reported that the dried tubers of *D. hatagirea* were sold @ Rs. 700 per kg (Selvam 2012). More recently, tubers of *D. hatagirea* were reported to fetch Rs. 7000-15000 per kg in Kashmir market (Dar *et al.,* 2017). According to another report, selling price of tubers of *D. hatagirea* could range from Rs. 1000-20000 per kg (https://www.indiamart.com/proddetail/salam-panja-*D.*-hatagirea-roots-raw-11822766362.html). In Nepal, despite ban on its collection from nature, tubers can be easily purchased in the markets @ NRs. 10,000-15,000 per kg (*https://en.wikipedia.org/wiki/ D. hatagirea*).

• Major Users

The tubers of *D. hatagirea* are in high demand by pharmaceutical industries which use them for preparation of various formulations (Singh, 1999; Selvam, 2012). The practitioners of Ayurveda, Unani, Homeopathy and Tibetan medicines use this plant in their medicinal preparations. Apart from this, natives in the high altitude areas use the plant to treat different diseases besides using the tubers as food material. The tuberous roots are collected on large scale and supplied to crude drug markets in Jammu and Kashmir (Kishtwar and Udhampur), Himachal Pradesh (Chamba and Kullu), Uttarakhand (Haridwar, Ramnagar, Tanakpur) and Sikkim Himalaya (Selvam, 2012; Ahuja *et al.,* 2015).

FUTURE DIRECTIONS FOR RESEARCH / VALUE ADDITION

D. hatagirea, a plant of therapeutic importance, is a threatened species due to its over-extraction from the wild. Unlike many other high elevations species, it has a rather restricted habitat in temperate/ alpine zone. In face of changing conditions of climate and/or increasing anthropogenic activity, the species ecological status can change drastically. *D. hatagirea* is reported to have significant variability among its wild populations, which opens a new area for conservation and genetic improvement of the species (Chauhan, *et al.*, 2014b).

The following areas for research and management are suggested:

- Generation of comprehensive information on ecological aspects (phytosociological data, phenology, functional traits), reproductive biology (to address inherent reproductive inefficiency, poor seed germination, high seedling mortality and low survival rates), changes in levels of disturbance, pollinator's behaviour, disease and pathogenic threats.
- Development of effective *in-situ* conservation measures (establishment of permanent plots for time series changes) and *ex-situ* conservation strategies (protocols for mass multiplication and gene-bank development).
- Establishment of germplasm centre for morphological, chemical and molecular characterization to identify the elite planting material.
- Biochemical and molecular characterization of drug to identify the elite population and active principles.
- Improved harvesting and post harvest techniques to minimize the loss of the materials harvested.
- Organized cultivation, followed by marketing network between growers, traders and herbal industries.

J. PATENTS

Om Prakash Chaurasia, Basant Ballabh B Raut, A medicated herbal tea and a process for the formulation thereof Indian Patent No. 242959; Application number 1053/DEL/2005

LITERATURES CITED

1. Agarwal A., Khokhar D., Vishwnath. A. 2008. Conservation through *in vitro* propagation of a critically endangered medicinal plant, *D. hatagirea* (*D.*Don) Soo. In: Reddy MV (ed). Wildlife Biodiversity Conservation. Daya Publishing House, 294-299.
2. Aggarwal S., Zettler L.W. 2010. Reintroduction of an endangered terrestrial orchid, *D. hatagirea* (D.Don) Soo, assisted by symbiotic seed germination: First report from the Indian subcontinent. *Nat. Sci.* 8(10):139-145.
3. Ahuja P.S. 2002. Current status of propagation of medicinal plants in Indian Himalaya. pp. 207-230. *In*: S. S. Samant, U. Dhar & L. M. S. Palni. (eds.). Himalayan medicinal plants. Potential and Prospects. Govind Ballabh Pant Institute of Himalayan Environment and Development, Almora.

4. Ahuja P.S., Lal B., Singh S. 2015. Medicinal and Aromatic Plants for Livelihood Enhancement In: V.L. Chopra (ed.) - Medicinal and Aromatic Plants for Livelihood Enhancement, New Scientific Publishers New Delhi, pp. 473-506.
5. Ali M. N., Prasad S. G. M., Gnanaraja R., Srivastava P., IbrahimM., Singh A. 2014. Assess the antioxidant activity of herbal ice cream prepared by using selected medicinal herbs. *The Pharma Innovation 3*(7, Part B): 57.
6. Anonymous, 1966. The wealth of India (Raw material). Council of Scientific and Industrial Research, New Delhi, India, pp. 4.
7. Anonymous, 2008. Non-Timber Forest Produce as Livelihood Option for Rural Communities of Mid Himalayas in Himachal Pradesh. Mid Himalayan Watershed Development Project, NTFP Study-Draft Report-1/July 2008.)
8. Anonymous, 2016. Agrotechniques of selected medicinal plants Vol.III 18-21 National Medicinal Plant Board (NMPB), Ministry of Ayush, GOI, New Delhi.
9. Anonymous, 2018. CSIR-IHBT Annual Report 2017-18, CSIR-IHBT Palampur, HP, pp.37-38.
10. Arditti J., Clements G., Fast G., Hadley G., Nishimura G., Ernst R. 1982. Orchid seed germination and seedling culture – A manual. In: Arditti J. (Ed). Orchid biology: Reviews and perspectives II. New York: Cornell University Press. p. 244-370.
11. Arditti J., Ernst R. 1992. Micropropagation of Orchids, New York: Wiley-Interscience Publication. 682p.
12. Aswal B.S., Mehrotra B. N. 1994. Flora of Lahaul-Spiti (A cold desert in north west Himalaya). Bishen Singh Mahendra Pal Singh, pp. 583-584.
13. Badola H.K., Aitken S. 2003. The Himalayas of India: A treasury of Medicinal plant under siege. *Biodiversity* 4: 3-13.
14. Badola H.K., Pal M. 2002. Endangered Medicinal plant in Himachal Pradesh. *Curr. Sci.* 83(7): 797-798.
15. Bancroft J. 2005. The endocrinology of sexual arousal. *J. Endocrinol.*, 186: 411–427.
16. Baral S.R., Kurmi P.P. 2006. A Compendium of Medicinal Plants in Nepal. Katmandu, Nepal: Central Bureau of Statistics, Government of Nepal.
17. Bennett S.S.R. 1985. Ethnobotanical studies in West Sikkim. *J. Econ. Tax. Bot.* 7: 321.
18. Bhatt A., Joshi S.K., Gairola S. 2005. *D. hatagirea* (D.Don) Soo-a west Himalayan Orchid in peril.*Curr. Sci.* 89: 610-612.
19. Bhattarai P., Pandey B., Gautam R.K., Chhetri R. 2014. Ecology and Conservation Status of Threatened Orchid *D. hatagirea* (*D.*Don) Soo in Manaslu Conservation Area, Central Nepal, *AJPS*. 5: 3483-3491.
20. Bhutya R.K. 2011. Ayurvedic medicinal plants of India,Scientific Publisher, Jodhpur. 2: 47-48.
21. Bubb P., May L., Miles I., Sayer J. 2004. Cloud Forest Agenda. Cambridge, UK: UNEP-WCMC.
22. Chauhan N.S. 1990. Medicinal orchids of Himachal Pradesh. *J Orchid Soc* India. 4: 99-105.
23. Chauhan N.S., Sharma V., Dixit V.K., Thakur M. 2014a. A review on plants used for improvement of sexual performance and virility. BioMed Research International volume 2014, Article ID 868062, 19 pages, Hindawi Publishing Corporation, http://dx.doi.org/10.1155/2014/868062.
24. Chauhan R.S., Nautiyal M.C., Vashistha R.K., Prasad P. 2014b. Morphobiochemical Variability and Selection Strategies for the Germplasm of *D. hatagirea* (D. Don) Soo: An Endangered Medicinal Orchid. *J Bot*. http://dx.doi.org /10.1155/2014/ 869167, pp 5.
25. Chauhan, N.S.1999. Medicinal and aromatic plants of H.P. Indus Publishing Company, New Delhi.
26. Chettri N., Sharma E., Lama S.D. 2005. Non-timber forest produces utilization, distribution and status in a trekking corridor of Sikkim, India.*Lyonia,* 8(1): 89-101.

27. Child B. 2002. The acceptable face of conservation. Nature 415: 581–582.
28. Dahal S., Sharma T.P. 2014. High Altitude medicinal plants of conservation concern in Sikkim ENVIS Newsletter on Medicinal Plants. 7 (1-4): 10-11.
29. Dar A.K., Hassan W.U., Lone A.H., Haji A., Manzoor N., Mir A.I. 2017. Study to assess high demand and high commercial value Medicinal Plants of Jammu and Kashmir India - with special focus on routes of procurement and identification. International Journal of Research and Development in Pharmacy & Life Science.6(2): 2576-2585.
30. Devi, U. and Thakur, Minakshi. 2011. Exploration of ethno botanical uses of some wild plants from cold desert of Himachal Pradesh. *Asian J. of Exp. Biol. Sci.* 2(2):362-366.
31. Dhar U., Rawal R.S.,Upreti J. 2000. Setting priorities for conservation of medicinal plants-A case study in the Indian Himalaya. *Biological Conservation* 95: 57-65.
32. Dhar U.,Kachroo P. 1983. Alpine flora of Kashmir Himalaya. Scientific Publishers, Jodhpur, India. pp 132
33. Dhyani P.P., Kala C.P. 2005. Current aspects on medicinal plants: five lesser known but valuable aspects, *Curr. Sci.* 88(3): 335.
34. Dutt, B., Sharma, S.S. Sharma, K.R., Gupta, A. and Singh, H. 2011. Ethnobotanical survey of plants used by Gaddi tribe of Bharmour area in Himachal Pradesh. ENVIS bulletin: Himalayan Ecology. 19.
35. Dutta I.C.,KarnA.K. 2007. Antibacterial Activities of Some Traditional Used Medicinal Plants of Daman, Nepal. Pokhara, Nepal.
36. Gaur R.D., Semwal J.K., Tiwari J.K.1983. A survey of high altitude medicinal plants of Garhwal Himalaya. *Bull. Medico-ethnobot. Res.* 4: 102 -116.
37. Giri D., Tamta S. 2010. A General Account on Traditional Medicinal Uses of *D. hatagirea* (D.Don) Soo. *New York Science Journal* 3: 78-79.
38. Giri D., Tamta S. 2012. Propagation and conservation of *D. hatagirea* (D.Don) Soo, an endangered alpine orchid. *Afr J.Biotech* 11: 12586–12594.
39. Guleria, V. and Vasishth., 2009. Ethnobotanical uses of wild mediicnal plants by Guddi and Gujjar tribes of Himachal Pradesh. Ethnobotanical Leaflets 13:1158-67.
40. Gupta R. 1966. *Orchis latifolia* L.: A little known plant from north western Himalaya. *Indian Forester* 92: 701-703.
41. Gupta R. 1981. Plants in folk medicine of the Himalayas. In SK Jain (ed.) *Glimpses of Indian Ethnobotany,* pp. 83-90.
42. Gupta, A.K., Vats, S.K. and Lal, B. 1998. How cheap can a medicinal plant species be? *Curr. Sci.* 74(7)565-566.
43. Gruenwald, J. 2000, European Herbal Market Update, ICMAP News, 7 (6), pp. 10-11
44. Hajra P.K., Balodi B. 1995. Plant Wealth of Nanda Devi Biosphere Reserve. Botanical Survey of India, Calcutta, India.
45. Haridasan K., Shukla G.P.,Deori M.L. 2002. Cultivation prospects of medicinal plants of Arunachal Pradesh-A Review. pp. 329-344. In: SS Samant, U Dhar and LMS Palni (eds.). *Himalayan medicinal plants. Potential and Prospects*. Govind Ballabh Pant Institute of Himalayan Environment and Development, Almora.
46. Hussain A. 1994, Essential oils plants and their cultivation. Central Institute of Medicinal and Aromatic Plants (CIMAP), Lucknow, UP
47. IUCN 2004. National Register of Medicinal and Aromatic Plants. International Union for Nature Conservation Nepal, Kathmandu, Nepal.
48. Jain S.K. 1991. Dictionary of Indian folk medicine and ethnobotany. Deep Publications New Delhi pp. 69.
49. Jain S.K. Mehortra A. 1984. A Preliminary Inventory of Orchidaceae in India, Botanical Survey of India, Calcutta.
50. Joshi K., Chavan P., Warude D., Patwardhan B. 2004. Molecular markers in herbal drug technology. *Curr. Sci.*, 87: 159-165.

51. Jakobsone G., Dapkûnienë S., Cepurîte B.M., Belogrudova I. 2007. The conservation possibilities of endangered orchid species of Latvia and Lithuania. Monographs of Botanical Gardens (European Botanic Gardens Together Towards the Implementation of Plant Conservation Strategies). *Baltic Botanic Gardens* 1: 65–68.
52. Kala C.P. 2000a. Commercial exploitation and conservation status of high value medicinal plants across the borderline ofIndia and Nepal in Pithoragarh. *Indian For.*129: 80–84.
53. Kala C.P. 2000b. Status and conservation of rare and endangered medicinal plants in the Indian Trans-Himalayas. *Biol. Cons*. 93(3): 371-79.
54. Kala C.P. 2015. Indigenous uses, population density and conservation of threatened medicinal plants in protected areas of Indian Himalayas, *Cons. Biol.* 19(2): 368-378.
55. Kala C.P., Rawat G.S., Uniyal V.K. 1998. Ecology and Conservation of the Valley of Flowers National Park, Himalaya, WII, Dehradun, p. 99.
56. Kaul, M.K. 1986. Weed flora of Kashmir Valley. Scientific Publishers, Jodhpur. India.
57. Kaul M.K. 1997. Weed flora of Kashmir valley, Indus publishing Co. New Delhi, pp.189.
58. Kaul M.K., Handa S.S. 2002. Medicinal plants on crossroads of western Himalaya. pp. 73-86. *In*: SS Samant, U Dhar and LMS Palni (eds.). *Himalayan medicinal plants. Potential and Prospects*. Govind Ballabh Pant Institute of Himalayan Environment and Development, Almora.
59. Kaul M.K.1997. Medicinal plants of Kashmir and Ladakh (Cold and Arid Himalaya), Indus publishing Co. New Delhi, pp.113.
60. Khadka C.B., Hammet A.L., Singh A., Balla M.K., TimilsinaY.P. 2016. Ecological status and diversity indices of Panchaule (*D. hatagirea*) and its associates in Lete village of Mustang district, Nepal.*Banko Janakari*. 26,1:45-52.
61. Kizu H., Kaneko E., Tommori T. 1999. Studies on Nepalese Crude Drugs. XXVI. Chemical Constituents of Panch Aunle, the Roots of *Dactylorhiza hatagirea* D. DON. *Chem. Pharm. Bull.* 47(11): 1618-1625.
62. Kumar S., Singh J., Shah N.C., Singh S. 1997. Indian Medicinal plants facing genetic erosion, CSIR-CIMAP, Lucknow. pp. 96-97.
63. Kunwar R.M. 2006. Non-timber forest product of Nepal: A sustainable management approach. Centre for Biological Conservation International Tropical Timber Organization, Kathmandu, Nepal, pp. 423.
64. Kunwar R.M., Nirmal A. 2005. Ethnomedicine of Dolpa district, Nepal: the plants, their vernacular names and uses.*Lyonia,* 8(1): 43-49.
65. Lal B., Negi H.R., Singh R.D.,Ahuja P.S. 2004. Medicinal Uses of *D. hatagirea* among the natives of higher altitudes in Western Himalaaya. *J. OrchiD.Soc.* India, 18 (1-2): 97-100.
66. Machaka-Houri N., Al-Zein M.S., Westbury D.B., Talhouk S.N. 2012. Reproductive success of the rare endemic *Orchis galilaea* (Orchidaceae) in Lebanon. *Turk J Bot* 36: 677–682.
67. Mishra M.K. 1998. Commercial trade in medicinal plants in India regulatory mechanism, trade enterprise, sensibilities and conservation concerns: an overview. Pages 142–157.
68. Molure S., Walker S. 1998. Report of the workshop on conservation assessment and management plan process (CAMP) (BCPP – Endangered Species Proejct), Zoo Outreach Organization, Conservation Breeding Specialist Group, India, Coimbatore, India, pp. 62.
69. Nautiyal M.C. 1995. Agrotechniques of some high altitude medicinal herbs. *In*: R.C. Sundriyal and E. Sharma (eds.) *Cultivation of medicinal plants and orchids in Sikkim Himalaya*. Himvikas Occasional Publication No. 7. Govind Ballabh Pant Institute of Himalayan Environment and Development, Almora, pp. 53-64.
70. Nautiyal M.C., B.P., Nautiyal. 2004. Agrotechniques for high altitude medicinal and aromatic plants. Bishen Singh Mahendra Pal Singh, Dehradun, pp 202.
71. Negi, V.S., Maikhuri, R.K. and Vashishtha, D.P. 2011. Traditional healthcare practices among the villages of Rawain valley, Uttarkashi, Uttarakhand, India. Indian Journal of Traditional knowledge 10(3):533-537.

72. Olsen C.S., Helles F. 1997. Medicinal plants, Markets and margins in the Nepal Himalaya: Trouble in Paradise. *Moun. Res. Dev.* 17(1): 363-374.
73. Phani K.G, Kumar R., Badere R., Singh S.B. 2010. Antibacterial and antioxidant activities of ethanol extracts from trans Himalayan medicinal plants. *Pharmacognosy J.* 2(17):66-69.
74. Rai L.K. Sharma E. 2002. Diversity and indigenous uses of medicinal plants of Sikkim. *In*: S. S. Samant, S.S., U. Dhar & L.M.S. Palni. 2002. *Himalayan medicinal Plants potential and prospects.* Himvikas occasional publication No. 14. Govind Ballabh Pant Institute of Himalayan Environment and Development, Almora, pp. 157-163.
75. Ranpal S. 2009. An assessment of status and antibacterial properties of *D. hatagirea* in Annapurna Conservation Area (A case study of Paplekharka, Lete VDC, Mustang). Master dissertation, Tribhuvan University Institute of Forestry, Nepal.
76. Rau M.A. 1975 High altitude flowering plants., BSI, Howrah.
77. Rawat J.K., Semwal J.K., Purohit A.N. 1985. Biossoming Garhwal Himalaya, HAPPRC, Garhwal University, Srinagar Garhwal, pp. 7.
78. Rawat, V.S. and Jalal, J.S. 2011. Sustainable utilization of medicinal plants by local community of Uttarkashi district of Garhwal Himalaya, India. European journal of medicinal plants 1(2):18-25.
79. Samal, P.K., Dhyani, P.P., Dollo, M., 2010. Indigenous medicinal practices of Bhotia tribal community in Indian Central Himalaya. Indian J. Tradit. Knowl. 9, 140–144.
80. Samant S.S., Dhar U. Rawal R.S.2001. Himalayan Medicinal Plants-Potential and Prospects (eds) Gyanodaya Prakashan, Nainital, India, 166-184.
81. Samant S.S., Dhar U., Palni L.M.S. 1998. Medicinal plants of Indian Himalayan: Diversity, Distribution Potential Values. HIMAVIKAS Publication. No.13, Gyanodaya Prakashan. Nainital, India.
82. Selvam, A.B.D. 2012. Pharmacognosy of Negative Listed Plants. Pp. 59-68.
83. Shabnam S.R. 1964. Medicinal plants of Chamba. *Indian Forester.* 50-63.
84. Sharma BM and Jamwal PS. 1998. Flora of Upper Liddar valley valleys of Kashmir Himalaya. Jodhpur, Raj., Scientific Publishers (India).
85. Sharma A.B. 2004. Global Medicinal Plants Demand May Touch $5 Trillion By 2050. Indian Express. Monday March 29, 2004.
86. Sharma, P.K., Chauhan, N.S., Lal, B., 2003. Commercially important medicinal and aromatic plants of Parvati valley, Himachal Pradesh. J. Econ. Taxon. Bot. 27 (4), 937-942.
87. Sharma, P.K., Thakur, S.K., Manuja, S., Rana, R.K., Kumar, P., Sharma, S., Chand, J., Singh, A., Katoch, K.K., 2011. Observations on traditional phytotherapy among the inhabitants of Lahaul valley through amchi system of medicine-A cold desert area of Himachal Pradesh in North Western Himalayas, India. Chin. Med. 2, 93-102.
88. Shrestha U.B., Shrestha S. 2004. Major Non-Timber Forest Products of Nepal. Bhundipuran Parkashan, Kathmandu, Nepal. 411p.
89. Shukla B.K. Sinha G.P. 2000. Salam panja, Parijaat (Hindi magazine) 5: 15.
90. Silori C.S., Badola R. 2000. Medicinal plants cultivation and sustainable development: A case study in buffer zone of the Nanda Devi Biosphere Reserve, Western Himalaya, India. Mt Res Dev 20: 272-279.
91. Singh B.P. 1999. Illustrated field guide to commercially important medicinal and aromatic plants of Himachal Pradesh (with special reference to Mandi district). Society for Herbal Medicine and Himalayan Biodiversity.
92. Singh R.S. 1983. *Vanaushadhi Nirdashika* (Ayurvedic Pharmacopoeia), Uttar Pradesh Hindi Sansthan, Lucknow, pp. 73.
93. Singh, A. and Duggal, S. 2009. Medicinal orchids: an overview. Ethnobotanical leaflets 13:351-63.

94. Singh, A., Lal, M. and Samant, S.S. 2009. Diversity, indigenous uses and conservation prioritization of medicinal plants in Lahaul valley, proposed Cold Desert Biosphere Reserve, India. International Journal of Biodiversity Science and Management. 5(3):132-154.
95. Singh, K.N., Lal, B., 2008. Ethnomedicines used against four common ailments by the tribal communities of Lahaul-Spiti in Western Himalaya. J. Ethnopharmacol. 115, 147-159.
96. Srivastava R.C., Mainera A.K. 1994. A note on *D. hatagirea* (D.Don) Soo: an important medicinal orchid of Sikkim, *Nat. AcaD.Sci. let.* (India) 17 (7-8): 129-130.
97. Stoutamire W.P. 1974. Terrestrial orchid seedlings. In: Withner CL (ed.). The Orchids-Scientific Studies. Wiley Interscience, New York.101-128.
98. Sundriyal R.C., Sharma E. 1995. Cultivation of medicinal plants and orchids in Sikkim Himalaya. Himvikas Occasional Publication No. 7. GBPIHED Almora.
99. Thakur M., Dixit V.K. 2007. Aphrodisiac activity of *D.hatagirea* (D.Don) Soo in male albino rats, J Evid Based Complementary Altern Med 4(1): 29–31.
100. Thakur, D., Rathore, N., Sharma, M. and Chawla, A. 2018. Enhanced reproductive success revealed key strategy for persistence of devastated populations in Himalayan food deceptive orchid, *Dactylorhiza hatagirea.* Plant species Biology
101. Uniyal S.K., Singh K.N., Jamwal P., Lal B. 2006. Traditional use of medicinal plants among the tribal communities of Chhota Bhangal, Western Himalaya. *J Ethnobiol Ethnomed* 2: 14.
102. UniyalS.K., Awasthi A., Rawat G.S. 2002. Current status and distribution of commercially exploited medicinal andaromatic plants in upper Gori valley, Kumaon Himalaya, Uttaranchal. *Current Science* 82(10): 1246-1252.
103. Vaasa A., RosenbergV. 2004. Preservation of the rare terrestrial orchid's in vitro. *Acta Univ. Latvi. Biol.* 676:243-246.
104. Ved D.K., Goraya G.S.. 2008. Demand and Supply of Medicinal Plants in India. Bangalore, India: BSMPS, D. Dun, and FRLHT.
105. Ved D.K., Kinhal G.A., Ravikumar K., Prabhakaran V., Ghate U., Sankar R.V., Indresha J.H. 2003. CAMP Report: Conservation assessment and Management prioritization for Medicinal plant of Jammu and Kashmir, Himachal Pradesh and Uttaranchal. *In*: Workshop at Shimla, Himachal Pradesh, FRLHT Bengalore, pp. 115-116.
106. Ved D.K., Tandon V. 1998. CAMP Report for high altitude medicinal plants of Jammu-Kashmir and Himachal Pradesh, FRLHT Bengalore, pp. 18.
107. Veyret Y. 1969. La stgructure des semences des Orchidaceae et leur aptitude a la germination *in vitro* en cultures pures. *Trava. Du. Lab. De. La. Jausnia*. 3:89-98.
108. Vij S.P., Nandi S.K., Palni L.M.S., Kumar A.2002. Orchids and Tissue Culture: Current Status, in Role of Plant Tissue Culture in Biodiversity Conservation and Economic Development. Nainital, India: Gyanodaya Prakashan, Nainital.
109. Vij S.P., Pathak P., Mahant K.C. 1995.Green pod culture of a therapeutically important species *D. hatagirea* (*D.*Don) Soo. *J. Orchid Soc. India*. 9:7-12.
110. Vij S.P., Srivastav R.C., Mainra A.K. 1992. On the occurrence of *D. hatagirea* (D.Don) Soo in Sikkim, *Orchid News* 8(9): 14-15.
111. Vij S.P., Verma J.C., Kumar S. 2013. *Orchids of Himachal Pradesh*, Bishen Sing Mahendrapal Sing, Dehradun: 65-67.
112. Warghat A.R.,Bajpai P.K., Srivastava R.B., Chaurasia O.P., Chauhan R.S., Sood H. 2014. In vitro protocorm development and mass multiplication of an endangered orchid, *D. hatagirea*. *Turk J Bot* 38: 737-746.
113. Xancó B.M., Aguilar J.V., Kenicer G.J., McHaffie H. 2012. Establishing *ex situ* conservation methods for *Dactylorhiza ebudensis* and *D. traunsteinerioides*, a combination of *in situ* turf removal and *in vitro* germinations. *J. Botanic Garden Horticulture*, No. 10.

6

Crepidium acuminatum and Allied Species

Crepidium acuminatum: Jeevak

Microstylis muscifera: Rishbhak

Akshay Nag and D. R. Nag

Crepidium acuminatum **(D. Don) Szlach.**
(A) Herbarium sheet, **(B)**Aerial plant parts, **(C)** Flowering stage
(D) Whole plant, **(E)** Crude drug material (Peseudobulbs)

INTRODUCTION

Himalayan region is known for the diversity of its cultural heritage and climate. In Indian part, this mountain range includes states of Uttarakhand, Himachal Pradesh and Jammu & Kashmir. The climatic variability of the region,from sub-tropical to extreme dry temperate, offers great attraction to naturalists, all across the globe. The forest formations here are the natural habitats of a large number of flora and fauna. The flora found in these forests varies with the altitudinal gradient and is quite distinct. In the sub-tropical Himalayas, for example,the predominant vegetation is of deciduous type consisting of trees, shrubs, scrubs, climbers and xerophytes. Man-made *Chir* pine forests can also be witnessed. At middle elevation, the climate is sub-temperate and supports a mix of broad leaves and pines forests.In the high reaches including Trans Himalayas, climate is typically wet or dry temperate. In the wet temperate climate, evergreen forests are the predominant vegetation, whereas, the dry temperate areas do not support any significant vegetation.This variation in climate and altitude has a great impact on the biodiversity found in the region, and represents about seven percent of the country's total biodiversity.

Among this diversity, many valuable and rare medicinal and aromatic plants grow wild in the region. The herbs which are traded from this region are of immense value for the preparation of Ayurvedic drugs in the form of monoherbal preparations or as multi-compound herbal formulations. *Charak Samhita* mentions that the herbs of the Himalayan forests excel in efficacy.

In the modern system of medicine too, plants have been accepted to be of curative value. *Taxus-baccata* and *Podophyllum hexandrum* are examples known to be effective in treating certain types of cancers. *Ephedra gerardiana* has shown promise for treating asthmatic disorders. *Berberis aristata* and its allied species yield berberine hydrochloride that is used to treat some eye diseases. Many plant species are considered as good tonic and are consumed as food supplements. In Ayurveda such plants are classified under term *"Rasayana"*, which in modern terminology can be co-related with nutraceutical. In Ayurveda, species of "*Ashtavarga*", a group of eight medicinal plants, are identified as "*Rasayana*" and believed to rejuvenate human body. Western Himalayan region is the natural habitat of most species of this group.

In this chapter an effort is made to deal with the species of genera which correlate with *Ayurvedic* nomenclature as *Jeevak* and *Rishbhak*. While *Crepidium acuminatum (*Synonyms*: Microstylis wallichii, Malaxis acuminata*) has been the focus of discussion, the other plants associated with *ashtavarga* as mentioned in the Ayurveda system of medicine have been listed (Table 1).

Table 1: *Ashtavarga* plants mentioned in the *Ayurveda* system of medicine

S. No.	*Ayurvedic*/ Sanskrit/ Common Name	Botanical name
1.	*Jeevak*	*Crepidium acuminatum* syn. *Microstylis wallichii*, *Malaxis acuminata*
2.	*Rishbhak*	*Microstylis muscifera* syn. *Malaxis muscifera*
3.	*Meda*	*Polygonatum verticellatum*
4.	*Mahameda*	*Polygonatum cirrhifolium* (Wall). Royle
5.	*Kakoli*	*Roscoea procera* Wall; *Fritillaria roylei* Hook.f
6.	*Kshirkakoli*	*Lilium polyphyllum* D.Don
7.	*Riddhi*	*Habenaria edgeworthii* H.f
8.	*Vriddhi*	*Habenaria intermedia* D.Don

A. BIOLOGICALASPECTS

The genus *Crepidium* belongs to Orchidaceae which is one of the most diverse and widespread families of the flowering plants.Orchidaceae comprises of around 450 genera and more than 5000 species distributed throughout the world from the high mountainous temperate regions to the tropical rain forests. Plants of this family are mostly herbaceous and shrubby in habit.

• Systematic classification

Kingdom : Plantae

Phylum : Magnoliophyta

Class : Liliopsida

Order : Asparagales

Family : Orchidaceae

Genus : *Crepidium*

Species : *acuminatum*

- **Synonyms:** (http://www.theplantlist.org/tpl/record/kew;http://biodiversity.bt/species/show)

a) *Crepidium acuminatum* (D.Don) Szlach.

Synonyms:*Corymborkis acuminata* (D.Don) M.R. Almeida, *Crepidium bilobum* (Lindl.) Szlach. ex Lucksom, *Malaxis acuminata* D.Don, *Malaxis acuminata* f. biloba (Lindl.) Tuyama, *Malaxis acuminata* var. *biloba* (Lindl.) Ames, *Malaxis allanii* S.Y.Hu & Barretto, *Malaxis biloba* (Lindl.) Ames, *Malaxis pierrei* (Finet) Tang & F.T.Wang, *Malaxis siamensis* (Rolfe ex

Downie) Seidenf. & Smitinand, *Malaxis wallichii* (Lindl.) Deb, *Microstylis biloba* Lindl., *Microstylis pierrei* Finet, *Microstylis pierrei* var. *rotundata* Finet, *Microstylis siamensis* Rolfe ex Downie, *Microstylis trigonocardia* Schltr. *Microstylis wallichii* Lindl., *Microstylis wallichii* var. *biloba* King & Pantl., *Microstylis wallichii* var. *brachycheila* Hook.f.

b) *Crepidium ridleyanum* (P.F.Hunt) M.A. Clem. & D.L. Jones Synonyms:*Microstylis acuminata* Ridl.

c) *Microstylis muscifera* (Lindl.) Ridl. Synonym: *Malaxis muscifera* (Lindl.) Kuntze

• Common names

Crepidium acuminatum: Most commonly used name is *Jeevak*; other names are:

Chiranjivi, Deerghaayu, Ksveda, Harsanga, Kurchashira.

Microstylis muscifera: *Rishbhak, Dheera, Durdhara, Bandhura, Gopti, Kakuda, Materka, Indraksa, Vasani, Virsa, Varisnabh.*

• Diversity within genus

More than 200 species of genus *Crepidium* are distributed in tropical and temperate regions of the world, across Indian sub-continent, Southeast Asia, China and northern regions of Australia. Within India, about 22 species are reported, and 18 species specifically from Himalayas (Gupta*et al.*, 2016). In Himachal Pradesh, three species viz. *Microstylis wallichii, M. muscifera* and *M. clyndrostachya* have been reported from the wild. These species can be differentiated from each other by the number of leaves and the size of flower, but more importantly by the elevation where these species grow naturally.

• Geographical distribution

C. acuminatum is a terrestrial orchid found in the sub-temperate to temperate locations, between elevations of 1400-2000 m amsl. In India, the species has been reported from Khasi and Jaintia hills, Western Ghats, Nilgiris hills, high offshore hills in Andaman and mountain slopes in Eastern Ghats (Adams *et al,.* 2018). Along the Himalayas, in Assam, Arunachal Pradesh, Himachal Pradesh, Manipur, Mizoram, Nagaland Tripura and Uttarakhand states of India, the species inhabits moist and humus-rich locations in shaded areas (Samant *et al.,* 1998; Chauhan, 1999; Singh, 2005, 2006; Dhyani *et al.,* 2010; Balkrishna *et al.,* 2012). The species is also reported from eroded cliffs, from the sloppy areas of oak forests, or growing as an under canopy herb.

• Habitat and ecology

Orchids can be classified as Terrestrial and Epiphytic. In the terrestrial habit, species grow on soil or sometimes on rocks, in moist, sub-tropical to extreme temperate conditions. Some species may grow in the marshy open fields also. Terrestrial species are usually succulent with tuberous or fibrous roots which annually produce erect flowering stem. On the contrary, epiphytic forms grow on trees and develop special aerial roots for absorbing moisture and nutrients from aerial habitat. *Crepidium* and *Microstylis* are genera of terrestrial and semi epiphytic orchids and are small herbs.Both the forms may develop pseudo bulbs which are borne on or above the soil. These thick swollen structures are actually a modified form of stem which stores water and nutrients.

• Taxonomic enumeration

Leaves are linear, leathery or fleshy, arise from stout stem, and are solitary or few, alternate or crowded, with sheathing base or sometimes reduced to scales.Flowers are unique with biseriate perianth. Perianth leaves are in two whorls and identical in colour. Out of the three petals, two are similar to sepals but the third one which forms the lower lip is usually much larger, entire or 2-4 lobed and is of contrasting conspicuous shape and colour, often with a spur which is a distinctive feature of family Orchidaceae. Stamens are united with the style to form a fleshy column in which neither stamen nor style is differentiated or visible. Pollens are in sacs or pollinia. Ovary is inferior and one chambered. Fruit is a capsule and seeds are minute.

C. acuminatum

It possesses a fibrous root system and horizontal rhizomes. The flowers grow in a spike-like raceme, which is short and bears numerous tiny flowers. There are three to five broadly lanceolate, acuminate leaves arising from the base of the stem. Perianth leaves are spreading, lip sessile, ovate, flat, and spur none. Anthers are sub-terminal, pollinia -4. Pseudobulbs are present, which remain above soil, are conspicuous, stout and horn shaped. Flowering stem is a spike like raceme which is 6-8 inch in width and is ribbed. This spike contains numerous flowers. Flowers have a short stalk of about one and half inch length. Flowers are yellowish green, tinged with purple, especially near the centre. Sepals are re-curved, two lateral rather shorter than the dorsal. Lip shield like, broadly ovate, tip notched, basal lobes straight and acute.

Microstylis muscifera: The species is commonly known as *Rishbhak*. Generally, the number of leaves in this species is two, which may be oblong or ovate, unequal and obtuse. Pseudobulbs are small and ovoid. Flowering stem is 6-18 inch long. Flowers are minute and half inch long, yellowish-green in colour.

Sepals are broadly lanceolate. Petals are linear. Lip ovate, abruptly pointed, margin thickened.

• Phenological period

Generally, these plants resume their growth after winter with rise in atmospheric temperature during spring months, blooms in summer, and flowers from July to September/October.

• Cytotypes

Crepidium acuminatum: 2n=30, 36, 42 (Vij *et al.*, 1986).

Microstylis khasiana (2n=42); (Rokaya et al., 2013).

B. REPRODUCTIVE BIOLOGY

• Reproductive phenology

In a two year study by Madhukar (2013), the rhizomes of *C. acuminatum* were planted in two different seasons, autumn (October 2011) and spring (March 2013). These rhizomes sprouted in the first week of June 2012 and last week of May 2013, respectively. The maximum sprouting ($\geq$ 80%) was achieved within one month from the day of planting. In both the cases, flowering commenced by the end of June and continued up to the second week of July. The development of fruit took place during August. Mostly, only a single fruit developed in each case, which ripened after third week of October. Thereafter, plants got into dormancy.

• Life Cycle

Tamta *et al.* (2015) studied propagation by raising psuedobulbs (3600 number of propagules) in nursery during the month of February-March. After shoot development, plants were transplanted in the forest field during the month of May. The plants were allowed to grow for the rest of the growing season, and during the nextyear as well without any harvesting. After two seasons of active growth, psuedobulbs were harvested during October. The yield during fourth year was 5-6 times higher compared to what was obtained during the second year. In nature, plants generate from underground dormant apical buds in the month of April. Generally, plants show vegetative growth for the first two years, and produce flowers by third year onwards (Sharma *et al.,* 2011).

• Pollination and pollinators

There is not much information regarding how pollination is achieved in *C. acuminatum*. However, as noticed for many orchids, nectar is the major attractant, besides pollens, oil, wax, etc. (Arditti, 1992). Nectaries, which produce nectar are well developed in orchids so as to be conspicuously visible to bees and insects that carry out pollination.

C. MEDICINAL ASPECTS

• Plant material used as crude drug

The crude drug, used in folk or Ayurvedic formulations, is derived from pseudobulbs of 2-3 years old plants (Adams *et al.*, 2018). Pseudobulbs, the aerial stem part of the plant, are horn shaped morphological entities, which are found attached to the mother plants just above the soil surface.

• General appearance

Pseudobulbs are traded in the herbal market, and can be identified as stout, succulent, greenish morphological entities. Each pseudobulb is covered within a whitish membrane. After harvesting, the pseudobulb remains green for a long period, and this property has led to the plant being commonly called as *Jeevak* in Sanskrit, which literally means – "*one which remains alive*".

• Organoleptic properties

Visually, the crude drug source can be easily recognized by its typical horn shaped greenish structure. It has a rough texture and a characteristic odor and taste. The pseudobulbs are sweet and refrigerant, and in Ayurvedic dynamics listed as sweet in taste, cold in potency, that pacifies *vata* and aggravate *kapha* (Singh, 2006).

• Microscopic characteristics

Transverse section of pseudobulb shows conspicuous barrel shaped epidermal cells that surround the inner ground tissue. Ground tissue consists of small parenchymatous cells in the outer layer just below epidermis, but is bigger in size, with clear inter-cellular spaces, towards the center. Scattered vascular bundles can be seen in ground tissue, in which phloem is encircled by xylem. Vascular elements show scalariform and spiral thickening (Sharma *et al.*, 2009).

• Powdered plant material

Dry powder of the material shows presence of calcium oxalate crystals of different shapes and sizes. The crushed product might consist of the parenchymatous cells containing brown yellowish pigments. Fibers, vessels and tracheids are also present (Sharma *et al.*, 2009).

• Major Chemical constituent

The pseudobulbs of *C. acuminatum* were reported tocontain alkaloid, glycoside, flavonoids and β-sitosterol (Chauhan, 1999).

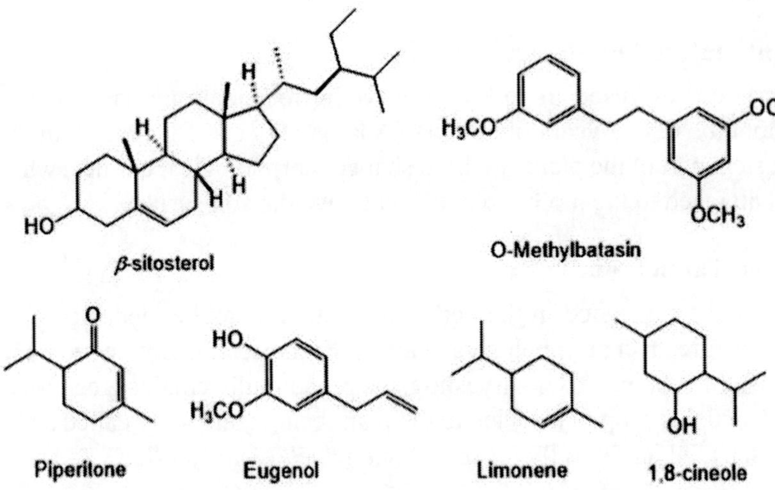

Chemical constituents of *Crepidium acuminatum*
(Chemical structure is contributed by Dinesh Kumar)

• Substitutes and adulterants

The demand for crude drug in market is greater than the existing available resources in nature. Terrestrial orchids like *C. acuminatum* are restricted in distribution, are found in small populations, in fragmented locations and are not easily accessible for collection in required volume. Consequently, the raw materials extracted from other species of the same genus are often used as substitutes for the real medicinal species (Roy *et al.*, 2013). Substitution of one species with another is a common practice in traditional pharmacopoeias (Khajuria *et al*., 2017). Dried tubers of "Vidarikand" (*Pueraria tuberosa*) are used as substitute drug material. Other substitutes are *Centaurea behen* L (*Safed behmen*), *Tinospora cordifolia* (Wild.) Miers. (*Guruchi*), *Malaxis mackinnoni* (Duthie) Ames and *M. cylindrostachya* (Lindl.) Kuntze (Balkrishna *et al.,* 2012).

• General identity tests

No information available

• Purity tests

• Foreign matter	Not more than 2 percent
• Total ash	Not more than three percent
• Acid soluble ash	Not more than 0.5 percent
• Alcohol soluble extractive	Not less than 4 percent
• Water soluble extractive	Not less than 12 percent
• Starch	Not less than 19 percent

• Medicinal usage

The somewhat sweet pseudobulbs are used as refrigerant, febrifuge and as tonic. These are also useful in haematemesis, fever, for seminal weakness, burning sensations, dyspepsia, emaciation, and tuberculosis. As stated earlier, the species represents an important component of *Ashtavarga* plants of *Ayurvedic* system of medicine. In *Ayurvedic* texts, almost all the indigenous plant species have been identified with some therapeutic use. Compilation of all such plants and their medicinal uses is available in a Medical classic text "*Bhav-prakash-nighantu*", an *Ayurvedic Materia Medica.*

Medicinal properties and the clinical uses of each plant species included in *Bhav-prakash-nighantu* have been described in detail and practiced to cure specific ailments. The classification of such plants is based on different criteria such as habit of the plants, its morphology, medicinal properties, therapeutic uses, etc. Depending upon the number of herbs used in the prescription, *Ayurveda* drugs are broadly classified as single or monoherbal preparations, as compound or poly herbal formulation. In monoherbal formulation, a single plant or its part is used for preparing the drug. In a compound formulation, a mixture of two or more crude drugs is made. Further, "*Vargikaran*" (grouping) of the plants was also done on the basis of medicinal properties where similar morphological parts of plats have common medicinal properties such as "***Dashmool***" (root part of ten different species), "***Panchtrin***" (five different grass species), "***Triphala***" (fruits of three different species) and so on. Among the various "Vargas" described, "*Ashtavarga* " (Table 1) is a group of eight different genera having common properties and therapeutic uses. Plants of the species of *Ashtavarga* are considered to be highly nutritive and widely used in various *Ayurvedic* formulations, such as "***Chyavanprash***", which is a well-known health supplement containing more than 30 crude drugs derived from different plant species. Regular use of *Chyavanprash* is claimed to enhance the immunity of the body.

• General uses

Ashtavarga plants including *Jeevak* and *Rishbhak* have been known to have antioxidant properties to overcome general weakness, since the time of Charaka (2nd century A.D), to whom the classic text of "*Charak-Samhita*" is attributed. *Charaka* described the medicinal uses and gave methodology of preparing these formulations. Some excerpts from this scripture on medicinal properties of *Jeevak* and *Rishbhak* are as follows:

जीवकर्षभकौ मेदामहामेदा काकोली क्षीरकाकोली मुद्ग्पर्णीं
माष्पण्यौं जीवन्ति मधुकमिति दशेमान जीवनीयानि भवन्ति।

Translation: "These drugs are called life giving and are called *Jeevak, Rishbhak, Meda, Mahameda, Kaakoli, Kshirkaakoli, Vanmoong, Maashparni (van urad), Jeevanti* and *Mulethi*"

जीवकर्षभकौ वीराम मेदाम्रिद्धिं शतावरीम्।
मधुकंचाश्वगन्धांच साधयेत्कुड्वोन्मिताम्॥

Jeevak, Rishbhak, Kaakoli, Meda, Ridhhi, Shatawar, Mulethi, Ashwagandha are the ingredients of "***Vaajikarann Ghrit"***, which are used for increasing sperm count in humans.

आजं घ्रितं क्षीरपात्रं मधुकंचोत्पलानि च्।
जीवकर्षभकौ चैवपिष्तवासर्पिर्विपाचयेत्।
सर्व नेत्राभिघाते तुसर्पिरितत्प्रशस्यते॥

"*Make a paste of Mulethi, Neel Kamal, Jeevak and Rishbhak in milk and add goat's ghee to this paste. This medicine is highly potent for all kinds of eye ailments*."

• Ethnobotanical uses

Plants have been used traditionally world over by different communities to treat different ailments and maintain good health. Species of *Microstylis* are taken to enhance sperm production and to address male fertility issues in humans. The paste of *C. acuminatum* is suggested to provide relief from insect bites and rheumatic pains. It is also prescribed to give relief in conditions of rheumatism, haemorrhage, dysentery, burning sensation and in general debility (Sharma *et al*., 2011). The paste of the crude drug is applied to treat diathesis, burning sensation, fever, sores and used as general tonic (Subedi *et al*., 2013).

D. CHEMISTRY ASPECT

• Chemical composition

Very little work has been done on phytochemical analysis of the plant. Chemical analysis done so far reveals that it contains a large number of alkaloids, glycosides, flavonoids and sterol (Sharma *et al.,* 2011). Bhatnagar *et al.* (1970) earlier reported presence of Piperitone, Citronellal, Euginol, Limonene, 1, 8-cineole, p-cymene, O-Methylebatatasin and cetyle alcohol. Lohani *et al.* (2013) reported a number of metals/nutrients, *viz.* Cu, Zn, Mn, Fe, K, Ca, Mg, Al, Ba, B, Mo, Cl, etc,. fatty acids (Linoleic acid, α -Linolenic acid, Oleic acid, Palmitic acid, Stearic acid, ã-Linolenic acid, Eicosanoic acid, Eicosenoic acid and Eicosadienoic acid), and Tocopherol and ã-Tocopherol from *C. acuminatum* to show that chemical contents qualitatively remained same in the wild and cultivated samples.

E. PHARMACOLOGICAL ASPECTS

Bioactivity

Orchids, in general, have been shown to possess a wide range of chemical compounds or metabolites with useful pharmacological activities (Pérez Gutiérrez, 2010). *C. acuminatum*, an important *Ashtavarga* plant, has been endowed with aphrodisiac and rejuvenating properties (Adams *et al.,* 2018; Gupta *et al*., 2016; Raturi *et al*., 2017). The following case studies depict some of the bioactive properties of this group of plants.

• Anti-microbial activity

Sharma *et al.* (2011b) studied antimicrobial activity of pseudobulbs of *C. acuminatum* and reported that it inhibited fungal inhibition better than bacteria inhibition. The antifungal activity was equivalent to that of reference fluconazole. The extract of pseudobulb of *C. acuminatum* showed potential antibacterial activity on strains of *Staphylococcus aureus, Escherichia coli, Bacillus subtilis* and *Pseudomonas aeruginosa* (Arora *et al.,* 2017).

• Anti-inflammatory

Sharma *et al.* (2007) reported that ethanolic extracts of pseudobulbs of *Microstylis wallichii* exhibited analgesic and anti-inflammatory activity in experimental rats. It was suggested that carrageenin-induced inflammation in rats resulted probably due to inhibition of inflammation and pain mediators.

• Antioxidant activity

The antioxidant activity of the butanol extract of *C. acuminatum* was studied by Garg *et al.* (2012), using different methods, including 1,1-diphenyl-2-picryl hydrazyl (DPPH) radicalscavenging activity, reduction capability by Fe3+ - Fe2+ transformation method andhydrogen peroxide scavenging method. Compared to the standard compounds, it was concluded that the species showed a goodantioxidant activity.

• Antiproliferative activity

Ethanolic extract of psuedobulbs of *C. acuminatum* was used to investigate anti-proliferative activity against four human cancer cell lines, such as A549 (non-small cell lung cancer cells), DU145 (human prostate carcinoma), DLD1 (human colorectal adenocarcinoma), and MCF-7 (human breast adenocarcinoma) using the sulphorhodamine B assay (Singh *et al.*, 2017). The study suggested the species to have a potent anti-proliferative activity.

Clinical Studies

• Role in management of specific ailments

Ashtavarga plants are considered in Ayurveda as source of good *rasayana* or containing beneficial compounds like antioxidants that strengthen immune system, cell regeneration capacity and overall vitality (Anon, 2006; Balkrishna *et al.*, 2012; Adams *et al.*, 2018).

Toxicity related information

All plant formulations may not be absolutely safe or free from health risks. Traditional use over a very long period of time has been one of the considered acceptable measure but critical evaluation with moderm tools of science have raised serious debate on the safety of many formulations or plant based crude drugs (De Smet, 2004). This situation is further questioned due to lack of reliable information or support from clinical studies on health safety issues for *Ashtavarga* plants. Incidentally, some of such plant based formulations do enjoy official approval. The crude drug (*Ashtavarga*) powder was fed to 40 rats for 41 days to assess its effect on liver function parameters (Hamid *et al.*, 2012). While all enzymes indicative of impaired liver function increased, the level of bilirubin, that reflects liver toxicity decreased,thereby indicating normal liver function.

• Formulations

The plants listed under *Ashtavarga*, including *C. acuminatum*, have been used to develop various formulations based on the knowledge of Ayurveda. Some of these are *Ashtavarga churna, Jevaniyo Dashko, Brahini gutika, Mahakshay* and *Chyawanprash rasayan, Chitrakali taila, Himvana agada, Vachadi taila, Mahakalyan ghrita,Mahamayura ghrita, Mahapadma taila, Jivania ghrita, Vajikaran ghrita,* which are sold in the market (Balkhrishna *et al.*, 2012; Chinmay *et al.*, 2011; Kaushik, 1983).

F. THREAT AND CONSERVATION ASPECT

Population status

The species that constitute *Ashtavarga* plants are distributed in sub-temperate to temperate wet zone of Himalayas, and mostly found growing in small formations, as an understory herb in oak forests.All these species, including *C. acuminatum*, are actively sought by the herbal industry.According to earlier studies, the plant did not figure as the major traded medicinal plants at national level, but were being extracted in apparently sustainable volumes from the wild. However, with increase in sale of *Chyavanprash* in recent times, the demand for *C. acuminatum* and *Microstylis* sp. is expected to increase by many folds. It is important to note that the slow regeneration potential of these species, in taking around 2-3 years for pseudobulbs to attain marketable size, may not quite meet their annual demand and extraction from wild.

• Threats

Nearly, all orchids are threatened species among flowering plants, and medicinal orchids face still greater threat due to over-extraction from wild and degradation and destruction of its habitat (Pant *et al.*, 2002). Many of these commercial species are sought for their medicinal or nutraceutical properties attributed to the roots or rhizomes. Consequently, the whole plant is uprooted to obtain the crude drug. *C. acuminatum* is one of such species that is regularly traded in bulk from Himalayas (Hinsley *et al.* 2017), and its unsustainable harvest from its rather restricted habitat is a major threat to its depleting natural status in wild. This valuable species has been listed as endangered in recent times (Ved *et al.*, 2003; Lohani *et al.*, 2013).

• Habitat loss/degradation/fragmentation

Oak forests, the favored habitat of *C. acuminatum* and species of *Malaxis*, has faced extensive degradation and large scale felling of oak trees for making high quality charcoal, has lead to loss of natural habitat of associated herbaceous species. The rising incidences of forest fires, depleting atmospheric humidity has further adversely affected this habitat. Together, these factors have lead to serious habitat destruction, poor regenerationand fragmentation of the size and frequency of its natural populations (Gupta, 2016).

• Over-exploitation

For commercial purpose,most medicinal species are collected from the wild by hired labourers. Such extractions are nearly always made beyond the sustainable limits, without consideration for proper time and stage for harvest. Since the marketable part of the plant remains attached with the mother plant, the whole plant is uprooted without leaving any vegetative part for regeneration. This is a major reason for loss of the natural resource in recent times. In the CAMP workshop in 1998, it was agreed by experts that during the preceding 10 years,more than 40% of the wild population of *Crepidium* and allied species declined in the Indian Himalayan region, thus listing the global status of the species as vulnerable.(1.http://dx.doi.org/10.2305/IUCN.UK.2015-; 2.RLTS.T50126625A50131390.en). Subedi *et al.* (2013) listed *C. acuminatum* as one of the 32 species overexploited for medicinal purpose trade in Nepal. Another report, that of 'Envis Centre on Medicinal Plants' (envis.frlht.org/ plantdetails/) stated that trade from eastern Himalaya (Sikkim) was much higher compared to its western part (Himachal Pradesh).

• Pollution and diseases

There are no reports regarding any disease or insect pest in this crop.

• Climate change

There are no specific studies on the impact of climate change on the species of *Ashtavarga*. However, considering their geographical distribution and habitat preference, some of the possible impacts are obvious. These include, warming related loss of atmospheric humidity in specific habitats, up-migration of species leading to renewed interaction with the existing vegetation, etc. The altered set of climatic variables in relation to loss of habitat specific advantage to species like orchids, interaction of climate change with habitat loss and habitat fragmentation, altered phenology thereby impacting its life cycle, can significantly impact species natural status (Barman and Devdas, 2013; Chen *et al.*, 2011; Root *et al.*, 2003). The multiple impact of climate change in long course of time for these species could be grave (Thomas *et al.*, 2004).

• Conservation strategies

As stated above, uprooting medicinal plant for the underground parts, can threaten status of mother stock of plant like *C. acuminatum* in its natural habitat. Effective conservation strategies must therefore focus on both *in-situ* and *ex-situ* measures. Under *ex-situ* conservation activities, new areas for cultivation of medicinal plants need to be identified and developed either within or outside forest areas. Under *in-situ* conservation, areas may be identified where the target species grows naturally,and from where extraction of the crude drug should be strictly prohibited. This is necessary to ensure that germplasm is conserved, and also where sizeable population remains available for further multiplication. Participation of local people in both conservation activities and cultivation is vital in this regard.

• *Ex-situ* and *in-vitro* conservation

Ex-situ and *in-vitro* measures for conservation have been suggested for some Himalayan orchids. Various options suggested for propagation of *C. acuminatum* by different workers can be employed to develop sizable population of the species for conservation and sustainable utilization. Arenmongla and Deb (2012) reported in vitro germination of immature seed on MS medium containing 3% sucrose and survival of 75% transplants in the field, to show a rapid way for multiplying *C. acuminatum* (*Malaxis acuminata*). Abraham (2010) reported adventitious shoot induction in *C. acuminatum* from cultured inter-nodal explants. Kaur and Bhutani (2009) studied micro propagation in the species (*C. acuminatum*) using pseudobulbs of different maturity age. Explants from relatively younger (<0.5cm in length) bulbs showed better results. *In vitro* multiple shoot induction system, using inter-nodal explants from sterile primary cultures was reported by Cheruvathur *et al.* (2010).

• *In situ* conservation

Most medicinal plants are extracted from forest areas as minor forest produce and the forest department normally allows its extraction, which nearly always exceeds the sustainable limits. While species status in its native habitat remains poorly defined for *C. acuminatum*, the sustainable extraction and maintenance of its valuable germplasm, as well as their possible diversity within, remains critically important for its long term existence. Extraction from wild therefore needs to be strictly regulated so that at least all known sites are ensured to maintain sustainable population structure, with minimal habitat degradation. Protection of habitat is far simpler and cheaper than habitat restoration.

G. CULTIVATION ASPECTS

Organized commercial cultivation is the need of the hour to replace the practice of gathering from the wild. Presently, many agencies working in the area, often claim to have developed cultivation protocols, but on ground these claims are hard to verify except for some cases of aromatic plants. The fact remains, that members of family Orchidaceae are very difficult to domesticate especially under *ex-situ* conditions, and like most orchids, seed germination is poor for *C. acuminatum* as well. Regeneration of the species in natural environment determines its life cycle, that begins after sprouting of vegetative parts, such as psuedobulbs, nodal segments of rhizomes, daughter plants, or tubers. While this offers clues for human intervention, better and more practical approaches are needed for successful commercial cultivation. Jadhav *et al.* (2016) compared propagation of *C. acuminatum* through pseudobulbs and rhizomes, and reported that plants produced through rhizomes were better, vigorous in growth, taller in height and higher in yield.

• Climate

C. acuminatum is the species that flourishes in temperate climatic conditions where moisture is optimum but rainfall is not heavy.

• Soil

The species prefers sandy loam soil with high organic matter.

• Nursery raising and planting

Seeds of most orchids have reduced endosperm and show very low germination in nature. It is difficult to germinate seeds under *ex-situ* and laboratory conditions, and therefore pseudobulbs are preferred for raising the crop. It has been estimated, in case of *M. muscifera,* that little over 1 lac bulbs (of two years' age) are required for planting one hectare area with a row spacing of 30x30 cm. For these species, no specific insect pest has been reported.

• Manure

Planting should be carried out in well prepared field by end of autumn, supplemented by 25 t/ha basal dose of farmyard manure.

• Irrigation

Light irrigation is recommended soon after transplanting. Weeding at an interval of one month needs to be carried out.

• Harvesting

By the end of autumn of second year of planting, the crop can be harvested, though relatively higher yields have been reported during third year onward. Harvested bulbs should be separated and washed well before storage. In case of *M. muscifera,* 60 kg/ha yield of fresh bulbs is expected. After drying, the produce is reduced to about one-third of the quantity. Under natural conditions, harvesting of about 10% of the produce (pseudobulbs) is recommended for commercial purpose, and the rest 60% should be left in the field for regeneration purpose.

H. COMMERCIAL ASPECTS

• Demand

The market demand for all the listed species under *Ashtavarga* is expected to increase in view of the renewed global interest in nutraceutical. With growing the awareness regarding the authenticity and traceability of standard formulations, adulteration is expected to be less acceptable. To maintain purity, the demand for crude drug is expected to increase manifolds.

• Market trends

Therefore, a major requirement is to generate raw or crude drug material in sufficient quantity. Currently, the market supplies are believed to be heavily adulterated/ substituted by related and other species of the same genus (Adams *et al.,* 2018; Balkrishna *et al.*, 2012).

• Trade data

Based on the available information, both *C. acuminatum* and *M. muscifera* are required in large quantities to meet the rising demand of products like *Chyawanprash* and other crude drug formulations. A fair quantitative assessment is hampered by the prevalent high level of adulteration. (http://dx.doi.org/10.2305/IUCN.UK.2015-2.RLTS.T50126625A50131390.en).

• Major users

Ashtavarga plant species are considered generally safe and have been consumed for hundreds of years, as part of formulations of Ayurvedic, Unani and Tibetan systems of medicine.

I. FUTURE DIRECTIONS FOR RESEARCH / VALUE ADDITION

Plant species of the *Ashtavarga* group, especially, *Jeevak* and *Rishbhak* have been considered as life giving herbs in *Ayurvedic* scriptures. This ancient literature is the only source of knowledge, based upon which these herbs are being used today. For general acceptance and scientific validation, proper data needs to be generated to evaluate and confirm the real potency of these crude drugs. Clinical trials and other relevant investigations can unravel the constituents and potential ingredients from these herbs that could prove beneficial for their plausible transformation into prescriptions of modern medicine in the future. Additionally, it is also high time now that stern conservation measures are devised. While the issues of safety are being evaluated, identification of better or elite populations is also suggested, in addition to the conservation of the species as discussed earlier. Equally important is planned, enlarged and sustained scientific research on biology, chemistry, pharmacology, cultivation and related cognate disciplines for validated and acceptable formulations of the traditional systems on medicine.

J. PATENTS

1. Riley, P. A. Modular system of dietary supplement compositions for optimizing health benefits and methods. US5976568A. 1999-11-02.

LITERATURES CITED

References

1. Abraham J. 2010. Adventitious shoot induction from cultured nodal explants of *Malaxis acuminata* D.Don,a valuable terrestrial medicinal orchid. *Int. J. Biodivers. Sci. Ecosys. Serv. & Manage.* 6(1-2).
2. Adams, S. J., Kumar, T. S., Muthuraman, G., Majeed, A. (2018). Distribution, morphology, anatomy and histochemistry of *Crepidium acuminatum*. *Modern Phytomorphology* 12: 15–32.
3. Anon. 2006.Ayurvedic Pharmacoepia of India, Part-I, Volume-V. 2006. Ministry of AYUSH, Government of India.
4. Arditti, J. 1992. Fundamental of orchid biology. New York: John Wiley & Sons, 704 pp.
5. Arenmongla, T. Deb C.R. 2012. Germination of immature embryos and multiplication of *Malaxis acuminata* D.Don, an endangered therapeutically important orchid, by asymbiotic culture *in vitro*. *Indian J. Biotechnol*.11:464-469.
6. Arora, M., Kaur, G., Kahlon, P., Mahajan, A., Sembi, J. 2017. Pharmacognostic Evaluation and Antimicrobial Activity of Endangered Ethnomedicinal Plant *Crepidium acuminatum* (D. Don) Szlach. *Pharmacognosy J.* 9. s56-s63. 10.5530/pj.2017.6s.158.
7. Balkrishna A., Shrivastava A., Mishra R., Patel S., Vashistha R., Singh A., Jadon V., Saxena P. 2012. *Astavarga* plants-threatened medicinal plants of the North-Western Himalaya. *Int. J. Med. Arom. Plants* 2: 2249–4340.
8. Barman, D., Devadas, R.2013. Climate change on orchid population and conservation strategies: A review. *J. Crop Weed*, 9(2):1-12.
9. Bhatnagar J.K., Handa S.S., Duggal, S.C. 1970. Chemical Investigation on *Microstylis wallichii.Planta Med.* 20:157-161.
10. Chauhan, N.S. 1999. Medicinal and aromatic plants of Himachal Pradesh. Indus Publishing Company, New Delhi.
11. Chen, I.C., Hill, J.K., Ohlemuller, R., Roy, D.B., Thomas, C.D. 2011. Rapid range shifts of species associated with high levels of climate warming. *Science*, 333: 1024–26.
12. Cheruvathur, M.K., Abraham, J., Mani, B., Thomas, T. D. 2010. Adventitious shoot induction from cultured internodal explants of *Malaxis acuminata* D. Don, a valuable terrestrial medicinal orchid. *Plant Cell Tiss. Organ Cult*. 101:163–170.
13. Chinmay R., Kumari S., Bishnupriya D, Mohanty RC, Dixit R.,Padhi M.M, Babu R. 2011.Phyto-Pharmacognostic studies of two endangered species of *Malaxis* (Jeevak and Rishibhak) *Pharmacognosy J.*3(26):77-85.
14. Ved, D. K., Kinhal, G.A., Ravikumar, K., Prabhakaran, V., Ghate, U., Vijayshanker, R., Indresha, J.H.2003. Conservation Assessment and Management Prioritization for the Medicinal Plants of Jammu and Kashmir, Himachal Pradesh and Uttarakhand, FRLHT, Bangalore.

15. De Smet, P. A. 2004. Health risks of herbal remedies: An update. *Clinical Pharmacolo. Therapeutics*, 76: 1-17. doi:10.1016/j.clpt.2004.03.005
16. Dhyani A., Nautiyal B.P., Nautiyal M.C. 2010. Importance of *Astavarga* plants in traditional systems of medicine in Garhwal, Indian Himalaya, *Int. J. Biodiversity Sci. Ecosyst. Serv. Manage.* 6 (1–2): 13–19.
17. Garg, P, P Aggarwal, P Sharma and S Sharma (2012) Antioxidant activity of the butanol extract of *Malaxis acuminata* (Jeevak). *J. Pharmacy Res.* 5(5),2888-2889.
18. Gupta A. 2016. Studies on *Malaxis acuminata* D. Don (= *Microstylis wallichii* Lindl.) - a medicinally important orchid. *J. Global Res. Comput. Sci. Technol.* 6: 1-11.
19. Hamid, K., Choudhuri, M., Kabir, S. 2015. Toxicological effect of mallasindura (msl) on liver function parameters of rat plasma after chronic administration. *Int. J. Pharmaceut. Sci. Res.* 6. 1958-1964.
20. Hinsley A., de Boer H.J., Fay M.F., Gale S.W. Gardiner L.M., Gunasekara R.S., Kumar P.,Masters S., Metusala D., Roberts D.L., Veldman S., Wong S., Jacob P. 2017. A review of the trade in orchids and its implications for conservation. *Bot. J. Linn. Soc.* box083. https://doi.org/10.1093/botlinnean/box083
21. Jadhav, R., Sharma, Y. P., Sarvade, S. 2016. Growth and yield *of Malaxis acuminata* D. Don as influenced by type of planting material. *Environ. Ecol.* 34(4A) 1935-1940.
22. Kaur S, Bhutani S.S. (2009). Micropropagation of *Malaxis acuminata* D. Don: A Rare Orchid of High Therapeutic Value. *J. Med. Aromatic Plants* 1(2): 29-33
23. Kaushik P. 1983. Ecological and anatomical marvels of the Himalayan orchids: 101–113. Today and Tomorrow's Printers and Publishers, New Delhi, India.
24. Khajuria AK, Kumar G, Bisht NS. 2017. Diversity with ethnomedicinal notes on orchids: a case study of Nagdev forest range, Pauri Garhwal, Uttarakhand, India. *J. Med. Plants* 5: 171–174
25. Lohani N, Tewari LM, Joshi GC, Kishor K, Kumar S, Tewari G, Joshi N. 2013. Chemical composition of *Microstylis wallichii* Lindl. from Western Himalaya. *J. Med. Plants Res.*7: 2289–2292.
26. Madhukar, JR 2013. Monographic studies on *Malaxis acuminata* D. Don Thesis submitted in partial fulfilment of the requirements for the degree of Master Of Science (Forestry) Medicinal And Aromatic Plants College Of Forestry Dr Yashwant Singh Parmar University of Horticulture and Forestry, Nauni, Solan (H.P.), INDIA
27. Pant, B., Chaudhary, R.P., Subedi A, Shakya, L.R. 2002. Nepalese Himalayan Orchids and the conservation priorities. In :Proceeding of International Seminar on Mountains. *Royal Nepal Acad. Sci. Technol.* pp. 485-495.
28. Pérez Gutiérrez RM. 2010. Orchids: A review of uses in traditional medicine, its phytochemistry and pharmacology *J Med Plants Res*;4:592–638.
29. Raturi, R., Kaur, G., Gupta, V., Maithani, M. and Singhal, RG, 2017. *Microstylis muscifera* (Jeevak): Highly Therapeutic and Endangered Orchid. *J. Biomed. Pharmaceut. Res.* 6, 6.
30. Root, T.L, Price, J.L., Hall, K.R., Schneider, S.H., Rosenzweig, C. and Pounds, J.A. 2003. Fingerprints of global warming on wild animals and plants. *Nature*, 421: 57-60.
31. Roy Amirian, Mallick Arindam, Kaour Amrinder. 2013. Adulteration and substitution, In Indian Medicinal Plants, (ISSN: 2277-8713); *IJPRBS*; 2(1): 208-218.
32. Samant S.S., Dhar U., Palni L.M.S. 1998. Medicinal plants of Indian Himalaya: Diversity, distribution, potential values. Gyanodaya Prakashan, Nainital.
33. Sharma A , Ressy G D, Kaushik A, Shanker K, Tiwari R K , Mukherjee R and Rao C V. 2007. Analgesic and anti-inflammatory activity of *Carissa carandas* Linn. Fruits and *Microstylis wallichii* Lindl. tubers. *Nat.l Product Sci.*13(1):6-10.
34. Rokaya, M.B., Raskoti, B.B., Timsina, B., Münzbergová, Z. 2012. An annotated checklist of the orchids of Nepal. *Nordic J. Bot.* 31: 511–550. doi: 10.1111/j.1756-1051.2013.01230.x,

35. Sharma Alok, Rao CV, Tiwari RK, Tyagi LK, Kori ML, Shankar *Kruna (2009).* Comparative Study on Physiochemical Variation of *Microstylis* wallichii: A Drug Used in Ayurveda. Acad. *J. Plant Sci.* 2(1):04-08.
36. Sharma P P, Dadhwal S, Singh G, Sharma D and Sharma S. 2011b. Antimicrobial activity of Butanol extract of *Malaxis acuminata. J.Pharma. Res.* 4(8):2703-2704.
37. Sharma P, Mahajan N, Garg P, Singh G, Dadhwal S, Sharma S. 2011a. *Malaxis acuminata* D. Don: A. Review. *IJRAP*. 2(2):422-425.
38. Singh, A.P. 2006. Raj Nighantu. Chaukhambha Orientalia, New Delhi, India.
39. Singh D., Kumar S., Pandey R., Hasanain M., Sarkar J., Kumar B. 2017. Bioguided chemical characterization of the antiproliferative fraction of edible pseudo bulbs of *Malaxis acuminata* D. Don by HPLC-ESI-QTOF-MS. *Med. Chem. Res.* 26 (12): 3307–3314. https://doi.org/10.1007/s00044-017-2023-6
40. Subedi, A., Kunwar, B., Choi, Y., Dai, Y., van Andel, T., Chaudhary, R.P., de Boer, H.J., Gravendeel, B., 2013. Collection and trade of wild-harvested orchids in Nepal. *J. Ethnobiol. Ethnomed.* 9, 64–74.
41. Tamta B. P., Sharma A. K., Puni L. and Singh A. 2015 Propagation and conservation of endangered orchid *Microstylis wallichii* Syn *Malaxis acuminata* (Jeevak) in its natural habitats of Uttarakhand Himalayas. *Int. J. Sci. Technol.* 4, 424–432.
42. Thomas,C.D.,Cameron,A.,Green,R.E.,Bakkenes, M., Beaumont, L.J., Collingham, Y C., Erasmus, B.F., De Siqueira, M.F., Gringer, A., Hannah, L., Huges,L., Huntley,B.Van Jarrsveld, A.S., Midgley, G.F., Miles,L., Ortega-Hueta, M.A., Petersen, A.T. Phillips, O.L. Williams, S.E. 2004. Extinction risk from climate change. *Nature*, 427: 145–48.
43. Vij, S.P., Shekhar, N. Kuthiala, R. (1986) Chromosome numbers in Indian Orchidaceae - a complete tabulation. In: Biology, Conservation, and Culture of Orchids (ed. S. P. Vij) pp. 221-291. Affiliated East-West Press, New Delhi.

7

Tribulus terrestris

English: *Caltrops fruit, Puncture-vine*

Hindi: *Gokhru*

A.K.S. Rawat and Sharad Srivastava

Tribulus terrestris **L.**

(A) Herbarium sheet, **(B)** *Tribulus* plant, **(C)** Flowering stage, **(D)** Fruit setting **(E)** Ripened fruits (crude drug material)

INTRODUCTION

Tribulus terrestris L. (Gokhru) is an important industrial herb, belonging to family Zygophyllaceae which includes 285 species representing around two dozen genera. Members of this family are distributed in tropical and warm climates. Most species are shrubs, but some grow as small trees or herbs. *T. terrestris* is a mat forming plant that grows extensively in warm and dry areas all over the world. In China, India, Greece and other parts of the world, the species finds wide application in folk medicine, in polyherbal formulations and as food supplement for physical rejuvenation, to enhance muscle strength, sexual potency, etc. Amidst many positive claims, clinical trials are recommended to evaluate its reported toxicity in animals.

A. BIOLOGICAL ASPECTS

• Systematic classification

- Kingdom : Plantae
- Division : Angiospermae
- Class : Dicotyledoneae
- Order : Geraniales
- Family : Zygophyllaceae
- Genus : *Tribulus*
- Species : *terrestris* L.

• Synonyms

T. languinosus L.

• Common names (in different Indian languages)

Sanskrit : *Svadamstra Goksuraka, Traikantaka, Trikatna*

Assamese : *Gokhurkata, Gokshura*

Bengali : *Gokhri, Gokshura*

English : *Caltrops fruit, Puncture-vine*

Gujrati : *Bethagokharu, Mithagokhru, Nanagokharu*

Hindi : *Gokhru*

Kannada : *Neggilamullu, Neggilu, Sannaneggilu*

Kashmiri : *Pakhda, Michikand*

Malayalam : *Nerinjil*

Marathi : *Gokharu, Sarate*

Oriya	:	*Gokhyura, Gukhura*
Punjabi	:	*Bhakhra, Gokhru*
Tamil	:	*Nerinjil, Nerunjil*
Telugu	:	*Palleru Kaya*
Urdu	:	*Khar-e-Khasak Khurd*

• Diversity within genus

Five species of *Tribulus* are distributed in different Indian phyto-geographical zones. These are:

1. *T. terrestris* Linn.
2. *T. pentandrus* Forssk.
3. *T. subramanyamii* P. Singh, G.S. Giri and V. Singh
4. *T. rajasthanensis* Bhandari and Sharma
5. *T. lanuginosus* Linn.

• Geographical distribution

T. terrestris, having wide distribution in the Old World, is now also widely naturalized over large areas of New World. The species is viewed as a noxious weed with the habit of sprawling over the ground and producing abundant hard, spiny nutlets. In India, it is well distributed in central states of India, such as Rajasthan, Madhya Pradesh, Uttar Pradesh, Gujarat, etc. It has also been reported from other areas, including Himalayas and from attitude up to 3000 m msl.

• Habitat and ecology

T. terrestris is native to warm temperate and tropical regions of southern Europe, southern and western Asia, throughout Africa, and Australia. It is widely distributed in Africa, Southern Europe, China, Japan, Korea and western parts of Asia. The species can grows over a wide range of soil types (Shah, 1978; Bapalal, 1982), but performs better in light textured soils. In India, it grows wild throughout the country, thriving particularly well in irrigated black soil.

• Taxonomic enumeration

Roots: Generally a deep taproot, slender, branched, often somewhat woody, with a network of fibrous roots.

Stems: The stems are green to reddish-brown in colour, highly branched, spreading radially from the crown along open ground and grow as prostrate branches up to 3-4 ft long. Stems can be more or less erect when shaded or competing with other plants. Individual stem is silky or hairy, sharply bristled to glabrous.

Stipules: These areleaf-like, subulate, and 2-3 mm long.

Leaves: Leaves are opposite, even-pinnate compound, approximately 1-5 cm long, with 3-7 leaflet pairs per leaf, and have a small extension at the rachis tip. Leaflets are elliptic or oblong, 3-15 mm long, with more or less oblique bases. Lower leaflet pair is unequal in size.

Inflorescence: Consists of peduncled flowers, solitary in axils of leaves.

Calyx/sepals: Thesepals are 5 (occasionally 4) in number, are caduceus, narrowly lanceolate, and 3-3.5 mm long.

Corolla/petals: Corolla/petals are bright yellow, 5-15 mm in diameter, about 5 mm in length, deciduous and 5 (occasionally 4) in number. The peduncle is reflexed and shorter than subtending leaves. Flowers are axillary, solitary; 5 glands occur between the stamens at the base of the ovary. Flowers are perfect, open in the mornings, close in the afternoons.

Gynoecium: Ovary has 5 carpels. Ovary chambers are twice the number of petals; a traverse partition separates seeds in each carpel. Style is deciduous and stigmas are 5 in number.

Androecium: Stamens are twice the number of petals, usually 10 that occur in 2 whorls, with 5 shorter stamens that are well below the level of the stigma, and grow opposite the sepals, each subtended by a small gland. The other whorl of 5 longer stamens has anthers at the same height of stigma, opposite petals.

Fruit: It is a schizocarp, gray to yellow-tan, hairy, approximately 1-1.8 cm in diameter, more or less flattened, and lobed that separates into 5 (occasionally 4) wedge-shaped, indehiscent nutlets (cocci), each with 2 stout dorsal spines, 4-7 mm long and spreading, with several prickles.

Seeds: Seeds are 2-5 per nutlets (cocci), remain enclosed within the burrs (coccus).

Key to the species

1 a. Fruits with spiny mericarps

b. Fruits with winged mericarps .... *T. pentandrus*

2 a. Intra-staminal glands not ciliate, style glabrous

b. Intra-staminal glands ciliate; style pubescent............. *T. subramanyamii*

3 a. Each mericarp with 4 spines, basal pair of spines sometimes reduced to tubercles

b. Each mericarp with 2 median and 20-30 dorsal spines, basal pair of spines absent ....*T. rajasthanensis*

4 a. Pedicels equal to or longer than subtending leaves; style longer than ovary .... *T. lanuginosus*

b. Pedicels shorter than subtending leaves; style shorter than ovary .... *T. terrestris*

• Phenological period

Phenological behaviour of *T. terrestris* is typical of a plant species that undergoes winter dormancy, resumes growth with rise of temperature by spring-summer months. Seed germination in wild is moisture dependent, and is generally observed within a week's time after shower. The process of seed germination preferentially takes place with availability of moisture, from spring to rainy season. Alam and Wiese (1985) reported that emergence of *T. terrestris* seedlings require 151 accumulated day degrees (heat units). Following growth, flowers appear by 15-40 days and the process lasts for a few months. Depending upon locations and climatic conditions, flowering has been observed from April to August. Fruit formation takes place after about two weeks' time and maturity attained by another two weeks. The process of fruiting completes anytime from August to November.

• Cytotypes

Intra-specific polyploids have been reported for *T. terrestris*. Gupta *et al.* (2016) reported three cytotypes of the species, *i.e.*, tetraploid, hexaploid and octaploid from Rajasthan in India. Morrison and Scott (1996) studied cytological diversity in 24 Australian and 24 overseas collections and reported it to range from diploid to octaploid (2n = 12, 24, 36, 48).

B. REPRODUCTIVE BIOLOGY

• Reproductive phenology

The high reproductive capacity of *T. terrestris* owes to high viability of pollen as well as embryo. Thespecies is strongly proterandrous, and reproduces sexually. The tetrasporangiate anthers produce microspore tetrads through microsporogenesis. The primary sporogenous cells, undergoing meiosis, mature and differentiate into microspore mother cells. After maturation, pollen grains are two celled, but three celled pollen grains have also been observed, though rarely. The syncarpous gynoecium is superior, with numerous ovules. The ovule is anatropous, with one-celled archesporium. Double fertilization leads to formation of embryo and endosperm. Mature embryo has two equal cotyledons. The fully mature seeds are non-endospermic (Tsvetkova *et al.,* 2014).

• Life Cycle

The species is a short lived perennial and behaves as annual where winter dormancy is pronounced. Life cycle starts from germination of seed, usually after the onset of warmer conditions during spring/ summer. The species can grow actively if moisture is available. Flowering takes place within months of germination, producing single flower in the axil of a smaller leaf. Flowers develop into fruit which is star shaped. Seeds, up to 5 in number, are contained in seedpod and are released when pods break apart at maturity. The ability of plants to develop deep root system enables it to become perennial.

• Pollination and pollinators

The bisexual flowers of *T. terrestris* are protandrous, and are favoured by high activity of pollinating insects. The result is prevalence of autogamy, geitonogamy and xenogamy. Under experimental as well as different environmental conditions, self-pollination has been shown to lead to 100% seed setting in *T. terrestris* (Ganaie, 2011). The closing petals push stamen inwards towards stigma and facilitate self-pollination. Cross pollination is mediated through insects that are attracted by pollen and nectar. Different species of *Apis, Pieris* and *Limenitis* are reported to be the dominant pollinators. Ants also play an important role because of their consistent presence (Reddi *et al.,* 1981; Ganaie, 2011). It has been reported that frequency of visits of insect pollinators in *T. terrestris* is highest by mid-noon and this may be due to flowering of the other plant species growing in the vicinity.

C. MEDICINAL ASPECTS

Plant material used as crude drug

Different parts of the plant have been used for its therapeutic values. Whole plant is useful in treating dysuria, vitiated conditions of *vata* and *pitta*, treatment of renal and vesicle calculi, anorexia, dyspepsia, helminthiasis, spermatorrhoea, anemia, and general weakness. Fruits are sweet, cooling, diuretic, aphrodisiac, emollient, appetizer, digestive, anthelmintic, expectorant, anti-inflammatory, laxative, cardio-tonic, styptic, lithotriptic and tonic. The seeds are astringent, strengthening and are useful in epistaxis, haemorrhages and ulcerative stomatitis. The ash of the whole plant is good for external application in rheumatic-arthritis.

General appearance

Fruit: Fruits are stalked, light or greenish yellow, five ribbed or angled, more or less spherical in structure and covered with short stiff or pubescent hairs, are about 1 cm in diameter with five pairs of prominent short stiff spines, pointed downwards. Spines are about 0.5 cm in length, form pentagonal frame work around fruit. Ripe fruit separates into five segments of cocci, each appearing as single-fruit. Each coccus is semi-lunar or plano-convex in structure with one chamber, and armed with a pair of spines, starting from its middle, containing four or more seeds, those are slightly astringent in taste.

• Organoleptic properties

The powdered plant material is brown in colour and may range from having a very faint odour to odourless condition. The taste of leaf, root and other parts has been described by many workers to vary from sweet, astringent to bitter

• Microscopic characteristics

Anatomical studies of *T. terrestris*

Cross-section of the fruit reveals five wedge-shaped cocci, lined externally by single layered epidermis, the cells of which protrude outwards to form a long, pointed, unicellular trichome. Major portion of the fruit wall is occupied by large thin-walled parenchymatous cell. On the inner side, fruit wall is limited by a distinct zone of sclerenchymatous cells which in turn enclose a cavity for seed. Outer layers of testa are more or less crushed against the endocarp. The inner epidermis is represented by a thin sheet of closely adherent cells having lignified inner and radial walls. Outer epidermis and mesophyll layers of tegument are crushed and persist as a sub lignified punctate layer.

• Powdered plant material

The powdered plant material is amorphous in nature, varying from light yellow green to brown in colour, having specific odour. It is fairly soluble in water and in 50% ethanol. Fibers are lignified and the bundles connected with stone cells, may contain calcium oxalate crystals of 10-20 μm diameter.

• Major chemical constituents

These comprise sterols, sapogenin with pyroketone ring (diosgenin), gitogenin and hecogenins, tribulosaponins A and B, tribulosin and terrestrosins A–K (WHO monograph, 2005).

• Substitutes and adulterants

In Ayurveda, *Tribulus* is known as *Gokshura* and many herbs available in herbal drug market are sold by the same vernacular name.Two most common kinds of *Gokshura* sold in the market are:

1. Small *Gokshura*: a spreading plant – *T. terrestris*
2. Large *Gokshura*: a big fruited variety and an erect plant – *Pedalium murex*

The following plant species are also source of adulteration

1. Different species of *Tribulus* (Zygophyllaceae)
2. *Pedalium murex* (Pedaliaceae)
3. *Acanthospermum hispidum (*Asteraceae)
4. *Xanthium strumarium* (Asteraceae)
5. *Martynia diandra* (Pedaliaceae)

These five species are described in the following section for general information.

1. *Tribulus* species

These are annual to perennial herbs, prostrate to semi-erect, branching, and hairy to glabrescent. Leaves are opposite, paripinnate, stipulate; leaflets 6-20 in number, opposite, sub-sessile to sessile. Flowers are solitary, axillary, pedicellate, usually yellow in colour. Sepals are 5, pubescent, deciduous or persistent, and imbricate. Petals are 5 in number, obovate to obovate oblong, and membranous. Disk is annular and 10-lobed. Stamens are 5-10 in number, with outer whorl of 5 slightly longer ones, lying opposite to the petals. Filaments are filiform and anthers oblong to ovoid. Each of inner whorls of 5 stamens subtend adaxially by intrastaminal gland and abaxially by extrastaminal gland. Ovary is globose, covered with bulbous-based stiff hairs, and is 5-locular. Ovules range from 1-5 in each cell. Style is simple, cylindrical and with 5-ridges. Stigma is 5-rayed. Schizocarps separate into 4-5, dorsally spiny, winged or tubercled, indehiscent, mericarps. Seeds are obliquely pendulous, endosperm is absent and cotyledons are ovoid (Chunekar, 2006; Pandita and Nighantu, 1998).

2. *Pedalium murex* (Pedaliaceae)

The species is herbaceous, simple or branched, erect or diffuse, pubescent. Leaves are 1-6.5 x 0.7-3.6 cm in dimension, are fleshy, glabrous, broadly ovate, elliptic-oblong or nearly orbicular. Flowers are axillary, solitary and 1.8-3.7 cm long. Drupes are 1.3-1.5 cm across, pyramidal-ovoid, with 4-gonous. The fruits have four lateral spines, and are generally called *Bada-gokharu*. *P. murex* has been advocated as *Gokshura* and sold as aphrodisiac, tonic, *rasayana* and for other *vatik* disorders (Bapalal, 1982).

3. *Acanthospermum hispidum* (Asteraceae)

It is an annual herb characterized by hairy stem, yellow-green florets, opposite leaves and fruits with hooked spines. Stems are erect and 20-80 cm long, diffusely branched, velvety with long hairs. Leaves are elliptic to ovate, 2-10 cm long, 1-7 cm wide, with glands on lower surface. Leaf margins are toothed to nearly entire, gradually tapering to the base, and are stalk less. Flower heads are yellow, 4-5 mm in diameter. Burs are wedge-shaped, strongly compressed, 4-7 mm long, the ribs bearing 1-2 rows of hooked prickles 1-2 mm long; the 2 apical prickles are stout, curved or straight and are 3-4 mm long (Theodore, 2006).

4. *Xanthium strumarium* (Asteraceae)

X. strumarium is an annual herb with short, stout, slightly branched stem, with short hairs. Leaves are numerous, broadly triangular-ovate or sub-orbicular, acute, and often 3-lobed, have hairs on both sides, are cordate and shortly cuneate at the base. Petiole is 1-3 inches long and hairy. Heads in terminal and axillary racemes; the barren heads rather numerous, crowded at the top of the stem, while the fertile heads are fewer, axillary. Involucre of fertile heads ovoid in fruit, about 5/8 inches long, with 2 erect mucronate beaks, pubescent, thickly clothed with usually hooked prickles, 2-celled, are hard and tough. Achenes are ½ in length, are oblong-ovoid, compressed, and glabrous (Shah, 1978).

5. *Martynia diandra* (Pedaliaceae)

The herb is native of Mexico, has become naturalized on rubbish-heaps and in waste places. Leaves are large, opposite, cordate, and minutely dentate, often covered with a glutinous dew-like substance. Flowers are diandrous, ill-smelling, rose-coloured, and are in racemes. The fruit is a green fleshy capsule, which contains a hard, black woody wrinkled nut, with two anterior hooks. Hooks are just like the scorpion's tail and therefore is known as *Vincchudo* – scorpion in Gujarat. This fruit is cleft shaped like a cow's hoof, having 3 hooks. So in the real sense of the term, this is the real *Gokshura* (Cow's hoof). It is a good diuretic and its fruits are used in medicine. A black coal-tar like oil is obtained and is used in eczema, ringworm, etc. (Bapalal, 1982).

• IDENTITY, PURITY AND STRENGTH (As per API)

Foreign matter	:	Not more than 1%.
Total Ash	:	Not more than 15%.
Acid-insoluble ash	:	Not more than 2%.
Alcohol-soluble extractive	:	Not less than 6%.
Water-soluble extractive	:	Not less than 10%.

• Medicinal usage

- **General uses:** *T. terrestris* is considered as an invigorating stimulant, aphrodisiac, nutritive herbal material (Oza and Mishra, 2004; Sharma and Sharma, 1979). It is used to treat dysuria, vitiated conditions of vata and pitta, renal and vesical calculi, anorexia, dyspepsia, helminthiasis, spermatorrhoea, anemia, and general weakness. Fruits are used as crude drug in Indian traditional systems of medicine and are considered as cooling, diuretic, emollient, appetizer, digestive, anthelmintic, expectorant, anti-inflammatory, laxative, cardiotonic, styptic, lithotriptic and tonic. The seeds are astringent, and are useful in epistaxis, haemorrhages and ulcerative stomatitis. The ash of the whole plant is good for external application in rheumatic-arthritis.
- **Ethnobotanical uses:** Several ethnobotanical uses of leaves and fruits by tribal have been reported from different parts of India and elsewhere for curing different ailments. Juice of fresh leaves of *T. terrestris* is given to treat colic and chronic cough of animals in Terai forest of western Nepal (Singh *et al.,* 2014). Ganzera *et al.* (2001) conducted survey in the mountainous region of districts Musakhel and Barkhan of Balochistan Province and reported use of *T. terrestris* to treat inflammation of urogenital system that helped relieve infection and uterine disorders. Whole plant infusion of the species has been suggested to remove kidney stones (Sajeev and Sasidharan, 1997). Similar use of dried seeds has been documented by natives of Palakkad district, Kerala (Kumar *et al.,* 2014). Katewa *et al.* (2004) reported that tribal of Rajasthan take the fruit powder with milk to increase sperms count.

D. CHEMISTRY ASPECT

• Chemical composition

T. terrestris has been well studied for its chemical composition and has been reported to contain various bioactive metabolites, including saponins, flavonoids, alkaloids, lignanamides and cinammic acid amides (Saleh *et al.,* 1982; Bourke *et al.,* 1992; Ren *et al.,* 1994; Wang *et al.,* 1997). The fruit and root contain pharmacologically important metabolites such as phytosteroids, flavonoids, alkaloids and glycosides (Wu *et al.,* 1996), and is also rich in iron. The diuretic property of the plant is due to nitrates. The freshly expressed aqueous extract of the whole plant contains inorganic nitrates, mostly potassium nitrate. Apart from this, the fruit contains minerals (14%), ascorbic acid, fats and essential oils. The content of ascorbic acid increases from roots to fruits in case of *T. terrestris*.The various saponins and their different derivatives identified in the leaf tissue are diosgenin, gitogenin and chlorogenin. The presence of spirostanol and furostanol saponins are characteristic feature of this plant, the latter is considered to be biogenetic precursors of their spiro analogs (Mahato *et al.,* 1982). Various derivatives of tigogenin, neotigogenin, gitogenin, neogitogenin, hecogenin, neohecogenin, diosgenin, ruscogenin, chlorogenin and sarsasapogenin are found. Four sulphated furo- and spiro saponins have also been isolated.

In general, saponins, flavonoids, alkaloids, lignanamides and cinnamic acid amides are of therapeutic value. Spirostanol and furostanol saponins are the main ingredient of these saponins (Kostova and Dinchev, 2005). Many pharmaceutical preparations or food supplements are available in the market based on these saponins.

Variation in saponin content with geographical regions: Interestingly, the saponin content and the saponin composition of *T. terrestris* growing in different geographic regions of the world differ (Kostova and Dinchev, 2005). The type of sapogenins, or the number ofsugars of the saponins and content of protodioscin also depends on the geographical region of collection. Saponins with a cis A/B-rings juncture (gitogenin/neogitogenin type) are found only in *T. terrestris* of Chinese origin (Kostova and Dinchev, 2005). The diosgenin type spiro saponins are reported to occur in *T. terrestris* of Bulgarian, Moldovian, Romanian, Turkish and South African origin (Tomova *et al.,* 1974; Miles *et al.,* 1993).Sulphated spirostanol and furostanol saponins were isolated only from plants of Bulgarian origin (Conrad *et al.,* 2004). The saponins analyzed from samples of India and China contained sugar units such as galactose, glucose, rhamnose and xylose, while those from Bulgaria had only glucose and rhamnose (Ganzera *et al.,* 2001).

Chemical markers: Estimation of biomarker compounds in *T. terrestris* germplasm collected from different phytogeographical zones of India

The author's group conducted a study under ICAR-NAIP sponsored project to estimate biomarker compounds in *T. terrestris* germplasm collected from different phytogeographical zones of India. Quantification of biomarkes *viz:* protodioscin, prototribestin, tribulosin and rutin in fruit of *T. terrestris* were analyzed by

HPLTC, and reverse phase HPLC–PDA. It was found that these biomarker compounds show significant variation in fruits of the species grown at different phytogeographical zones of India (Table-1).

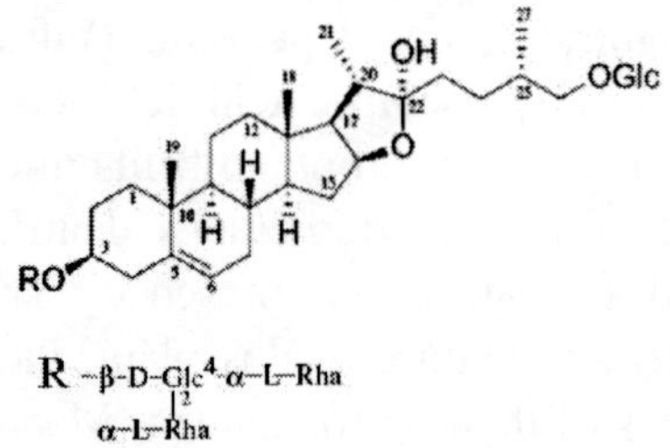

R –β-D-Glc[4]-α-L-Rha; Glc[2]: α-L-Rha

Protodioscin
Aphrodisiac, Increases the levels of the hormones like testosterone, dihydrotestosterone and dehydroepiandro sterone
Prototribestin

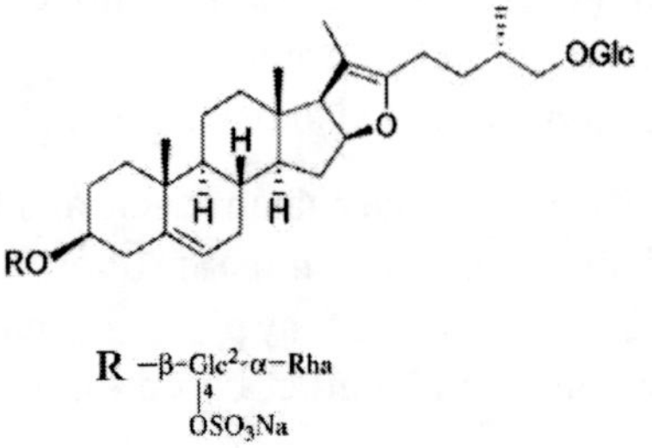

R –β-Glc[2]-α-Rha; Glc[4]: OSO_3Na

Proerectile effect
Increases the levels of the hormones like testosterone, dihydrotestosterone and dehydroepiandro sterone

R -β-D-Gal[4]-β-D-Glc[3]-β-D-Xyl; Gal[2]: α-L-Rha; Glc[2]: β-D-Xyl

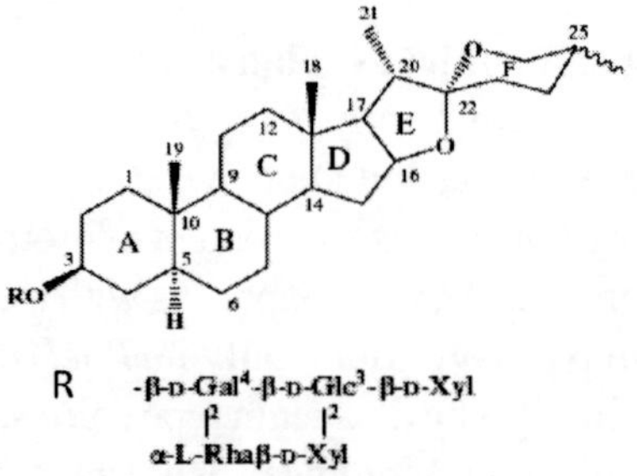

Tribulosin
Testosterone, dihydrotestosterone hormones level enhancer

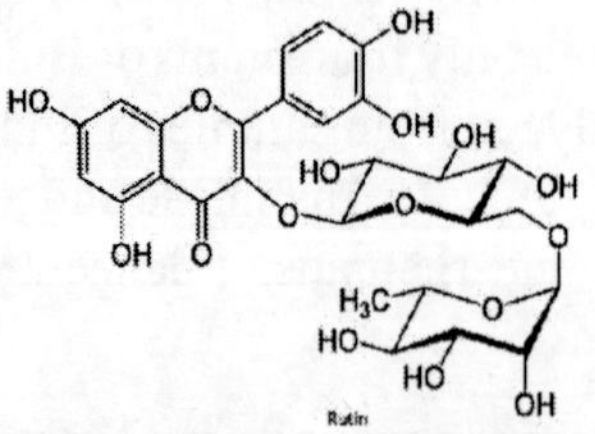

Rutin
Antioxidant, anti-inflammatory, Inhibits the aldose reductase activity which reduce the risk of heart disease.

HPTLC conditions
Mobile phase: Butanol: Glacial acetic acid: water [8.0:0.6: 2.0 v/v/v].
Stationary phase: HPTLC normal phase silica gel glass plate (20 x 10 cm)
Seperation time: 45 minutes at room temperature (25 ± 2°C).
Detection reagent: p-anisaldehyde solution
HPTLC scanning: Camag TLC Scanner 3 at λ_{450} nm in the UV absorbance mode operated by Win CATS Software [Version 3.2.1].
Statistical Analysis: Graphpad prism

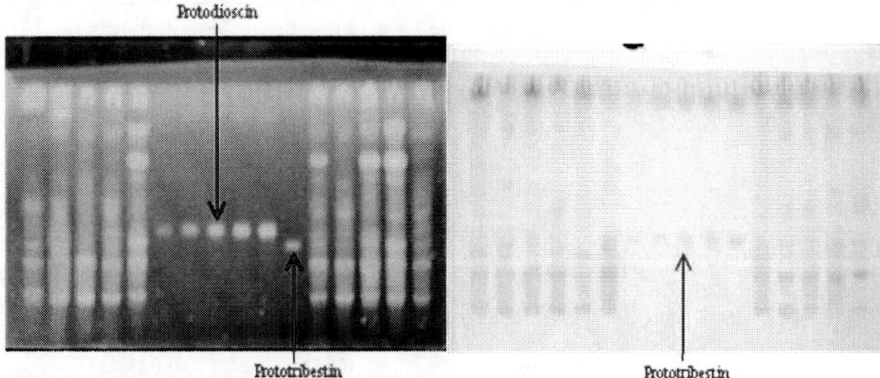

Figure 1. HPTLC plate of *Tribulus terrestris* fruits indicatingProtodioscin and Prototribestin biomarkers

Results

Of the different accessions of *T. terrestris* analysed, NBT-06 (samples from arid zone, Rajasthan) contained maximum content of Prototribestin (0.636 %). Protodioscin content was maximum (0.317 %) in NBT-03 (Eastern Ghat, Tamilnadu), but not detected in NBT-03 (Tamilnadu), NBT-05 (U.P.), NBT-08 (U.P.) and NBT-34 (Madhya.Pradesh). Protodioscin was not detected in NBT-01 (Gujrat) and NBT-31 (Uttar Pradesh) analysed through HPTLC. The result of HPTLC analysis are tabulated in Table 1.

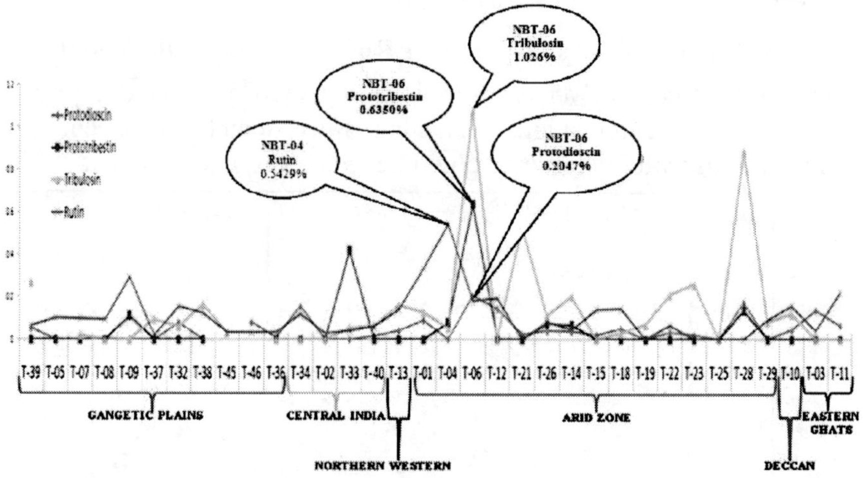

Figure 2. HPTLC quantification of biomarkers in fruits of *Tribulus terrestris*

Table 1. HPTLC analyses of fruits of *Tribulus terrestris*

Phyto-geographical zone	Sample code	Protodioscin %	Prototribestin %
GANGETIC PLAINS	NBT-39	0.035	0.052
	NBT-5,TT-7	0.090 , 0.051	NQ, 0.041
	NBT-8, NBT-31	0.057, NQ	NQ , 0.100
	NBT-09	0.053	0.416
	NBT-37	0.143	0.260
	NBT-32	0.02)	0.165
	NBT-38	0.018	0.059
	NBT-36	0.098	0.200
CENTRAL INDIA	NBT-34	0.047	NQ
	NBT-02	0.067	0.407
	NBT-33	0.053	0.023
NORTHERN WESTERN GHATS	NBT-13	0.113	0.129
ARID ZONE	NBT-1	NQ	0.193
	NBT-4	0.046	0.121
	NBT-6	0.072	0.636 (max)
	NBT-12	0.075	0.108
	NBT-21	0.033	0.110
	NBT-26	0.017	0.041
DECCAN	NBT-10, NBT35	0.026, 0.223	0.214, 0.314
EASTERN GHAT	NBT-03	0.317 (max)	NQ
	NBT-11	0.274	0.154

Conclusion: Following elite accessions of *T. terrestris* species have been identified-
Chemotype-1L Protodioscin rich- (0.20472%) NBT-06 (Arid zone)
Chemotype-1: Prototribestin rich-(0.635076%) NBT-06 (Arid zone)
Chemotype-1: Tribulosin rich- (1.07618%) NBT-06 (Arid zone)
Chemotype-2: Rutin rich- (0.5429%) NBT-04 (Arid zone)

RP-HPLC Chromatographic study for Rutin (from PDA detector)

The quantity of Rutin in *T. terrestris* fruit was analyzed by Reverse phase high performance liquid chromatographic method. It was found that concentration of Rutin in fruits varies in different phyto-geographical zones of India.

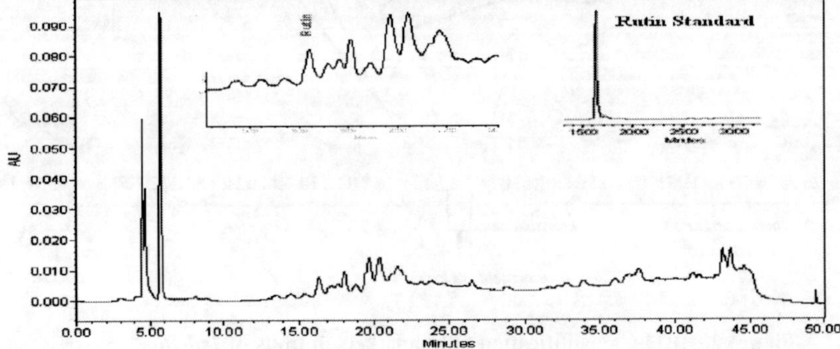

Figure 3. HPLC chromatogram of *Tribulus terrestris* fruits with Rutin standard (Rt-15.5)

HPLC conditions

HPLC system: Water HPLC with PDA detector and autosampler

Mobile phase: 0.025 acetic acid in water /acetonitrile gradient as a mobile phase.

Stationary phase: Phenomenex Luna C-18 column (250 x 4 mm, 5ìm)

Detection: at λ_{256} nm, Separation time: 45 mins at 30 ± 2°C temperature.

HPLC Software: Empower for waters HPLC, Statistical Analysis: Graphpad prism

Results

The RP-HPLC analysis showed considerable variation (0.0292 % to 0.542%) of rutin concentration in fruits of *T. terrestris* (Figure 3). It was observed that maximum percentage of rutin (0.542%) was found in NBT-04 (Gujarat) accession from arid zone.

E. PHARMACOLOGICALASPECTS

T. terrestris has been used since time immemorial in India and China, and can be traced back to as far as 5,000 years. In Ayurveda, the plant is known for its diuretic, aphrodisiac and anti-urolithiatic properties (Deepak *et al.,* 2002), while its fruits have been used for treatment of eye ailments, abdominal distention, emission, edema, morbid leucorrhea, sexual dysfunction and veiling in traditional Chinese medicine, (Xu *et al.,* 2000; Cai *et al.,* 2001). In South Africa, the species is recommended as herbal tonic for diarrhea, and diseases of throat and eyes (Drewes *et al.,* 2003). In the Bulgarian folk medicine, *T. terrestris* is used for blood purification and for treating hemorrhoids (Stoyanov, 1973). Several workers have carried out extensive studies on this plant for targeting following diseases/disorders:

Bioactivity

• Sexual disorders

T. terrestris is known for its aphrodisiac properties and as a traditional medicine for treating male infertility (Gauthaman *et al.,* 2002). A formulation of the saponins from *T. terrestris* was developed for veterinary application in Bulgaria (Tomova *et al.,* 1966; Tomova and Gjulemetova, 1978b). This formulation was effective in stimulating the sexual system (spermatogenesis, libido sexualis). Using *T. terrestris* extracts, an increase in sexual function of rats was demonstrated and attributed to an increase in testosterone, dihydrotestosterone, and dehydroepiandrosterone (Gauthaman *et al.,* 2002). The Bulgarian formulation with the name of *Tribestan* was optimized and widely used for treatment of infertility and libido disorders in men and women (Tomova *et al.*, 1978a; Protich *et al.*, 1981; Viktorov *et al.*, 1982). The furastanol saponins (protodioscin, protogracilin) extracted from *T. terrestris* were reported to have stimulating effect on spermatogenesis *via* Luteinizing Hormone (LH) that stimulates the secretion of male hormone, testosterone. Testosterone in turn significantly improved the quality and quantity of sperms (Tomova *et al.*, 1981; Brown *et al.*, 2002). Libilov is another formulation of the saponin fraction of *T. terrestris* that is reported for similar activities. Protodioscin, the main component of product Libilov, proved to be effective for treating male infertility by increasing the level of dehydroepiandrosterone (DHEA) in infertile men. Protodioscin obtained from *T. terrestris* was suggested to be DHEA precursor in patients with low serum level of this hormone (Adimoelja and Adaikan, 1997; Adimoelja, 2000). Different formulations containing *T. terrestris* extracts are marketed in USA and Europe as food supplements for stimulation, vigor (De Combarieu *et al.*, 2003; Mulinacci *et al.*, 2003) and for several other ailments.

• Antiurolithiatic activity

In a preliminary study, the diuretic effect of water extracts of *T. terrestris* and *Hygrophila spinosa* was evaluated in albino rats. The diuretic effect was attributed to the presence of potassium salts in high concentration (Adaay and Mosa, 2012). The diuretic activity related to *T. terrestris* with minimal side effect was reported in albino rats (Adimoelja, 2000). In order to evaluate the therapeutic use of *T. terrestris* in various urinary disorders including urolithiasis, ethanol extract was tested for activity against artificially induced urolithiasis in albino rats. The extract was administered orally at 25, 50 and 100mg/kg daily for 4 months. It exhibited dose dependent anti-urolithiatic activity and almost completely inhibited stone formation. Other biochemical parameters in urine and serum which were altered during the process of stone formation were normalized by the plant extract in a dose dependent manner (Angelova *et al.*, 2013).

The effect of an aqueous extract of *T. terrestris*, administered orally at a dose of 5g/kg body weight, was studied in rats with induced hyperoxaluria (Intraperitoneal injection of 4-OH proline at a dose of 2.5g/kg body weight for three successive days) and maintained by sodium glycolate. After twenty four hours, urine was collected and analyzed for creatinine and oxalate. The oxalate excretion reversed to normal from 1.97 + 0.314 to 0.144 + 0.004 mg/mg creatinine ($P<0.01$) within 21 days of administration of *T. terrestris* extract and remained so until 15 days after withdrawal of extract and sodium glycolate (Bayati and Al-mola, 2008).

To confirm the earlier results, the effect of administering aqueous extract of *T. terrestris* on the metabolism of oxalate in rats fed with sodium glycolate was evaluated. Glycolate feeding resulted in hyperoxaluria as well as increased activities of oxalate synthesizing enzymes of the liver namely glycolate oxidase (GAO), glycolate dehydrogenase (GAD) and lactate dehydrogenase (LDH) and decreased kidney LDH activity. *T. terrestris* fed rats exhibited a significant decrease in urinary oxalate excretion and a significant increase in urinary glycoxylate excretion, as compared to sodium glycolate fed animals. The supplementation of *T. terrestris* extract also caused a reduction in liver GAO and GAD activities, where as liver LDH activity remained unaltered. The isoenzyme pattern of kidney LDH revealed that normalization of kidney LDH by *T. terrestris* was mainly due to an increase in the LDH5 fraction (Bhattacharjee, 2004).

Activity guided fractionation was conducted on fruits of *T. terrestris*. The ethanolic extract showed significant dose dependent protection against uroliths induced by glass bead implantation in albino rats. On subsequent fractionation of the ethanolic extract, maximum activity was localized in the 10% aqueous methanol fraction which provided significant protection against deposition of calculogenic material around the glass bead and which also protected against leucocytosis and elevation in serum urea levels. Further fractionation lead to decreased activity which was attributed to the loss of active compounds during fractionation or due to combined effect of several constituents in the methanolic fraction (Andersen, 2010).

• Cardiac diseases

Studies revealed that saponins obtained from *T. terrestris* play a role in dilating coronary artery and improving coronary circulation. The results of a clinical trial on 406 patients, when treated with saponin of *T. terrestris,* showed that the total efficacious rate of remission of angina pectoris was 82.3% and the total effective rate of ECG improvement was 52.7% higher.

Many research studies revealed that aqueous extract of *T. terrestris* fruits have some antihypertensive effect (Yang *et al.,* 1991; Chui *et al.,* 1992). These beneficial effects have partly been attributed to its ability to increase nitric oxide (NO) release from the endothelium and nitrergic nerve endings (Adaikan *et al.,* 2000). Antihypertensive mechanism of *T. terrestris* was studied in 2K1C hypertensive rats by measuring circulatory and local angiotensin-converting enzyme (ACE) activity in aorta, heart, kidney and lung. The systolic blood pressure and ACE of *T. terrestris* fed hypertensive rats was significantly decreased compared to hypertensive rats indicating a negative correlation between consumption of *T. terrestris* and ACE activity in serum and different tissues in 2K1C rats (Sharifi *et al.,* 2003). Another research study indicated that dietary intake of *T. terrestris* can significantly lower serum lipid profiles, decrease endothelial cellular surface damage and rupture, and may partially repair the endothelial dysfunction resulting from hyperlipidemia in New Zealand rabbits fed a cholesterol-rich diet (Altug *et al.,* 2009).

• Antimicrobial activity

The methanolic or ethananolic extracts of different parts (fruits, roots and stems with leaves) of Iranian, Indian or Turkish *T. terrestris* inhibited the growth of different tested microorganisms (Bedir and Khan, 2000; Ali *et al.,* 2001; Kianbakht and Jahaniani, 2003). Among the seven different saponins from *T. terrestris* tested, only the spirostanol saponins showed antifungal activity against *C. albicans* and *Cryptococcus neoformans*, while none of the furostanol derivatives exhibited inhibitory activity. Further, these compounds were not effective against *S. aureus, Aspergillus fumigatus, P. aeruginosa* and *Mycobacterium intracellulare* (Bedir *et al.,* 2002). Antimicrobial activity of organic and aqueous extracts from fruits, leaves and roots of *T. terrestris* from Iraq was examined against 11 species of pathogenic and non-pathogenic microorganisms. All the extracts from the different parts of the plant showed antimicrobial activity against most tested microorganisms. The most active extract was ethanol extract from the fruits with a minimal inhibitory concentration of 0.15 mg/ml against different bacteria and fungi (Al-Bayati and Al-Mola, 2008).

• Anti-cancerous activity

Among the saponins analyzed from stem and fruit of *T. terrestris* collected from different regions, only spiro compounds exhibited remarkable activity (Kostova and Dinchev, 2005). The most active spirostanol glycoside (Hecogenin) exhibited a broad range of anticancer activity against cell lines SKMEL, KB, BT-549 and SKOV- 3 (Bedir and Khan, 2000). Dioscin and prosapogenin A of dioscin, showed anti-cancerous activity against the cancer cell line K562 in vitro (Hu *et al.,* 1996). Protodioscin proved to be cytotoxic against cell lines from leukemia and solid tumors and particularly against one leukemia line (MOLT-4), one NSCLC line (A549/ATCC), two colon cancer lines (HCT-116 and SW-620), one CNS cancer line (SNB-75), one melanoma line (LOX IMVI), and one renal cancer line (Hu and Yao, 2002). Based on computer analysis of methylprotodioscin as a seed compound, a potential novel mechanism was suggested for its anti-cancer action (Hu and Yao, 2003). The saponin mixture from Chinese plant had potent inhibitory effects in a concentration dependent manner on Bcap37 breast cancer cells (Sun *et al.,* 2003).

• Anthelmintic activity

Anthelmintic activity was observed only in the 50% methanol extracts of *T. terrestris* (whole plant) using the nematode *Caenorhabditis elegans*. Further, investigation revealed tribulosin and sitosterol glucoside as active components responsible for anthelmintic activity (Deepak *et al.*, 2002).

• Miscellaneous uses

Lin *et al.* (1999) investigated the use of *T. terrestris* fruits for treatment of vitiligo with positive results. The lyophilized saponin mixture of the plant caused a significant decrease on peristaltic movements of isolated sheep ureter and rabbit jejunum preparations and it was proposed that the saponin mixture may be useful for some smooth muscle spasms or colic pains (Arcasoy *et al.*, 1998). The preventive and therapeutic effects of saponins from *T. terrestris* on diet induced hyperlipidemia in mice have been studied. It was shown that the saponins could lower the levels of serum TC, LDLC and liver TC, TG and increase the activities of superoxide-dismutase in liver (Chu *et al.*, 2003). Hypoglicemic effect of saponins from *T. terrestris* was investigated and the saponins were found to reduce the level of serum glucose significantly (Li *et al.*, 2002).

Clinical Studies

Some clinical studies conducted in relation to various ailments have been discussed in the above section. In recent times, *T. terrestris*, believed to be a testosterone booster, has been considered as a nutritional supplement for enhancing human vitality, especially of athletes. Several clinical trials, conducted to verify this have not shown strong evidence to support safe usage or usefulness of the species as sport related nutritional supplement (Pokrywka *et al.*, 2014). One of the affects that may place athletes at risk of a positive drug test is the increase in urinary testosterone/epitestosterone (T/E) ratio. Rogerson *et al.* (2007) reported that *T. terrestris* had little effect on the urinary T/E ratio. On the contrary, dietary supplement that contained *T. terrestris*, significantly increased muscular power and blood testosterone in men (Milasius *et al.*, 2009).

Some clinical studies, that evaluated effect of higher doses of *T. terrestris* in animal system, reported impairment of cardiac muscles, or liver or kidney when fed to sheep and goats (Aslani *et al.*, 2003; Aslani *et al.*, 2004). Severe ill affects to kidney and liver that relate to exceptionally higher intake of *T. terrestris* have been reported in humans (Talazas *et al.*, 2010), though such reports are very rare.

IMPORTANT FORMULATIONS

Goksuradi, Guggulu, Traikanaka Ghrta, Draksadi Churna

Dose

- 3-6 g of the crude drug in powder form.
- 20-30 g of the crude drug for decoction.

F. THREAT AND CONSERVATION ASPECT

Despite its high demand for different indigenous schools of medicine, *T. terrestris* seems to be under no threat as regards its status in the wild. In fact, there are reports where physical, chemical and biological methods of its removal have been sought and given (https://en.wikipedia.org/wiki/Tribulus_terrestris). Its seeds are preferred byweevil beetle (*Microlarinus lareynii*, Jacquelin du Val) and stem and crown by *M. lypriformis*, Wollaston). In India, herbivore threat has been listed from *Ephysteris subdiminutella, Tegostoma comparalis* and *Eriophyes tribuli* (Ramaseshiah, 1976; Sankaran and Ramaseshiah, 1981).

- **Threats and Conservation strategies:** While the species faces no immediate threat of any kind, the high genetic diversity reported within the species and lack of its ecological understanding with possibility of erosion of important germplasm lines are cause of concern. Considering that demand of all plants sought by pharmaceutical industries will increase in future, protection and conservation of elite germplasm is highly recommended. Desirable perhaps would be the introduction and cultivation of its diverse lines from fertile to dry regions for conserving its genetic diversity.

G. CULTIVATION ASPECTS

Since *T. terrestris* is considered as a weed, there is not much information on its commercial cultivation. The plant has been reported to become invasive in warmer climate in Australia and United States. The capacity of the species to adapt to a wide range of soil type and considerable changes of temperate conditions has enabled it to spread extensively. It has been listed as weed in about 37 countries for more than 20 crops (Pacanoski *et al.*, 2014). In countries, such as Bulgaria, where demand of the industrial species exists, organized cultivation however has been emphasized.

Cultivation Technology

The plant is propagated through seeds. Fresh seeds are not quite useful for cultivation owing to prevalent seed dormancy. Seeds however can remain viable for years. In fields, where bare patches are high, germinated populations can be used to fill gaps to enhance crop density (Petkov, 2011). The species has large root volume and deep root system (about 2 m or more) that helps it to survive adversity of drought. Water requirement for *T. terrestris* is low such that for production of 1 kg of dry matter, 96 kg of water is required compared to 840 and 300 kg water for alfalfa and sorghum, respectively (Davis and Wiese, 1964). The plant grows fast and within 3-4 months the whole plant can be harvested and sold for commerce.

• Climate

T. terrestris can be cultivated under warmer and open sun conditions across a wide range of climatic conditions, ranging from tropical, subtropical and semiarid climate. The open space for growing requires low rainfall, and free from being water logged.

• Soil

Being an invasive weed in nature, the species thrives in most soil types, from clay-loam to light sandy-loam, but does not prefer highly alkaline soil. A well drained sandy soil is ideal for commercial cultivation.

• Nursery raising and planting

For nursery raising or crop cultivation, direct sowing of seeds is recommended, one seed sowed per cup or in nursery beds at a depth of 4-5 mm. For better results, seeds are germinated and then planted in rows in the field. Subsequently, seeds raised nursery plants, grown to the size of 10-15 cm long, are transplanted at a distance of 15x15 cm during summer-rain months. Germination is favoured at high temperature. For stratification, seeds are placed in moist sand for 1-3 months, and planted after the sign of swelling or sprouting are noticed. These sprouted seeds can be grown in plug trays or pots before transferring to field. For quicker results, seeds may be soaked overnight, and treated for 48 hours with 20 ppm gibberellic acid, before sowing in nursery or field. Seeds are sown at the rate of 1-2 kg seeds per ha.

• Manure

Generally, the crop may not require fertilizers of any kind, but for commercial cultivation application of manure at the rate of 10 t/ ha in soil before sowing is recommended. Two application of N, P and K at 60:40:40 kg/ha, with a gap of 60 days are recommended. Weeding may be carried out to keep the field weed-free.

• Irrigation

Initial irrigation is required but after establishment and during later stages of growth, much water is not required. Irrigation is recommended during long dry spells.

• Harvesting

The crop is harvested after about 240-250 days in the field. Crop is uprooted after the seeds ripen. Seeds and roots harvested, separated and dried under shade . Storage should be under dry cool conditions. A reported total yield per ha is 2.88 tonnes (fresh weight) or 0.738 tonnes per ha (dry weight) (http://vikaspedia.in/agriculture/crop-production/package-of-practices/ medicinal-and-aromatic-plants/tribulus- terrestris-2#section-5).

• Crop rotation

There is no information on crop rotation for *T. terrestris*. There are also no significant pests or diseases reported, except a rare mention of some weevils (*Microlarinus lareynii* and *M. lypriformis*), attacking seeds and stems. These are also known to have been used as biocontrol agents.

H. COMMERCIAL ASPECTS

There is no reliable data to document the quantity of collection of *T. terrestris* from the wild. The export data from the country, however, show large scale sale of products containing the plant material, wholly or in parts. Concern has been raised by different State Forest Departments (SFDs) and Indian Subcontinent Specialist Group (1998) about this deficiency and it has been suggested to conduct detailed study to evaluate the situation and to provide protection to about ten plant species of which *T. terrestris* is one.

I. FUTURE DIRECTIONS FOR RESEARCH / VALUE ADDITION

For a species that has a tag of being a 'weed', it does not attract attention for conservation, except that knowledge on diversity within species could be of academic interest. While considerable genetic diversity has been suggested with the species, our understanding of the same in ecological perspective is poor. Modern tools like SSR markers can be handy techniques for genetic characterization of the diverse germplasm of the species. Variation in amount and quality of active constituents in different cytotypes of medicinal plants has been highlighted (Berkov, 2001), and calls for further understanding for identification of better chemotype/ cytotype of *T. terrestris.*

J. PATENTS

There are over 200 patents granted or applied for T. terrestris. The following section contains some patents those are granted. For details please refer to (https://patents.google.com/patent/US20110020443A1/en?q=Tribulus +terrestris &q= granted&q=patents&oq=Tribulus+terrestris+granted+patents)

1. Sakai, Y., Yokoo, Y. Process for producing a plant extract containing plant powder US20040161524A1 2009-04-21.
2. Lillard, J. W., Singh, R., Singh, S. Nanoparticles for delivery of active agents. US8568784B2. 2013-10-29
3. Liu, G., Mao, F. Slow release magnesium composition and uses thereof. US20110020443A1. 2013-02-19.
4. Pushpangadan, P., Prakash, D. Herbal health protective and promotive nutraceutical formulation for diabetics and process for preparing the same. US20030185913A1. 2006-03-21.
5. Young, D. C. Systemic herbicides and methods of use. US5288692A. 1994-02-22
6. Hasegawa, J., Okamoto, T., Tanaka, K. NF-êB activation inhibitor. JP2005194246A. 2004-01-09.
7. Koike, K., Kojima, H., Tomono, N., Yoshida, S. Hair cosmetic. JP2008024628A. 2011-11-16.
8. Anzai, T., Hori, M., Ito, K., Kojima. H. Fibroblast proliferation promoting agent. JPH1036283A. 2007-08-29.
9. Hasegawa, J., Ito, K., Ueno, H. Cosmetic composition and beauty and healt food composition. JP2003206225A. 2008-06-11.
10. Arunasiri, I., Ito, K., Naba, Y. Activator of peroxisome proliferator-activated receptor. JP2007119430A. 2011-07-27.

ACKNOWLEDGEMENT

The authors are thankful to ICAR-NAIP to have funded some of the activities under report and Director, CSIR-NBRI for the necessary facilities and research team of the NAIP project.

References

1. Adaay M.H., A.A.R. Mosa. 2012. Evaluation of the effect of aqueous extract of *Tribulus terrestris* on some reproductive parameters in female mice. *J Mater Environ Sci.*, 3(6): 1153-1162.
2. Adaikan, P.G., K. Gauthman, R.N. Prasad and S.C. Ng. 2000. Proerectile pharmacological effects of *Tribulus terrestris* on the rabit corpus cavernosum. *Ann. Acad. Med., Singapore.* 29: 22-26.
3. Adimoelja, A. 2000. Phytochemicals and the breakthrough of traditional herb in the management of sexual disfunction. *Int. J. Androl.,* 23: 82-84.
4. Adimoelja, A. and P.G. Adaikan. 1997. Protodioscin from herbal plant *Tribulus terrestris* L. improves the male sexual functions, probably via DHEA. *Int. J. Impotence Res.*, 9: 1-15.
5. Alam, L.R. and Wiese, A. F. 1985. Effect of degree days on weed emergence. *Proc. South. Weed Sci. Soc.* 38:448.
6. Al-Bayati F.A. and H.F. Al-Mola. 2008. Antibacterial and antifungal activities of different parts of *Tribulus terrestris* L. growing in Iraq. *J Zhejiang Uni. Sci. B.*, 9: 154-159.
7. Ali NA, Julich WD, Kusnick C Lindequist U (2001) Screening of Yemeni medicinal plants for antibacterial andcytotoxic activities. *J. Ethnopharmacol.* 74: 173–9.

8. Altug T.M., B. Yaymacib, L.Satic, C. Sevil, A. Goksemin, T. Altuge and D. Ramazan. 2009. Influence of *Tribulus terrestris* extract on lipid profile and endothelial structure in developing atherosclerotic lesions in the aorta of rabbits on a high-cholesterol diet. *Acta Histochemica.* 111: 488-500.
9. Andersen M.M.L., Hachul H, Tufik S. 2010. Medicinal plants as alternative treatments for female sexual dysfunction: utopian vision or possible treatment in climacteric women? *J. Sex Med*,; 7(11):3695-714.
10. Angelova S, Z. Gospodinova, M. Krasteva, G. Antov, Lozanov V, T. Markov *et al.,* 2013. Antitumor activity of Bulgarian herb *Tribulus terrestris* L. on human breast cancer cells. *J Bio Sci Biotech*., 2(1): 25-32.
11. Arcasoy, H.B., A. Erenmemisoglu, Y. Tekol, S. Kurucu and M. Kartal. 1998. Effect of *Tribulus terrestris* L., saponin mixture on smooth muscle preparations: a preliminary study. *Boll. Chim. Farm*., 137: 473-475.
12. Aslani MR, Movassaghi AR, Mohri M, Pedram M, Abavisani A. 2004. Experimental *Tribulus terrestris* poisoning in goats. *Small Rumin Res*,51: 261-267.
13. Aslani MR, Movassaghi AR, Mohri M, Pedram M, Abavisani, A. 2003. Experimental *Tribulus terrestris* poisoning in sheep: clinical laboratory and pathological findings. *Vet Res Commun*,; 27: 53-62.
14. Bapalal, V. 1982. Some controversial drugs in Indian Medicine, Ed. 1st, , Varanasi: *Chaukhambha Orientalia*. 76-77.
15. Bayati FAA, Al-mola HF. Antibacterial and antifungal activities of different parts of *Tribulus terrestris* L. growing in Iraq. 2008. *Zhejiang Univ Sci*., B 9(2): 154-159.
16. Bedir, E. and I.A. Khan. 2000. New steroidal glycosides from the fruits of *Tribulus terrestris*. *J. Nat. Prod.,* 63: 1699- 1701.
17. Berkov S. 2001. Size and alkaloid content of seeds in induced autotetraploids of *Datura innoxia, Datura stramonium* and *Hyocyamus niger. Pharmaceutical Biology,* 39:329-331.
18. Bhattacharjee S.K., 2004. Handbook of Medicinal plants. Jaipur, Pointer Publishers:, pp.353.
19. Bourke, C.A., G.R. Stevens and M.J. Carrigan. 1992. Locomotor effects in sheep of alkaloids identified in Australian *Tribulus terrestris*. *Aust. Vet. J.,* 69: 163-165.
20. Brown, A.G., M.D. Vukovich, E.R. Martini, M.L. Kohut, W.D. Franke, D.A. Jackson and D.S. King. 2002. Endocrine and lipid responses to chronic androstenediolherbal supplementation in 30 to 58 year old men. *J. Am. Coll. Nutr.,* 20: 520-528.
21. Cai, L., Y. Wu, J. Zhang, F. Pei, Y. Xu and S. Xie. 2001. Steroidal saponins from *Tribulus terrestris*. *Planta Medica*, 67:196-198.
22. Chu, S., W. Qu, X. Pang, B. Sun and X. Huang. 2003. Effect of saponin from *Tribulus terrestris* on hyperlipidemia. *Zhong Yao Cai.,* 26: 341-344.
23. Chui H.C., J.I. Victoroff, D. Margolin, W. Jagust, R. Shankle R. Katzman. 1992. Criteria for the diagnosis of ischemic vascular dementia proposed by the state of california alzheimer's disease diagnostic and treatment centers. *Neurology,* 42: 473-80.
24. Chunekar, K. C. 2006. In *Bhavamishra, Bhavaprakasha Nighantu*. Dr. G.S. Pandey, Ed. Reprint, Chaukhambha Bharati Academy, Varanasi,; 292
25. Conrad J., D. Dinchev, I. Klaiber, S. Mika, I. Kostova W. Kraus. 2004. A novel furostanol saponin from *Tribulus terrestris* of Bulgarian origin. *Fitoterapia.,* 75: 117-122.
26. Das S., Vasudeva, N. Sharma, S 2017. Pharmacognostical evaluation of *Tribulus terrestris* L. *Int. J. Pharmaceutical Sci. Res.* 8(3): 1393-1400.
27. Davis R, Wiese A, 1964. Weed root growth pattern in the field. Proceedings of the 17th Southern Weed Conference. Knoxville, USA: University of Tennessee at Knoxville, 367-368.

28. De Combarieu, E. N. Fuzzati, M. Lovati, E. Mercalli. 2003. Furostanol saponins from *Tribulus terrestris*. *Fitoterapia.*,74: 583-591.
29. Deepak M., G. Dipankar, D. Prasanth, M.K. Asha, A. Amit and B.V. Venkatraman. 2002. Tribulosin and â-sitosterol-Dglucoside, the anthelmintic principles of *Tribulus terrestris.Phytomedicine*, 9: 753-756.
30. Drewes E.S., J. George and F. Khan. 2003. Recent findings on natural products with erectile-dysfunction activity. *Phytochem*. 62: 1019-1025.
31. Ganaie S.A (2011). Mechanism of pollination in *Tribulus terrestris* L. (Zygophyllaceae). *Int. J. Pharma Bio Sci.*. 2(2):316-320.
32. Ganzera M., E. Beddir and I.A. Khan. 2001. Determination of steroidal saponins in *Tribulus terrestris* by reversed phase high-performance liquid chromatography and evaporative light scattering detection. *J. Pharma Sci.*, 90: 1752-1758.
33. Gauthaman K., P.G. Adaikan and R.N.V. Prasad. 2002. Aphrodisiac properties of *Tribulus terrestris* extract (protodioscin) in normal and castrated rats. *Life Sci.*, 71: 1385-1396.
34. Gupta RC, Kaur K, Kataria V (2016). Meiotic chromosomal studies in family Zygophyllaceae R. Br. from Rajasthan. Ind J Genet Plant Breed 76(1):111–115.
35. Hu K, Dong A, Yao X, Kobayashi H & Iwasaki S (1996) Antineoplastic agents: I. Three spirostanol glycosides from rhizomes of *Dioscorea collettii* var. *hypoglauca* . *Planta Med.*62: 573–575
36. Hu K. Yao, X. 2003. The cytotoxicity of methyl protodioscin against human cancer cell lines in vitro. *Cancer Invest.,* 21: 389-393.
37. Hu, K,. Yao. X. 2002. Protodioscin (NSC-698 796): its spectrum of cytotoxicity against sixty human cancer cell lines in an anticancer drug screen panel. *Planta Med.,* 68: 297-301.
38. Katewa, S.S., Chaudhary, B.L., Jain, A.. 2004. Folk herbal medicines from tribal area of Rajsthan, India. *J. Ethnopharmacol.*, 92:1, 41-46.
39. Kianbakht S. and F. Jahaniani. 2003. Evaluation of Antibacterial Activity of *Tribulus terrestris* L., growing in Iran. *Iranian J. Pharmacol. Therapeutics*, 2: 22-24.
40. Kostova I. and D. Dinchev. 2005. Saponins in *Tribulus terrestris*-chemistry and bioactivity. *Phytochem. Rev.,* 4: 111-137.
41. Kumar S.P., L.K. Abdul, A.B. Remashree. 2014. Ethnobotanical survey of diuretic and antilithiatic medicinal plants used by the traditional practioners of Palakkad district. *Int. J. Herbal Medicine*. 2(2), 52-56.
42. Li M., W. Qu, Y. Wang, H. Wan and C. Tian. 2002. Hypoglycemic effect of saponin from *Tribulus terrestris*. *Zhong Yao Cai.*, 25: 420-422.
43. Lin Z.X., J.R. Hoult and A. Raman. 1999. Sulphorhodamine B assay for measuring proliferation of a pigmented melanocyte cell line and its application to the evaluation of crude drugs used in the treatment of vitiligo. *J. Ethnopharmacol.*, 66: 141-150.
44. Mahato S.B., A.N. Ganguly and N.P. Sahu. 1982. Steroid saponins. *Phytochem*. 21: 959-978.
45. Milasius K, Dadeliene R, Skernevicius J. 2009. The influence of the Tribulus terrestris extract on the parameters of the functional preparedness and athletes' organism homeostasis. Fiziol Zh,55: 89-96
46. Miles C.O., A.L. Willkins, S.C. Munday, A. Flaønen, P.T. Holland and B.L. Smith. 1993. Identification of insoluble salts of the b-D-glucuronides of episarsasapogenin and epismilagenin in the bile of lambs with alveld and examination of *Nartecium ossifragum*, *Tribulus terrestris*, and *Panicum miliaceum* for sapogenins. *J. Agric. Food Chem.,* 41: 914-917.
47. Morrison SM, Scott JK (1996) Variation in populations of Tribulus terrestris (Zygophyllaceae). 2. Chromosome numbers. *Aust J Bot* 44:191–199.

48. Mulinacci N., P. Vignolini, G. Marca, G. Pieraccani, M. Innocenti and F.F. Vincieri. 2003. Food supplements of *Tribulus terrestris* L. an HPLC-ESI-MS method for an estimation of the saponin content. *Chromatographia*. 57: 581-592.
49. Oza, Z. and Mishra, D. 2004. Dhanvantari Nighantu with Hindi Gunakarmatmaka commentary, edited by Dr. Zarkhand Oza and Dr. Umapati Mishra, Chaukhamba Surabharati Prakashan, Varanasi,; 35.
50. Pacanoski Z., Tyr, S. Veres, T. (2014). Puncturevine (*Tribulus terrestris*) L.): noxious weed or powerful medical herb. *Journal of Central European Agriculture*, 15:11-23.
51. Pandita, N. and Nighantu, R. 1998. Hindi commentary *Dravya*guna prakashika, edited by Dr. Indradeva Tripathi, Ed. 2nd, Chaukhambha Krishnadas Academy, Varanasi, 69.
52. Petkov G. (2011). Enhancement of *Tribulus terrestris* L. yield by supplement of green house seedlings. *Biotechnology and Biotechnology.* DOI: 10.5504/bbeq.2011.0039.
53. Pokrywka A., Obminski, Z., Malczewska-Lenczowska, J., Fijalek, Z., Turek-Lepa, E., Grucza, R. 2014.Insights into supplements with Tribulus terrestris used by athletes. J Hum Kinet.41:99–105.
54. Protich M., D. Tsvetkov, B. Nalbanski, R. Stanislavov and M. Katsarova. 1981. Clinical trial of Tribestan on infertile males. (*Pharmachim, Bulgaria - Scientific-technical report*).
55. Ramaseshiah G, 1976. Survey for natural enemies of Tribulus terrestris. Report of work carried out during 1975. Trinidad: Commonwealth Institute of Biological Control, CAB International.
56. Reddi S. C., Reddi, E. U. B. and Reddi, N. S., (1981) Breeding struc-ture and pollination ecology of *Tribulus terrestris* Proc. Indian Natl. Sci. Acad Part B, 47 , 185 -193.
57. Ren, Y.J., Chen, H.S, Yang,. G.J., Zhu, H. 1994. Isolation and identification of a new derivative of cinnamic amide from *Tribulus terrestris*. *Acta. Pharm. Sin.,* 29: 204-206.
58. Rogerson S, Riches CJ, Jennings C, Weatherby RP, Meir RA, Marshall-Gradisnik SM (2007) The effect of five weeks of Tribulus terrestris supplementation on muscle strength and body composition during preseason training in elite rugby league players. *J. Strength Cond. Res.* 21(2):348-53.
59. Sajeev K.K., Sasidharan, N. 1997. Ethnobotanical observations on the tribals of Chinnar wildlife sanctuary. *Ancient Sci.Llife.* (16)4, 284-292.
60. Saleh, N.A.M., Ahmed A.A., Abdalla, M.F. 1982. Flavonoid Glycosides of *Tribulus pentandrus* and *Tribulus terrestris*. *Phytochemistry.* 21: 1995-2000.
61. Sankaran T, Ramaseshiah G, 1981. Studies on some natural enemies of puncturevine *Tribulus terrestris* occurring in Karnataka State, India. Proceedings of the 5th International Symposium on Biological Control of Weeds. Australia: CSIRO, 153-160.
62. Shah G.L., Flora of Gujarat State. Part I, Ed. 1st; February 1978, Vallabh Vidyanagar: Sardar Patel University.
63. Sharifi A.M., R. Darabi and N. Akbarloo. 2003. Study of antihypertensive mechanism of *Tribulus terrestris* in 2K1C hypertensive rats: role of tissue ACE activity. *Life Sci.,* 73 (23): 2963-71.
64. Sharma, P.V. and Sharma, G. 1979. Kaiyadeva Kaiyadeva Nighantu, edited by P.V. Sharma and Guruprasad Sharma, Ed. 1st, Chaukhambha Orientalia, Varanasi, 1979; 16
65. Singh A.G., Kumar, A. and Tiwari, D.D. 2014. *Asian Pacific J. Trop. Biomed.* 4(1), 460-467.
66. Stoyanov N. 1973. Our Medicinal Plants 2. Nauka *and* Izkustvo, Sofia, 454-455.
67. Sun B., W. Qu and Z. Bai. 2003. The inhibitory effect of saponins from *Tribulus terrestris* on Bcap-37 breast cancer line in vitro. *Zhong Yao Cai.,* 26: 104-106.
68. Talazas AH, Abbasi MR, Abkhiz S, Dashti-Khavidaki S. 2010.Tribulus terrestris-induced severe nephrotoxity in a young healthy male. *Nephrol. Dial. Transplant*, 25: 3792-3793.

69. Theodore, C. 2006. The Flora of the presidency of Bombay. Vol 2:37, Bishen Singh Mahendra Pal Singh, Dehradun.

70. Tomova M. and R. Gyulemetova. 1978a. Bulg. Patent (11) 26221, 2(51) A 1K 35/1978.

71. Tomova M., D. Panova, S. Zarkova and V. Dikova. 1966. Bulg. Patent (11) 11450, 30 1/ 02 A61K/1966.

72. Tomova M., R. Gyulemetova, S. Zarkova, S. Peeva, T. Pangarova and M. Simova. 1981. Steroidal saponins from *Tribulus terrestris* L. with a stimulating action on the sexual functions. *First Intern. Conf. Chem. Biotechnol. Biol. Active Nat. Prod., Proc.*, Varna, Sept 3: 289-291.

73. Tomova M.P. and R. Gyulemetova. 1978a. Steroidsaponins and steroidsapogenins. VI. Furostanol bisglycoside from *Tribulus terrestris* L. *Planta Med.,* 34: 188-191.

74. Tomova, M., D. Panova and N.S. Wulfson. 1974. Steroid saponins and sapogenins IV. Saponins from *Tribulus terrestris*. *Planta Med.,* 25: 231-237.

75. Tsvetkova, Y.E., Semerdjieva, I., Baldjiey, G. and Grancharova, Y.T. (2014) On the reproductive biology of tribulus terrestris L. (Zygophyllaceae): embryological features; pollen and seed viability. *Biotechnology Biotechnological Equipment*. 25: 2383-2387.

76. Viktorov I., D. Kaloyanov, A.I. Lilov, L. Zlatanova and V. Kasabov. 1982. Clinical investigation on Tribestan in males with disorders in the sexual function. *Med. Biol. Inform.*, (Pharmachim, Bulgaria - Company documentation).

77. Wang Y., Ohtani, K., Kasai, R. Yamasaki, K. 1997. Steroidal saponins from fruits of *Tribulus terrestris. Phytochem.* 45: 811-817.

78. WHO monographs on selected medicinal plants. Vol. 4. 1. Plants, Medicinal. 2. Angiosperms. 3. Medicine, Traditional. I. WHO Consultation on Selected Medicinal Plants (4th: 2005: Salerno-Paestum, Italy) II. World Health Organization.

79. Wu G., Jiang S., Jiang F., Zhu D., Wu H., Jiang, S. 1996. Steroidal glycosides from *Tribulus terrestris*. *Phytochem.* 42: 1677-1681.

80. Xu Y.X., Chen, H.S., H.Q. Liang, Z.B. Gu, W.Y. Liu, W.N. Leung and T.J. Li. 2000. Three new saponins from *Tribulus terrestris. Planta Med.,* 66: 545-550.

81 Yang S.S., H.L. Chang, C.B. Wei and H.C. Lin. 1991. Reduce waste production in the Kjeldahl methods. *J. Biomass Energy Soc. China,* 10: 147-155.

8

Ephedra species

Hindi: *Ain, Khanta, Somlata, Torgatha, Tutgautha*

S.S. Samant and Manohar Lal

***Ephedra* species**
(A) Herbarium sheet, **(B)** Plants in natural habitat
(C) Flowering stage, **(D)** Fruiting stage

INTRODUCTION

In India, medicinal plants have traditionally occupied an important place in the socio-cultural, spiritual and health related aspects of peoples' life. As much as 34% (ca. 6000) of total plant wealth have some known medicinal value (Ved, 2008). The genus *Ephedra* is one such important group of medicinal plants, from which the drug Ephedrine and a range of medicinally active compounds such as ephedrine, pseudoephedrine, norephedrine, norpseudoephedrine, methyl ephedrine, methyl pseudoephedrine, alkaloids, phenols, terpenoids are obtained. Owing to its medicinal properties, genus *Ephedra* is in high demand by pharmaceutical industries. It is currently facing heavy extraction pressure from its restricted habitat in the wild, and utterly lacks effective proper conservation and management efforts. For long-term conservation management of this genus, a better and comprehensive understanding of the species remains a top priority, for which detailed studies on its biology, habitat ecology, and agro-technology are required.

A. BIOLOGICAL ASPECTS

The genus *Ephedra* belonging to family Ephedraceae, is a group of plants that inhabit temperate regions in Eurasia, Northern Africa, South-western North America, and western South America. Its preferred habitats include cold deserts; dry rocky slopes; grasslands, and maritime areas with a Mediterranean climate. It is a drought and frost resistant species with xerophytic characters that make it apt for habitation in colder as well as semiarid and arid regions. It is represented by 50–65 species of shrubs, vines, or rarely small trees (Stevenson, 1993; Price, 1996; Sharma and Uniyal, 2008a, b; Sharma *et al.*, 2010). In India, its distribution extends from Sikkim in Eastern Himalaya to North-West in Uttarakhand, Himachal Pradesh and Jammu & Kashmir.

• Systematic classification

Kingdom	:	Plantae
Division	:	Gnetophyta
Phylum	:	Tracheophyta
Class	:	Gnetopsida
Order	:	Ephedrales
Family	:	Ephedraceae
Genus	:	*Ephedra*

Synonyms: No synonyms are recorded for this species.

Common names

Hindi	:	*Ain, Khanta, Somlata, Torgatha, Tutgautha*
Sanskrit	:	*Som, Soma*

• Diversity within Genus

Ephedra genus has eleven species that occur in the Indian sub-continent, mostly distributed in higher elevations of Himalaya preferring alkaline soils. Different species that occur in various regions of country are markedly different from each other in term of their habitat preferences that promote diversity within the genus. The regional spread of various species (Sharma and Uniyal, 2008a, b; Sharma *et al.*, 2010) is as follows:

E. foliata **Boiss.**: Punjab and Rajasthan

E. gerardiana **Wall. ex Stapf**: Semiarid and temperate Himalaya from Kashmir to Kumaon

E. intermedia **Schr. & Meyer**: Kashmir, Himachal and Garhwal

E. nebrodensis **Tineo**: Kashmir and Himachal Pradesh

E. pachyclada **Boiss.** : Jammu and Kashmir

E. saxatilis **(Stapf.) Royle ex Florin:** Uttarakhand

E. saxatilis **(Stapf.) Royle ex Florin var.** ***sikkimensis*** **(Stapf.) Florin:** Sikkim

E. regeliana **Florin:** Ladakh

E. przewalskii **Stapf. :** Nanga Parbat in Kashmir and in the bed of Shyok River in Ladakh.

E. sumlingensis **Sharma & Uniyal:** Lahaul and Spiti

E. khurikensis **Sharma & Uniyal:** Lahaul and Spiti

E. kardangensis **Sharma & Uniyal:** Lahaul and Spiti

Geographical Distribution

- ***E. gerardiana*****:** This species is found scattered in the drier regions of temperate and alpine Himalayas from Kashmir to Sikkim at altitudinal range of 2100-4800 m amsl and similar heights at Pangi (Chamba), Lahaul & Spiti, Chini and Kilba Kailash ranges of Kanawar (Kinnaur), Shali hills (Shimla), Kashmir and Ladakh.
- ***E. intermedia*****:** Species recorded in Lahaul & Spiti and Kinnaur districts of Himachal Pradesh. Four varieties of this species have been recorded, of which ***E. intermedia*** **var.** ***tibetica*** occurs in India.

- ***E. pachyclada***: It is a closely allied species found in Chitral, Baluchistan and Afghanistan.
- ***E. kardangensis***: Species recorded from the Western Himalayas (Himachal Pradesh) in India where it grows on gentle slope in Kardang (District Lahaul & Spiti) above 3,000 m elevation.
- ***E. khurikensis***: Species has been reported from Khurik (Spiti) in Himachal Pradesh.
- ***E. przewalskii***: It is distributed in Pakistan, Tajikistan, Uzbekistan, N Gansu, Nei Mongol, Ningxia, N Qinghai, Xinjiang Kazakstan, Kyrgyzstan.
- ***E. saxatilis***: It is distributed in Sikkim, Bhutan, Nepal, S Xizang and N Yunnan.
- ***E. regeliana***: It is distributed in Xinjiang (China), Afghanistan, N India, Kazakstan, Kyrgyzstan, Pakistan, Tajikistan, Uzbekistan.

Habitat and Ecology

- *Ephedra* is dioecious, shrubby plant with jointed ribbed green stems differentiated into nodes and internodes and small leaves. It is commonly known as Jointed fir genus, closely related to conifers. It is however quite different from conifers in many botanical aspects. It is resistant to drought and frost, shows various xerophytic characters, which make it suitable for habitation in colder as well as semiarid and arid regions. It also resembles *Equisetum*, a pteridophytic genus as its green stem is ribbed, branched, hollow inside and differentiated into nodes and internodes.

The habitat preference of different species is given below:

- ***E. gerardiana*:** It prefers stony, slopes and rock-strewn terraces, in open/ sunny and driers area of Himalayas.
- ***E. intermedia***: The habitat preference is similar to that of *E. gerardiana.*
- ***E. kardangensis***: **I**t grows on gentle slope in Kardang (District Lahaul) above 3,000 m elevation in the western Himalaya.
- ***E. khurikensis***: It grows luxuriantly in open ground in dry gravel soil of Khurik (Spiti, HP).
- ***E. przewalskii***: It prefers dry and sandy places.
- ***E. saxatilis***: It occurs on mountain slopes and sandy places.
- ***E. regeliana***: It prefers rocky slopes, flood lands and sandy places.

• Taxonomic Enumeration

The various species of *Ephedra* differs from each other in taxonomical enumerations, which are as follows:

E. gerardiana Wallich ex C. A. Meyer

An erect shrub of varying sizes but usually not more than a few inches in height; bears cylindrical, striated, often curved branches arising in whorls; dark green in colour. The internodes of branchlet measure 1-4 cm in length and 1-2 mm in width are red in colour, sweet in taste and edible. The ovoid fruits contain 1-2 seeds covered by luscious bracts. The rhizomes have large knobs, size of a football. *E. gerardiana*var. *saxalitis* Stapf. is taller and ascending and occurs in Garhwal and Kumaon. *E. gerardiana* var. *sikkimensis* Stapf. is erect, robust but soft, andoccurs in Sikkim.

E. intermedia Schrenk & Meyer.

Synonyms: - *E. ferganensis* V.A. Nikitin, *E. intermedia* var. *persica* Stapf., *E. intermedia* var. *schrenkii* Stapf., *E. intermedia* var. *tibetica* Stapf., *E. microsperma* V.A. Nikitin, *E. persica* (Stapf.) V.A. Nikitin, *E. valida* V.A. Nikitin, *E. vulgaris* var. *submonostachys* Boiss. & Buhse

It is a densely branched, erect or prostrate shrub. Four varieties of *E. intermedia* have been recorded, of which var. *tibetica* occurs in India. *E. pachyclada* Boiss. is a closely allied species found in Chitral, Baluchistan and Afghanistan.

E. major Host

Synonyms: *Chaetocladus monostachys* J. Nelson, *E. atlantica* Andr., *E. equisetiformis* Webb & Berthel., *E. graeca* C.A.Mey, *E. major* ssp. *major*, *E. major* var. *nebrodensis* (Tineo) Hayek, *E.major* sub-sp. *villarsii* (Gren. & Godr.) P. Fourn., *E. nebrodensis* Tineo, *E. nebrodensis* var. *scoparia* (Lange) Nyman, *E. scoparia* Lange, *E. villarsii* Gren. & Godr., *E. vulgaris* Willk.

It is an upright, rarely ascending, densely branched shrub, up to 6 ft. high, of which var. *procera* (Fisch. & Mey.) Aschers. & Graebn. is reported from Lahaul. The twigs of this species closely resemble those of *E. gerardiana.*

E. kardangensis P. Sharma & P. L. Uniyal

Synonyms: No synonyms are recorded for this name.

It is a dioecious, bushy, erect shrub that grows upto1m in height with brownish black bark, and dark green branchlets, having ridges and furrows, 0.5–2 mm in diameter and internodes measuring about 1–2 cm. Scaly leaves showing opposite and decussate arrangement are usually two per node, opposite and decussate, turn brown at maturity, triangular with sharp tips, 1–1.5 mm in length, mostly 1/2 connate, membranous covering, later fissured. Male strobilus, broadly ovoid, 3×2.5 mm with three pairs of flowers is stalked. Female strobili elliptic, 7×5 mm, stalked, three pairs of bracts, bracts mucronulate. Seed ovate to triangular at the apical region, dull maroon in colour, *ca.* 3×2 mm, two enveloped, micropylar tube straight, 2–4 mm long, exerted with round opening.

E. khurikensis P. Sharma & P. L. Uniyal

Synonyms: No synonyms are recorded for this name.

The species is a medium erect shrub, dioecious, 40-50 cm in height, bark is brownish and branchlets are ridged and furrowed 0.5-3 mm in diameter, and internodes 2.5-3 cm long. Leaves scaly, usually two per node, opposite and decussate, turning brown at maturity, triangular with acuminate tips, 3-3.5 mm long, mostly 3/4 connate, membranous covering, later fissured with chocolaty base. The stalked male strobilus is elliptic, 4×2.5 mm, with four pairs of flowers, bracts binate in four pairs, mucronulate, lower and middle pairs 1/4 connate, upper ½-3/4 connate. Female strobili elliptic, 9×5 mm stalked, three pairs of bracts, bracts mucronulate, connate. Seed elliptic, brown - black, 4×2 mm, two enveloped, micropylar tube coiled, fringed with oblique opening, 2 mm long.

E. przewalskii Stapf.

Synonyms:*E. kaschgarica* B. Fedtsch. & Bobrov, *E.przewalskii* var. *przewalskii, E.przewalskii* var. *kaschgarica* (B. Fedtsch. & Bobrov) C.Y. Cheng

The species is a branched ascending shrub that grows up to 2.4 m in height. Young branchlets are green to pale brownish, with 2.5–5 cm long internodes. Triangular leaves in 3 whorls with acute or acuminate apex, rarely opposite. Cones are often sessile with opposite or whorled at node. Bracts of pollen cones with central, herbaceous keel are surrounded by broad membranous margin, in whorls of 3-4, rarely opposite, connate; shortly stipulate anthers 5–8 in number. Subglobose seed cones with connate, light brown bracts in 4- 5 whorls of 3, rarely opposite, almost completely free and membranous at maturity, abruptly narrowed toward base, apical whorl of bracts with female flowers; integument tube usuallystraight, 1.5–2 mm long. Elongate-ovoid seeds 2-3, 4–5 × 2–2.5 mm, covered by scarious bracts.

E. gerardiana var. *sikkimensis* Stapf.

Synonyms: *E. saxatilis* (Stapf.) Royle ex Florin, *E. gerardiana* var. *congesta* C.Y. Cheng, *E. gerardiana* var. *saxatalis* Stapf., *E. saxatilis* var. *sikkimensis* (Stapf.) Florin

It is a shrub growing up to 60 cm in height, with erect or ascending greyish brown or greyish yellow stout stems. Green branchlets radially arranged, clustered at nodes having internodes 2–4 cm long, notably furrowed when dry, smooth. Leaves measure about 2–6 mm, connate for 1/2–3/4 their length with opposite arrangement. Pollen cones 1–3 in number, sessile or shortly pedunculate, usually opposite at nodes. Bracts in pairs of 3-5 or 6-7, with broad margins, membranous; staminal columns with 6–8 sessile anthers exerted at apex. Seed cones are sessile or shortly pedunculate, solitary or opposite sometimes at subsequent nodes, bracts in 2-3 pairs, connate for 1/5–2/3 their length, narrowly membranous margins, turning red and fleshy at maturity; integument tube about 0.5 mm, straight. Greyish black, pruinose, ovoid to sub-cylindric seeds often 2 ca. 6 × 3 mm, ridged, usually exerted.

E. regeliana Florin

Synonyms: *E. monosperma* var. *disperma* Regel, *E. pulvinaris* V.A. Nikitin

Shrubs up to 8–15 cm in height, only with several woody basal branches 1–2 cm long, lacking prominent woody stems above ground; main branches usually subtle; slender and short branchlets ca. 1 mm in diameter, clustered at nodes andradially arranged with internodes 1–3 cm, apex slightly curved. Leaves are acute or slightly blunt apex, connate for 1/2–2/3 their length, oppositely arranged. At nodes of branchlets pollen cones are usually solitary or 2-3, ellipsoid measuring about 2–6 mm; bracts in 4–8 pairs; staminal column with 6 to 8 shortly stipitate anthers exerted at apex. Seed cones, ovoid or broadly ovoid, ca. 5×3–4 mm at maturity opposite at nodes or clustered at apex of branchlets; bracts usually in 3 pairs, apical pair connate for ca. 5/6 their length or more, red and fleshy at maturity; integument tube 1–2.5 mm, straight. Seeds: 1-2, shiny, narrowly ovoid, 3–4.5 × 1.5–2 mm, covered by bracts.

• Phenological Period

E. gerardiana resumes active growth with warming after winter, and show flowering by May-June.The strobili are seen on species from July to August. These mature into fruits, then seeds in August-September.

• Cytology

Different chromosome counts have been reported in *E. gerardiana* such as 2n=56 (Kawatani *et al.*, 1959); 2n= 28 (Chouhdry, 1984); 2n =28 (Leitch, 2001); 2n= 14, 28, 56 (e flora). All seem to be multiples of basic number 14.

B. REPRODUCTIVE BIOLOGY

Reproductive Phenology

- *Ephedra* shows alternation of generation within its lifecycle and the dominant phase is sporophyte. Sporophytic phase of *Ephedra* plant is represented by small woody much branched shrub, bearing minute scale like stem. Plant produces two types of small strobili as diminutive spikes at the nodes.

- **Male strobili**: Male strobili occur in whorls at nodes of branches, each strobilus possesses 2-8 pairs of bracts borne on canal axis where lower 1-2 bracts may be sterile. Male flowers arise from axils of fertile bracts. Axis at apex bears 1-8 microsporangia (anthers), each of which possess 2-3 loci and each loculus opens by terminal slit to release microspores (pollen grains). Boat shaped unwinged microspore possess an inner thin wall (intine) and outer thick wall (exine).

- **Female strobili:** Female strobili arise in the axis of leaves at node. Each female strobilus is an elongated structure pointed at the apex and, consists of 2-4 more pairs of bracts borne on central axis. Only apical 1-2 bracts are fertile and bear single female flower in the axil. Each female flower has a short stalk bearing an ovule at the apex. The ovule has two integuments, inner integument has 2 bracts fused at the base and the outer has four basal bracts along with four vascular bundles. At the time of pollination, inner integument becomes hard and forms long micropyler tube at the top. Nucleus has mass of sterile tissue that overlies embryo sac. Pollen chamber is funnel shaped and extends to the embryo sac thereby exposing the neck of the archegonium so that pollen grains come in contact with archegonium during pollination.

- **Pollination:** Anemophilous type of pollination is recorded for *Ephedra* spp. where wind acts as pollination agent. Wind carries pollen grains that have five nuclei to female gametophyte. Microspores, lodged in pollination drop, are sucked into pollen chamber and germination takes place within a few hours as exine ruptures and pollen tube is formed.

- **Fertilization:** The pollen tube bursts apically after forcing its way by the neck of archegonium and releases tube nucleus, stalk cell and two male nuclei into cytoplasm of mature egg, which at the time of fertilization, is surrounded by cytoplasmic sheath. The fusion of male and egg nuclei results in formation of diploid zygote.

C. MEDICINAL ASPECTS

• Plant material used as Crude Drug

Ephedra spp. has been used to treat different ailments since 5000 BC. This genus is generally used as a folk medicine for curing allergies, bronchial asthma, chills, colds, coughs, oedema, fever, flu, headaches and nasal congestion. In Russia, a decoction of the stems and roots is used to treat rheumatism and syphilis.The juice of the berries is used to treat respiratory afflictions. The stems are used in Tibetan medicine system as febrifuge, tonic and vulnerary in spite of it having bitter taste. They are used in the treatment of severe bleeding and chronic fevers (Anonymous, 1993; Bown, 1995; Tsarong, 1994).

- **General appearance:** The sporophytic plant body of genus *Ephedra* is divided into roots, stem and leaves. The stems and roots are of pharmacological importance. In crude form, it is available as dried stems in American markets and is sold as teas, tablets, capsules and other forms (www.allnatural.net/herbpages/ephedra.shmtl). The stem is ribbed, branched, hollow inside and differentiated into nodes and internodes.

- **Organoleptic properties:** These are properties experienced via senses including those of sight, smell, taste and touch. The stems of *Ephedra* are ribbed with ridges and furrows distinguished into nodes and internodes. They are therefore rough in touch. Ephedrine extracted from genus *Ephedra* in pure form iswhite microcrystalline powder. It is waxy solid, granules, crystals or granules (O'Neil, 2013). Its colour ranges from white to colourless (Lewis, 2007).

- **Microscopic characteristics:** The stem shows wavy outline in transverse section. The epidermis is heavily cutinized with stomata present in pits in region of furrows. Below the epidermis in the region of each ridge, group of sclerenchymatous cells are present. The cortex is formed of parenchymatous cells with chloroplasts distinguished into two layers, with outer palisade and inner spongy tissue. The stele is surrounded by endodermis with pith in the centre. The vascular bundles are collateral and endarch and vascular bundles form a continuous ring due to narrowness of medullary rays (www. Biologyboom.com).

- **Powdered plant material:** Microscopic analysis helps in rapid and accurate identification of powdered drug. Crude drug powder of *E. gerardiana* is reported to be greyish green in colour, has presence of calcium oxalate crystals, reticulate vessels, bordered pitted tracheids, fragments of epidermal cells with sunken stomata, palisade-like chlorenchymatous cells and reddish brown masses of tannin (Friedrich and Wiedemeyer, 1976). The powder on analysis was astringent in taste with mild aroma. Fluorescence analysis of the crude drug, which is helpful in fulfilling the inadequacy of physical and chemical methods for identification have been given for *E. gerardiana* by Rungsung *et al.* (2015).

• Major chemical constituents

The aerial parts of plant contain ephedrine alkaloid as the main constituent and smaller quantities of pseudoephedrine, methylephedrine, methyl-pseudoephedrine, norephedrine, and nor-pseudoephedrine (Ibragic and Sofiæ, 2015).

• Substituent and adulterants

The genus*Ephedra* resembles pteridophytic genus *Equisetum.* In crude drug market, *Equisetum* could be used as adulterants and substitute, though no such report or literature is available to authenticate it.

• General identity tests

Physico-chemical analysis of powdered drug is helpful in finding its purity. Incinerated drug samples across a range for minerals and ash values are fairly indicative, characteristics to a species. The following values have been given for *E. gerardiana* by Rungsung *et al.* (2015).

- Total ash : 6.50-7.20%
- Acid-insoluble ash : 2.50-3.40%
- Water-soluble ash : 0.90-1.30%
- Alcohol-soluble extractive : 23.20-25.20%
- Water-soluble extractive : 18.90-20.70%

• Purity identity test

Schaneberg *et al.* (2003) suggested that chemical fingerprinting could be employed to authenticate the presence of ephedrine in ground plant material. This will prevent spiking of genuine material with adulterants, including other species of *Ephedra* known to be devoid of the ephedrine alkaloids or synthetic ephedrine. Chemical fingerprinting is the only available method of ensuring quality of dietary supplements.

• Medical usage

- **General uses:** Ephedrine drug obtained from *Ephedra* spp. isused during spinal anaesthesia to avert low blood pressure. However, using this drug is not a preferred practice. The drug has also been used for treatment of asthma, narcolepsy, and obesity. Among various modes of taking the drug, intravenous delivery gives fast results, intramuscular injection takes about 20 minutes, and oral intake can take an hour to show effect. When given by injection, the effect lasts for about an hour and when taken by mouth the effect can last for up to four hours.A 5cc liquid extract of *Ephedra*, standardized to contain 0.5 grain of the total alkaloids as such, has been practiced, or in combination with other mixtures for controlling asthmatic paroxysms. For heart malfunction, resulting from pneumonia and diphtheria, Ephedra tincture is a valuable cardiac stimulant.

- **Ethnobotanical uses:** Traditionally, the Chinese have used Ephedra to treat asthma and other respiratory problems. Ephedra has been used as folk medicine for curing allergies, bronchial asthma, chills, colds, coughs, oedema, fever, flu, headaches and nasal congestion.

The genus Ephedra has been used for other purposes too by tribal communities where it is eaten as fruits in high altitude areas. It is also utilised as fuel by the tribal communities of the Cold Desert region. Aerial part is also used for washing utensils. Tribal people of cold desert region also use this plant for religious purposes.

D. CHEMISTRY ASPECTS

• Chemical composition

Ephedrine and its isomers were isolated by the Japanese chemist Nagai Nagayoshi (https://en.wikipedia.org/wiki/Ephedra) in 1881 from *E.dystachia*. His work promoted cultivation of three species (*E. sinica*, *E. vulgaris* and *E. equisetina)* in mainland China. These species are used as source for natural ephedrine and its isomers for pharmaceutical use. *E. sinica* and *E. vulgaris* have six optically active phenylethylamines, mostly ephedrine and pseudoephedrine (Freudenberg *et al.,* 1932).

About 45 *Ephedra* species exist, having varying alkaloid contents. Indian and Chinese species are known to contain higher active alkaloid content in comparison to American, Chilean, and European species. Based on HPLC analysis at industrial scale, the concentrations of total alkaloids in dried *Ephedra* ranged between 1 to 4%, and in some cases up to 6% (Chopra *et al.*, 1931). Among the Indian species, *E. major* is the richest source of ephedrine. The plants collected from Lahaul contained over 2.5 per cent total alkaloids, of which nearly three fourths is ephedrine. Total alkaloid content in *E. gerardiana* ranges from 1.0 to 2.5%, but below 1.0% for *E. sinensis*. The total alkaloid content of the green stems of *E. intermedia* ranges from 0.7 to 2.33 per cent of which only about one-tenth is ephedrine, the rest being pseudo-ephedrine. Other species, namely, *E. equisetina*, *E. sinica*, *E. distachya*, *E. gerardiana,* also yield ephedrine (Anonymous, 1960). Ephedrine is the major component of the total alkaloid content, while the rest are pseudoephedrine, N-methyl ephedrine, norephedrine and related compounds.

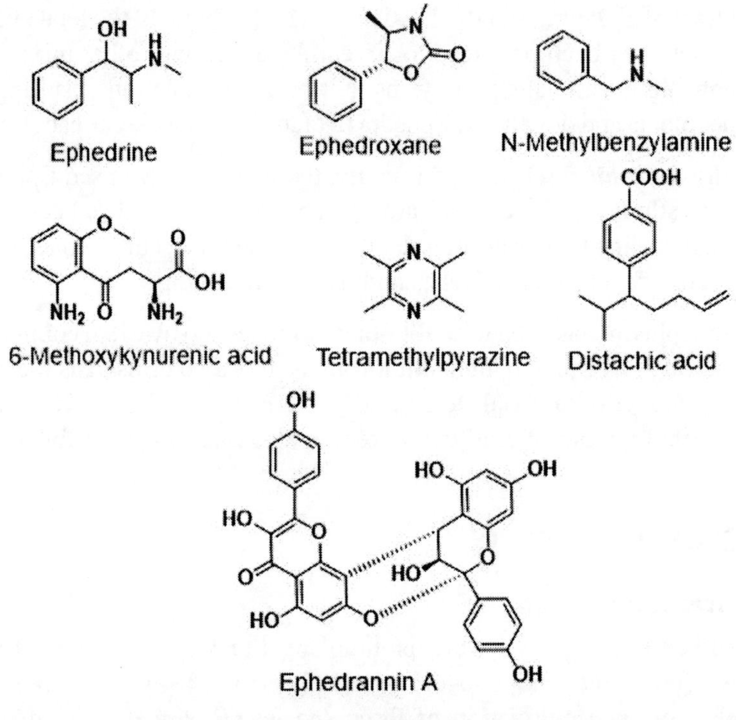

Ephedrine Ephedroxane N-Methylbenzylamine

6-Methoxykynurenic acid Tetramethylpyrazine Distachic acid

Ephedrannin A

Chemical constituents of *Ephedra gerardiana*
(Chemical structure is contributed by Dinesh Kumar)

The ratio of ephedrine to d-pseudo-ephedrine seems to vary in different species, the real value of the herb being determined by high γ-ephedrine content. The alkaloid ephedrine can exist in no less than six forms, *i.e.* γ-ephedrine, d-ephedrine, d γ-ephedrine, γ-pseudo-ephedrine, d-pseudo-ephedrine and dγ-pseudo-ephedrine. Ephedrine, $C_{10}H_{15}ON$, is a colourless crystalline substance with a melting point of 41-42°C. The hydrochloride forms colourless needles, melting point 216°C; specific rotation in water is -34.20 and in absolute alcohol -6.810. The platinic chloride of the base crystallizes in colourless needles, melting point 186°C. Pseudoephedrine, or iso-ephederine ($C_{10}H_{15}ON$), is also reported in addition to ephedrine from *E. gerardiana* and *E. intermedia.* Chemically, it is formed by heating ephedrine with hydrochloric acid. It is a dextrorotatory isomer of ephedrine with specific rotation of +50 in absolute alcohol and crystallises from ether (m.p. 118°C).

Various factors that affect alkaloid content of the plant, are its age and physiological state, its harvested part, climate as well as season. The alkaloid content decreases from May to August and thereafter gradually increases till it acquire its maximum value in October-November. Berries and roots contain hardly any alkaloid. The green twigs have significantly more alkaloid content than the woody stems. The apt time to collect the green twigs, which constitute the crude drug of *Ephedra* of commercial importance, is when the plants are 4 years old and are in blossom. The alkaloid content increases with the age of the plant. The twigs must be harvested after the last rainfall, and before the winter frost sets in, as rainfall has a marked adverse effect. Natural drying of twigs in the sun is most appropriate. Artificial drying at high temperatures must be avoided because it may decrease the alkaloid content. Humid conditions also have adverse effect on alkaloid content of the twigs. It is therefore essential to store dried twigs under complete dry conditions. Air dried drug stored in dry closed containers, protected from light, retains the activity without loss for a long period (Chopra *et al.*, 1931).

A number of workers have studied alkaloids of *Ephedra* in both plant material and clinical trials (urine samples) using gas chromatography, chiral-gas chromatography, gas chromatography-mass spectrometry and high-performance liquid chromatography (Cui *et al.*, 1991; Betz *et al.*, 1997; Li *et al.*, 2001; Spyridaki *et al.*, 2001; Jian *et al.*, 1988; Hurlbut *et al.*, 1998; Gmeiner *et al.*, 2002; Hong *et al.* 2011) and through liquid chromatography-mass spectrometry method for dietary supplements (Gay *et al.*, 2001). Carbon-13 nuclear magnetic resonance also has been used to qualitatively and quantitatively analyze ephedra alkaloids (Yamasaki and Fujita, 1979).

Four species of *Ephedra*, viz., *E. sinica*, *E. intermedia, E. equisetina,* and *E. distachya,* in 22 herbal samples collected from markets in Taiwan were studied by Liu *et al.* (1993) to analyze status of six ephedrine alkaloids (ephedrine, pseudoephedrine, methylephedrine, methyl-pseudoephedrine, norephedrine, and nor-pseudoephedrine).It was reported that ephedrine and pseudoephedrine constituted the major proportion of alkaloids, and its content was highest in *E. sinica*.

Other pharmacologically active compounds isolated from Ephedra are tetramethylpyrazine (Spyridaki *et al.*, 2001; Lv *et al.*, 2000; Zhao *et al.*, 2009), feruloylhistamine (Hikino *et al.*, 1984), and ephedradines A-D (Hikino *et al.*, 1983; Tamada *et al.*, 1979) from roots. The roots of ephedra have yielded a variety of hypotensive compounds, including the flavonoid ephedrannin A (Hikino *et al.*, 1982), feruloylhistamine (Hikino *et al.*, 1984) and the spermine alkaloids ephedradines A-D (Hikino *et al.*, 1983) and polysaccharide ephedrans A-E from stems of Ephedra (Konno *et al.*, 1985).

In addition to alkaloids, non-alkaloid contents also show pharmacological potential. While analgesic effect of species of the genus is attributed to pseudoephedrine, Hyuga *et al.* (2016) evaluated ephedrine-free Ephedra Herb extract (EFE) and found it to be safer alternative with comparable analgesic, anticancer, and anti-influenza activities. EFE was suggested to suppress hepatocyte growth factor (HGF)-induced cancer cell motility by preventing both HGF-induced phosphorylation of c-Met and its tyrosine kinase activity.

• Chemical markers

Ephedrine alkaloids and herbacetin, a flavonoid aglycon and potential putative marker for EFE quality control (Oshima *et al.*, 2016)

E. PHARMACOLOGICAL ASPECTS

Bioactivity

A number of medicinally active compounds, such as, ephedrine, pseudoephedrine, norephedrine, norpseudoephedrine, methyl ephedrine, and methyl pseudoephedrine, other alkaloids, and phenolic, terpenoid compounds are found in members of this genus. These compounds are known to possess various bioactivities like anti-arthritic, anti-bacterial, anti-cancerous, anti-inflammatory, antioxidant, anti-parasitic, anti-ulcer, antivirus and anti-obesity.

- **Therapeutic activity**: The therapeutic activity of *Ephedra* is due to the presence of the alkaloids, ephedrine and pseudo- ephedrine. Ephedrine is the principal alkaloid of *E. gerardiana* and *E. major*. It is soluble in water. On boiling with 25% hydrochloric acid, it is partially converted to pseudo- ephedrine. The conversion of pseudo- ephedrine to the useful l-ephedrine, however, is not so easily accomplished. Ephedrine has been synthesized and the synthetic product is marketed under the name, Enex-ephedrine.
- **Pharmacological action:** Pharmacologically, ephedrine is similar to adrenaline in action. It is more stable and can be given by mouth, unlike adrenaline. It strengthens heart action by stimulating respiratory centre and increasing the depth of respiration. It is also used in treatment of bronchitis as it dilates the bronchi.It also causes pupil dilation and contraction of uterus (Chopra *et al.*, 1931).
- **Analeptic action:** It also possesses analeptic action. Its central nervous stimulation, gives the basis for its use in the treatment of anxiety and for relief of narcolepsy. Ephedrine (methedrine) and emphetamine seem to have some advantages. It is used in vasomotor rhinitis, congestion of the

mucous membrane, acute sinusitis and hay fever. Ephedrine imparts light local anaesthetic action and this property seems to be associated in the l-,d-, and dl-forms of cinnamyl ephedrine. However, in case of allergic rhinitis, extreme caution has been advised for prescription of ephedrine and pseudoephedrine, which are highly efficient amines for relief of nasal congestion (Laccourreye *et al.,* 2015).

- **Anti-oxidant activity:** Antioxidant activities, using ferric reducing antioxidant power (FRAP) and 2, 2 diphenyl-1-picrylhydrazyl (DPPH) free radical scavenging assays, studied in *E. sarcocarpa* (Rustaiyan *et al.,* 2011a) and *E. laristanica* (Rustaiyan *et al.,* 2011b) showed a high potential. A strong free radical scavenging activity in the aqueous, ethyl acetate, and n-butanol fractions of stem of *E. gerardiana* was reported by Khan *et al.* (2017).

- **Anti-microbial activity:** Parsaeimehr *et al.* (2010) studied antimicrobial activities in three *Ephedra* species (*E. strobiliacea, E. procera*, and *E. pachyclada*). Of these, *E. strobiliacea* showed the highest activity against *Pseudomonas aeruginosa* (gram negative bacteria) and *Staphylococcus aureus* (gram positive bacteria). *E. procera* proved most effective against *Aspergillus niger* (fungus), while methanolic extract of *E. pachyclada* showed activity against *Klebsiella pnemoniae* (gram negative bacteria) and *Bacillus subtilis* (gram positive bacteria). Other studies have reported biological activities of *E. major* to inhibit growth of *Aspergillus parasiticus* and reduction in its aflatoxin production (Bagheri and Bigdeli, 2009), and antibacterial activity of *E. alata* against *Staphylococcus aureus, Bacillus anthracis, B. diphtheriae, B. dysenteriae, B. typhosus* and *Pseudomonas aeruginosa* (Soltan and Zaki, 2009). Caveney *et al.* (2001) reported antimicrobial compound cis-3,4-methanoproline in large amounts in the stems and seeds of many *Ephedra* species.

- **Anti-inflammatory**: Extract of *E. sinica*, studied for its anti-inflammatory affect, showed capacity for functional recovery in case of injured spinal cord in experimental rodents (Li *et al*., 2009). It likely acted by reducing inflammation and improving motor function in injured rats by inhibiting complement activation. Wu *et al*., (2014) studied affect of pseudoephedrine and ephedrine against lipopolysaccharide-induced acute liver failure in D-galactosamine-sensitised male rats, and reported a potent anti-inflammatory affect resulting from inhibition of tumour necrosis factor-α (TNF-α). Ephedrannin A and B were also reported to suppress the transcription of TNF-α and interleukin-1 beta in macrophages and in induced hepatic failure in mice (Kim *et al.,* 2010; Yamada *et al*., 2008).

- **Anti-arthritic potential**: Aqueous, aqueous-ethanol, n-butanol and ethyl acetate extract of *E. gerardiana* were reported to have anti-arthritic activity (Uttra and Alamgeer, 2017). In animals, protein denaturation is the principal cause of inflammatory and arthritic disorders that result in rheumatic disorders. Therefore, substances which can avert denaturation of protein are considered valuable for developing antiarthritic drugs. The study showed that hydro-alcoholic extract and fractions of *E. gerardiana* markedly subdued protein denaturation, and provide protection against arthritis

Clinical Studies

Role in management of specific ailments

Ephedrine is the important therapeutic ingredient associated with *Ephedra* species. As a drug, it produces physiological effects characteristic of sympathetic nervous system. The compound acts as a bronchodilator and decongestant owing to its sympathomimetic properties, which stimulate á-, â1-, and â2-adrenergic receptors through direct and indirect effects. When used in therapeutic doses, ephedrine primarily renders relief in respiratory and cardiovascular ailments (Andraws *et al.,* 2005). Other effects of ephedrine include CNS stimulation, relaxation of the smooth muscle of the gastro-intestinal tract, urinary retention, glycogenolysis, increased oxygen consumption and accelerated metabolic rate (Tsarong, 1994). Compared to ephedrine, d-pseudo-ephedrine is less toxic as well as cheaper, and has shown efficacious results in asthma patients (Duke, 2018).

Combined with a number of other herbs, *Ephedra* is also used for treating a wide range of complaints. A combination of ephedrine and a caffeine-containing supplement has been most frequently used for weight loss (Greenway, 2001; Vukovich *et al.*, 2005), and reported to increase endurance in running and cycling experiments (Dhar *et al.*, 2005; Shekelle *et al.*, 2003).

• Toxicity

The reported cytotoxicity to cultivated cells by *Ephedra* extracts is not primarily caused by ephedrine (Joyal, 2004; Lee *et al.*, 2000). However, N-nitrosamines of ephedrine and pseudoephedrine have been reported as carcinogens (Alwan *et al.*, 1986; Tricker *et al.*, 1987). A report from The United States Food and Drug Administration (USFDA) highlighted 15 deaths and 395 adverse effects due to products containing ephedrine (Theoharides, 1997), the principal active alkaloid found in *Ephedra*, and first banned the sale of all dietary supplements containing it in 2004, but later allowed the products containing quantities up to 10 mg or less. However, in 2007, the ban continued for dietary supplements containing adulterated ephedrine alkaloids (USFDA, 2007). Ephedrine causes headache, insomnia, nausea, nervousness, palpitation, sweating, vertigo, vomiting, occasionally precordial pain and sometimes dermatitis if taken in excessive doses. Ephedrine is the major component responsible for toxicity of *Ephedra* along with pseudo-ephedrine (Lee *et al.*, 2000). Both active constituents stimulate the adrenergic system and the clinical presentation of toxicity reflects the pharmacologic activity of these agents. Consumption of doses of ephedrine, taken 2 to 3 times and pseudoephedrine 4 to 5 times, above its therapeutic range, is likely to cause noticeable toxicity within 1 hour of ingestion. The clinical toxicity of *Ephedra* crude drug primarily affects cardiovascular system and the central nervous system (CNS). Adverse CNS effects associated with ephedrine toxicity cause anxiety, insomnia, restlessness, psychosis and seizures. Additional signs of ephedrine toxicity include nausea, vomiting, headache, flushing, difficulty in micturation and precordial pain. Patients with sympathomimetic toxicity are at an increased risk of rhabdomyolysis.

• Formulations

Traditional preparation: Dry stalks of *E. gerardiana* are boiled for ten minutes in water to prepare Tea. Dried herbage (about 6 grams) is a medicinal dose, while up to 30 grams may be taken to induce feelings of euphoria. However, alkaloid content may vary widely from plant to plant, so dosages must also vary (von Reis and Lipp, 1982). Usage of Ephedra is advised under medical supervision. Over dosage may lead to side effects like restlessness, anxiety, headaches, insomnia, vomiting, nausea and urinary problems. Pregnant women, lactating mothers, persons who take medications for depression and high blood pressure should not use *Ephedra* (www.alwaysayurveda.com/ephedra-gerardiana/).

F. THREAT AND CONSERVATION ASPECTS

Population status

The natural stands of different species vary in different regions of its distribution. *E. gerardiana* has been defined as threatened status by IUCN (Walter and Gillet, 1998), while *E. major* is listed asof least concern by IUCN UK (2011), but listed in the Red Data Books of Cyprus and Croatia as Vulnerable and Near Threatened, respectively (Nikolic *et al.,* 2007; Tsintides *et al.*, 2007). Based on distribution studies of *E. gerardiana* in Indian Himalayas, the species was listed as critically endangered (Samant and Pant, 2006). In Jordan, *E. alte* has been referred to as highly competitive and a threat to different fruit and forestry species (Qasem, 2012). *E. intermedia* is sparsely distributed in Himalaya, and its status is not well studied. *E. ciliata* prefers sandy habitat and has threatened status in India (Pandey *et al.*, 1983). *E. foliata* is a species of desert of Rajasthan in India and its status has also not been well defined (Bhandari, 1990).

Threats

• Habitat loss/degradation (human induced)

The natural distribution locations of the species in Himalayas does not face much threat in term of human settlement since, that continue to remain sparsely populated. However, trampling and overgrazing, destructive harvesting of the plant to the extent of leaving little regenerative source are some of the human induced threats.

• Over-exploitation

In Nepal, *E. gerardiana* has been categorized as threatened through over-collection for the export trade (http://www.floraofnepal.org/). Conservation status of *E. gerardiana,* studied in Mankial valley of Hindukush Range, Pakistan, revealed high extraction rates leading to reduction in its population size by more than 70% over the last 10 years/three generations (Ullah and Rashid, 2014). It was listed as endangered species.

• Introduction and spread of exotic/alien species

The species is naturally distributed in areas, such as dry temperate or cold desert, which is not conducive for growth of many species. Owing to lack of water availability and poor precipitation during the active seasons of growth, no significant competition is faced by *Ephedra.*

• Pollution and diseases

The area of distribution of species is not reported to be under any risk of pollution and no specific diseases are known.

• Climate change

The dry temperate or cold desert areas are particularly sensitive to warming and changing climate. The expected rise in temperature may prove beneficial by extending the growing period available for the species. However, warming induced up-migration of other plant species can pose competition for the limited resources, especially scarce availability of water. How would warming impact secondary metabolite profile remains the other important area for future investigation.

Conservation strategies

- ***Ex-situ* conservation:** A sound conservation strategy is required for all medicinal plants which are sought in large quantities and are collected for commercial use from wild. In addition to a well developed propagation protocols, comprehensive approach requires identification of quality planting material from the variety of its natural populations, large scale production capacity at functional nurseries, facilities for seed storage and breeding for better varieties, etc. *Ex-situ* production of seeds, developing seed/ pollen bank, conserving pollen using tissue culture and cryopreservation, are the other suggested measures. Developing plantlets in nurseries using conventional and *in vitro* propagation methods and distributing the germplasm among various stakeholders like farmers, NGOs etc. interested in its cultivation, are important steps for its effective conservation.

- ***In situ* conservation:***In-situ* conservation of *E. gerardiana* can be done through natural habitat management of species by directly manipulating populations of *E. gerardiana*, establishment of botanical garden in suitable places within natural habitats of species.

Germplasm conservation through in vitro route

Micro propagation techniques have been standardized for a number of species of *Ephedra* for mass multiplication. O'Dowd *et al.* (1993) reported successful micro propagation for *E. equisitina, E. gerardiana, E. minima* and *E. saxatilis* using half strength Murashige and Skoog (MS) medium, supplemented with 1% sucrose and Indole-3-Acetic Acid). *In vitro* regeneration of *E. gerardiana* was standardized by Rautela *et al.* (2018) in which shoot induction and elongation was achieved for shoot and root development.

Efforts have been made to transfer tissue culture raised plants of *E. gerardiana* to field. Rooted plants were hardened and grown in glass house before shifting to soil beds in forest nurseries in central Himalaya (http://www.ucost.in/update/rnd/field-establishment.html).

G. CULTIVATION ASPECTS

Climate

The members of genus *Ephedra,* characterized by xerophytic characteristics, thrive well in dry-temperate to semi-arid and arid desert-like climatic conditions. These species are well represented in drier areas of Himalaya. Plants are hardy and able to tolerate low temperature stress.

Soil

Ephedra is drought resistant and tolerant lime deposits. It requires a well-drained loamy or sandy soil. However, detailed characterization of desired soil as well cultivation requirements have not been worked out.

Nursery raising

For commercial cultivation, propagation through seed is suggested. The seeds should be sown as soon as they are ripe, at a distance of 5cm with a row distance of one meter in a greenhouse/polyhouse during autumn. Seeds can also be sown during spring in sandy compost inside a greenhouse. Young plantlets must be shifted to individual pots as soon as they are large enough to handle. During this period, proper watering and weeding is necessary. Plants should be allowed to grow in individual pots, still in greenhouse, at least for their first winter. The plantlets should be taken out in the spring or in early summer for field cultivation.

E. gerardiana can be cultivated across an altitudinal range of 2500-3500m, where annual rainfall should not exceed 50cm. It can also be propagated by layers or divisions of the root stock. Detailed information on different aspects of cultivation, such as, fertilizers, irrigation, weeding, harvesting, crop rotation, per hectare yield, etc., have not been worked out.

H. COMMERCIAL ASPECTS

China has dominated the commercial supply of *Ephedra* in the world market. Spain also contributes significantly to the export market for the species. However, demand for Indian *Ephedra* in international markets has risen in recent past because of hindered supplies from China and Spain. The preferred demand for Indian species is due to its higher percentages of ephedrine. Recent studies have shown that the crude drug from Sikkim material is also rich in alkaloids (up to 1.607% ephedrine hydrochloride). Baluchistan, at one stage was the principal source of supply of *Ephedra* from this region for this reason. Efforts are now being made to tap sources within India. *Ephedra* from Himachal Pradesh and Kashmir is suitable for commercial exploitation.

Demand

With resurgence of global interest in plant based therapeutics, industrial demand for most species known to have medicinal value, have increased. This holds true for *Ephedra* as well. The present trade is based on collections made from the wild where different species of *Ephedra*, which often occur together, are collected in bulk. The practice makes it difficult for collectors to distinguish one species from another. While this can be overcome by employing trained workers who can distinguish between species and collect the drug source from suitable localities at the proper time, the possibility of adulteration of the best specimens with poor grades is ever present.

To obviate the difficulty arising from variation in the alkaloid content of commercial consignments, the Forest Research Institute, Dehradun, has recommended the preparation and marketing of *Ephedra*. The standardized extract must contain 18-20 percent of alkaloids and represent about 5 percent of the weight of the crude drug. The loss of alkaloids during extraction should not exceed 7-8 percent. The suggested onsite extraction does not involve the use of any complicated equipment and it can be carried out with advantage in localities where the plants occur naturally, thereby minimizing the costs of handling and transport. The dry extract can be used for the preparation of pure ephedrine. It is possible to undertake commercial cultivation of selected species of *Ephedra* for ensuring regular supplies for the manufacture of ephedrine in India.

Market trends

The medicinal plants sold as crude drugs in Indian market lack stringent control for purity of samples. The possibility of adulteration with material of lower grade remains high and is detected at any stage. Indian *Ephedra* thus face stiff competition from drug obtained from Chinese or other sources in the foreign market.

Keeping in mind the demand of current market trends, it is therefore important that quality plant material should be deployed for production process, for pharmaceutical industry, in suitable habitat areas, and to maintain strict quality control. *Ephedra* from Lahaul, Pangi and Kashmir is suitable for commercial exploitation. Recent studies have shown that the drug from Sikkim is rich in alkaloids.

Trade data

According to All India Trade Survey of Prioritized Medicinal Plants, demand for some high-value medicinal plants has increased 50%, but availability has declined by 26% (Samant and Pal, 2003). *Ephedra* containing more than 1% ephedrine is demanded by foreign markets and a few consignments from India satisfy this requirement. However, authentic information regarding trade of the different species of *Ephedra* is not available.

Major users

Traditionally, *E. gerardiana* has been used as medicine as well as for religious purposes by tribal people. On the international platform, owing to the presence of ephedrine, an amphetamine-like compound closely related to adrenaline, it has a new set of users such as athletes (Shekelle *et al.,* 2003). The recent publicity (Nazeri *et al*., 2009) about its ability to aid weight loss (Dwyer *et al*., 2005; Joyal, 2004; Pittler and Ernst, 2005) and increase energy and alertness has caught fancy of average people to use *Ephedra*. The claims made that dietary supplements containing *Ephedra* or ephedrine can enhance athletic performance or help with long-term weight loss, has ushered in another set of users, and creation of products of different kinds (Gurley *et al.*, 2000; Avois *et al*., 2006; Dulloo, 2002). However, lack of authentic data and evidence (Greenway, 2001; Dhar *et al.,* 2005), made FDA ban *Ephedra* from Olympics, and all other competitive sports at all levels, and by the US military at all its bases worldwide.

FUTURE DIRECTIONS FOR RESEARCH

Safety issues related to use of *Ephedra* are an area which demand high research attention in global context. Until the early 2000s, *Ephedra* was presumed to be'safe' under the Dietary Supplement Health and Education Act of 1994, US, and was sold as an herbal supplement, increasingly used by athletes for its stimulant properties and was touted as a dietary supplement (appetite suppressant) for weight loss (Abourashed *et al*., 2003). However, it has been the subject of much controversy and is deemed too risky by the US FDA. Abuse of *Ephedra*led to many incidents of adverse cardiovascular outcomes such as heart attack, stroke, and even death. The International Olympic Committee, the National Football League, the National Collegiate Athletic Association, and the US Armed Forces have banned products containing *Ephedra*. Consumption of Ephedra alkaloids is prohibited in competition by the World Anti Doping Agency (WADA) (Chang *et al*., 2018).

The second important issue demanding immediate attention points to its conservation, more importantly in the national context. *Ephedra* is one of the important genera of Trans Himalaya, the species of which are sought by various indigenous schools of medicines in the country. It also serves as important traditional medicine for people residing in high altitude areas. The species from which drug ephedrine is obtained have restricted distribution in arid and semi-arid regions. In spite of its high importance from conservation point of view, nothing has been done towards its proper conservation and management. Following are some suggestions for its sustainable use, conservation and management:

- Assessment and monitoring of populations of this high value medicinal genus using standard ecological methods, and detailed assessment of population biology for their effective conservation.

- Assessment of suitable habitat using ecological niche modeling to identify places for its reintroduction is required.
- Identification of various threats to natural population of its genera and counter strategies must be developed.
- Establishment of seed/ gene bank and botanical gardens in native places of the genus.
- In-depth research studies on its phenology and reproductive biology must be undertaken.

PATENTS

- Gregory Pogue, Guy Della-Cioppa, Gershon Wolfe, Wenjin Zheng. Methods of creating dwarf phenotypes in plants.Application number-US20020194646A1.2002-12-19.
- Omer Y. Katzenelson, Ileana L. Katzenelson.Systems and methods to perform inhibition diagnostic testing.Application number-US20090310839A1.2009-12-17.
- Deeksha Dixit, Palpu Pushpangadan, Vinod Kumar Kochhar, Sunita Kochhar, Chandana Venketeshwara Rao. Isozyme of autoclavable superoxide dismutase (sod) derived from curcuma longa l. Application number-WO2005017134A2. 2005-02-24.

LITERATURES CITED

1. Abourashed E.A., El-Alfy A.T., Khan I.A., Walker L. 2003. *Ephedra* in perspective-a current review. *Phytother Res* 17(7):703–712.
2. Alwan S.M., Al-Hindawi M.K., Abdul-Rahman S.K., Al-Sarraj S. 1986. Production of nitrosamines from ephedrine, pseudoephedrine and extracts of *Ephedrafoliata* under physiological conditions. *Cancer. Lett.* 31(2):221-226.
3. Andraws R., Chawla P., Brown D.L. 2005. Cardiovascular effects of ephedra alkaloids: a comprehensive review. *Prog Cardiovasc Dis* .47(4):217-225.
4. Anonymous, 1960. Medicinal Plants of the Arid Zones, United Nations Educational, Scientific and Cultural Organization.
5. Anonymous, 1993. Medicinal Plants of Nepal Dept. of Medicinal Plants. Nepal.
6. Avois L., Robinson N., Saudan C., Baume N., Mangin P., Saugy M. 2006. Central nervous system stimulants and sport practice. *Br. J. Sports Med.* 40(suppl1): i16-i20.
7. Bagheri G. and Bigdeli M. 2009. Inhibitory Effects of Ephedra major Host on Aspergillus parasiticus. Growth and Aflatoxin Production. *Mycopathologia.* 168: 249–255.
8. Betz J.M., Gay M.L., Mossoba M.M., Adams S., Portz, B.S. 1997. Chiral gas chromatographic determination of ephedrine-type alkaloids in dietary supplements containing Má Huáng. *J. AOAC International.* 80(2):303-315.
9. Bhandari M.M. 1990. Flora of the Indian desert. M.P.S. Repros., Rajasthan.
10. Bown D. 1995.Encyclopaedia of Herbs and their Uses. Dorling Kindersley, London.

11. Caveney S. , Charlet D.A., Freitag H., Maier-StolteM., Starratt A.N. 2001. New observations on the secondary chemistry of world Ephedra (Ephedraceae). *Am. J. Bot.* 88: 1199–1208.
12. Chang C.W., Hsu S.Y., Huang G.Q., Hsu M.C. 2018. *Ephedra* alkaloid contents of Chinese herbal formulae sold in Taiwan. *Drug Test. Ana.* 10(2): 350-356.
13. Chopra R.N., Krishna S., Ghose T.P. 1931. Indian Ephedras: Their chemistry and pharmacology. *Indian J. Med. Res.* 19:177-219.
14. Chouhdry A.S. 1984. Karyomorphological and cytological studies in *Ephedra*. *J. Sci. Hiroshima Univ. Ser. B.* 19: 57–109.
15. Cui J.F., Zhou T.H., Zhang, J.S., Lou Z.C.1991. Analysis of alkaloids in Chinese Ephedra species by gas chromatographic methods. *Phytochem. Anal.*2(3):116-119.
16. Dhar R., Stout W., Link M.S., Homoud M.K., Weinstock J., Estes N.A. 2005. Cardiovascular toxicities of performance-enhancing substances in sports. *Mayo. Clin. Proc.* 80(10): 1307-1315.
17. Dulloo A.G. 2002.Herbal simulation of ephedrine and caffeine in treatment of obesity. *Int. J. Obes. Relat. Metab. Disord.* 26(5): 590-592.
18. Duke J.A. 2018. Handbook of Medicinal Herbs: Herbal Reference Library, CRC Press, 689 pages.
19. Dwyer J.T., Allison D.B., Coates P.M. 2005. Dietary supplements in weight reduction. *J. Am. Diet. Assoc.* 105(5 suppl1): S80- S86.
20. Freudenberg K., Schoeffel E., Braun E. 1932. Study on the configuration of ephedrine. *JAm Chem Soc*.54: 234-236.
21. Friedrich H. and Wiedemeyer H. 1976. Quantitative determination of the tannin-precursors and the tannins in *Ephedra helvetica. Planta Med.*30(3):223-231.
22. Gay M.L., White K.D., Obermeyer W.R., Betz J.M., Musser, S.M. 2001. Determination of ephedrine-type alkaloids in dietary supplements by LC/MS using a stable-isotope labeled internal standard. *J. AOAC Int.* 84(3):761-769.
23. Gmeiner G., Geisendorfer T., Kainzbauer J., Nikolajevic M.,Tausch H. 2002. Quantification of ephedrines in urine by column-switching high-performance liquid chromatography. *J. Chromatogr. B.* 768 (2):215-221.
24. Greenway F.L. 2001. The safety and efficacy of pharmaceutical and herbal caffeine and ephedrine use as a weight loss agent. *Obes. Rev.* 2(3): 199-211.
25. Gurley B.J., Gardner S.F., Hubbard, M.A. 2000. Content versus label claims in ephedra-containing dietary supplements. *Am. J. Health Syst. Pharm.* 57(10): 963-969.
26. Hikino H., Kiso Y., Ogata M., Konno C., Aisaka K., Kubota H., Hirose N., Ishihara T. 1984. Pharmacological Actions of Analogues of Feruloylhistamine, an Imidazole Alkaloid of *Ephedra* Roots. *Planta Med.* 50(06): 478-480.
27. Hikino H., Ogata K., Konno C., Sato S. 1983. Hypotensive actions of ephedradines, macrocyclic spermine alkaloids of *Ephedra* roots. *Planta Med. 48*(08): 290-293.
28. Hikino H., Takahashi M., Konno, C. 1982. Structure of ephedrannin A, a hypotensive principle of *Ephedra* roots. *Tetrahedron Lett.*23(6): 673-676.
29. Hong H., Chen H.B., Yang D.H., Shang M.Y., Wang X., Cai S.Q., Mikage M. 2011. Comparison of contents of five ephedrine alkaloids in three official origins of *Ephedra* herb in China by high-performance liquid chromatography *J. Nat. Med.* 65(3-4):623-628.
30. Hurlbut J.A., Carr J.R., Singleton E.R., Faul K.C., Madson M.R., Storey J.M. Thomas T.L., 1998. Solid-phase extraction cleanup and liquid chromatography with ultraviolet detection of ephedrine alkaloids in herbal products. *J. AOAC Int.* 81(6):1121-1127.
31. Hyuga S., Hyuga M., Oshima N., Maruyama T., KamakuraH., Yamashita T., Yoshimura M., Amakra Y., Hakamatsuka T., Odaguchi H.,Goda Y.2016. Ephedrine alkaloids-free Ephedra Herb extract: a safer alternative to ephedra with comparable analgesic, anticancer, and anti-influenza activities. *J. Nat. Med.* 70(3): 571-583.

32. Ibragic S. and Sofiæ E. 2015. Chemical composition of various *Ephedra* species. *BosnianJ Basic Med.* 15(3): 21.
33. IUCN UK 2011. http://dx.doi.org/10.2305/IUCN.UK.20112.RLTS. T201716A9172369
34. Jian Z., Zhen T., Zhi-cen, L.1988. Simultaneous determination of six alkaloids in Ephedrae Herba by high performance liquid chromatography. *Planta Med.* 54(01): 69-70.
35. Joyal S.V. 2004. A perspective on the current strategies for the treatment of obesity. Curr Drug Targets CNS. *Neurol. Disord.*3(5): 341-356.
36. Kawatani T., Fujita S., Ohno T., Kuboki N., Hoshizaki K. 1959. On the alkaloidal content of Ephedras cultivated at Kasukabe. *J. Pharmaceutical Soc. Jap.* 79(3): 392-393.
37. Khan A.,Jan G.,Khan A.,Gul Jan F., BahadurA., Danish M. 2017. *In vitro* Antioxidant and Antimicrobial Activities of *Ephedra gerardiana* (Root and Stem) Crude Extract and Fractions. *Evid. Based Complement. Alternat. Med.*doi:10.1155/2017/4040254.
38. Kim I.S., Park Y.J., Yoon S.J., Lee H.B. 2010. Ephedrannin A and B from roots of Ephedra sinica inhibit lipopolysaccharide-induced inflammatory mediators by suppressing nuclear factor-kB activation in RAW 264.7 macrophages. *Int. Immunopharmacol.* 10(12): 1616-1625.
39. Konno C., Mizuno T., Hikino H. 1985. Isolation and hypoglycemic activity of ephedrans A, B, C, D and E, glycans of *Ephedra distachya* herbs. *Planta Med.* 51(2): 162-163.
40. Laccourreye O., Werner A., Giroud J.P., CouloignerV., Bonfils P., Bondon-Guitton E. 2015. Benefits, limits and danger of ephedrine and pseudoephedrine as nasal decongestants. *Eur Ann Otorhinolaryngol Head Neck Dis.*, 132 (1): 31-34.
41. Lee M.K., Cheng B.W., Che C.T., Hsieh D.P. 2000.Cytotoxicity assessment of ma-huang (Ephedra) under different conditions of preparation. *Toxicol Sci.* 56(2):424-430.
42. Leitch I.J., Hanson L., Winfield M., Parker J., Bennett M.D. 2001. Nuclear DNA C values complete familial representation in gymnosperms. *Ann. Bot.*: 88, 843–849.
43. Lewis R.J. Sr. (2007). Hawley's Condensed Chemical Dictionary 15th Edition. John Wiley & Sons, Inc. New York, p. 505.
44. Li H.X., Ding M.Y., Lv K., Yu J.Y. 2001. Separation and determination of ephedrine alkaloids and tetramethylpyrazine in *Ephedra sinica* Stapf by gas chromatography-mass spectrometry. *J. Chromatogr. Sci.* 39(9): 370-374.
45. Li L., Li J., Zhu Y., Fan G. 2009. *Ephedra sinica* inhibits complement activation and improves the motor functions after spinal cord injury in rats. *Brain Res. Bull.* 78(4-5):261-266.
46. Liu Y.M., Sheu S.J., Chiou S.H., Chang, H.C., Chen Y.P. 1993. A comparative study of commercial samples of Ephedrae Herba. *Planta Med.* 59(04):376-378.
47. Lv K., Li H., Ding, M. 2000. Analysis of tetramethylpyrazine in *Ephedrae* herba by gas chromatography–mass spectrometry and high-performance liquid chromatography. *J. Chromatogr. A.* 878(1): 147-152.
48. Nazeri A., Massumi A., Wilson J.M. *et al.* 2009. Arrhythmogenicity of weight-loss supplements marketed on the Internet. *Heart Rhythm.* 6(5):658-662.
49. O'Dowd N.A. and Richardson D. H. S. 1993 In vitro micropropagation of Ephedra. *J. Hort. Sci.* 68(6): 1013-1020.
50. Nikoliæ, T. and Topiæ, J. (eds). 2007. Red Book of Vascular Flora of Croatia. Ministry of Culture, State Institution for Nature Protection, Zagreb.
51. O'Neil M.J. 2013. The Merck Index - An Encyclopedia of Chemicals, Drugs, and Biologicals. Cambridge, UK: Royal Society of Chemistry.
52. Oshima N., Yamashita T., Hyuga S., Hyuga M., Kamakura H., Yoshimura M., Maruyama T., Hakamatsuka T., Amakura Y., Hanawa T. Goda Y. 2016. Efficiently prepared ephedrine alkaloids-free *Ephedra* Herb extract: a putative marker and antiproliferative effects.*J. Nat. Med.* 70: 554–562.

53. Pandey R.P., Shetty B. V., Malhotra S. K. 1983. A Preliminary census of rare and threatened plants of Rajasthan. In: S.K. Jain & R.R. Rao. (eds.). An assessment of threatened plants of India. Naba Mudran Private Limited, Calcutta, India: Director, Botanical Survey of India.
54. Parsaeimehr A., Sargsyan E., Javidnia K.2010. A comparative study of the antibacterial, antifungal and antioxidant activity and total content of phenolic compounds of cell cultures and wild plants of three endemic species of *Ephedra*. *Molecules* 15(3): 1668-1678.
55. Pittler M.H. and Ernst E. 2005. Complementary therapies for reducing body weight: a systematic review. *Int. J. Obes*. 29(9): 1030-1038.
56. Price R.A. 1996 Systematics of the Gnetales: a review of morphological and molecular evidence. *Int. J. Plant Sci.* 157: S40-S49.
57. Qasem, J.R. 2012. *Ephedra alte* (Joint Pine): An Invasive, Problematic Weedy Species in Forestry and Fruit Tree Orchards in Jordan, *The Scientific World J.*, Article ID 971903, 10 pages, 2012. https://doi.org/10.1100/2012/971903.
58. Rautela I., Dhiman M., Sharma M.D., Misra P. 2018. In vitro regeneration of medicinal plant *Ephedra gerardiana*. *Int. J. Pharmaceut. Sci. Res.* 9 (3): 1183-1188.
59. Rungsung W., Dutta S., Ratha K. K., Mondal D. N., Hazra, J.(2015). Pharmacognostical and Phytochemical Study on the Stem of *Ephedra gerardiana*.*J. Internat. Res. Med. Pharmaceut. Sci.* 2(3): 80–85.
60. Rustaiyan A., Javidnia K., Farjam M. H., Aboee-Mehrizi F., Ezzatzadeh E. 2011a. Antimicrobial and antioxidant activity of the Ephedra sarcocarpa growing in Iran. *J. Med. Plants Res.* 517: 4251–4255.
61. Rustaiyan A., Javidnia K., Farjam M. H., Mohammadi M. K., Mohammadi N. 2011b. Total phenols, antioxidant potential and antimicrobial activity of the methanolic extracts of *Ephedra laristanica*. *J. Med. Plants Res.* 524 : 5713–5717.
62. Samant S.S. and Pal M. 2003. Diversity and conservation status of medicinal plants in Uttaranchal State. *Ind. For.* 129 (9): 1090-1108.
63. Samant S.S. and Pant S. 2006. Diversity, distribution pattern and conservation status of the plants used in liver diseases/ailments in Indian Himalayan Region *J. Mountain Sci.* 3:28-47.
64. Schaneberg B.T., Crockett S., Bedir E., Khan I.A. 2003. The role of chemical fingerprinting: application to *Ephedra*. *Phytochem*. 62: 911-918.
65. Sharma P. and Uniyal P.L. 2008a. Biodiversity Characterization of *Ephedra* in Lahaul & Spiti (India). International conference: Biodiversity research safeguarding the future, Bonn, Germany, Abstract, P I.21, 59-60.
66. Sharma P. and Uniyal P.L. 2008b. *Ephedra sumlingensis* (Ephedraceae - A new species from Himachal Pradesh, India. *Bull. Bot. Surv. India* 50: 179-182.
67. Sharma P., Uniyal P.L., Hammer Ø. 2010. Two new species of *Ephedra* (Ephedraceae) from the Western Himalaya. *Syst. Bot*. 35(4): 730-735.
68. Shekelle P.G., Hardy M.L., Morton S.C., Maglione M., Mojica W.A., Suttorp M.J., Rhodes S.L., Jungvig L., Gagné J. 2003. Efficacy and safety of ephedra and ephedrine for weight loss and athletic performance: a meta-analysis. *JAMA*. 289(12): 1537-1545.
69. Soltan M.M.and Zaki A.K. 2009. Antiviral screening of forty-two Egyptian medicinal plants. *J. Ethnopharmacol*. 126(1): 102-107.
70. Spyridaki M.H.E., Tsitsimpikou C.J., Siskos P.A., Georgakopoulos C.G. 2001. Determination of ephedrines in urine by gas chromatography–mass spectrometry. *J. Chrom. B. Biomed. Sci. Appl.* 758(2): 311-314.
71. Stevenson D.W. 1993. Ephedraceae. In: Flora of North America Editorial Committee. Flora of North America, New York: Oxford Univ. Press, pp 428-434.

72. Tamada M., Endo K., Hikino H., Kabuto C. 1979. Structure of ephedradine A, a hypotensive principle of *Ephedra* roots. *Tetrahedron Lett.* 20(10): 873-876.
73. Theoharides, T.C. 1997. Sudden Death of a Healthy College Student Related to Ephedrine Toxicity from a Ma Huang-Containing Drink. *J. Clin. Psychopharmacol.* 17(5):437-439.
74. Tricker A.R., Wacker C.D., Preussmann R. 1987. Nitrosation products from the plant *Ephedra altissima* and their potential endogenous formation. *Cancer. Lett.* 35(2): 199-206.
75. Tsarong T.J. 1994.Tibetan Medicinal Plants Tibetan Medical Publications, India.
76. Tsintides, T., Christodoulou, C.S., Delipetrou, P., Georghiou, K. (eds). 2007. The Red Data Book of the Flora of Cyprus. Cyprus Forestry Association, Lefkosia.
77. Ullah A. and Rashid A. 2014. Conservation status of threatened medicinal plants of Mankial Valley Hindukush Range, Pakistan. *Int. J. Biodiver.Conser.* 6: 59-70.
78. UttraA.M. 2017. Assessment of anti-arthritic potential of *Ephedragerardiana* by in vitro and in vivo methods. Bangladesh *J. Pharmacol.* 12(4): 403-409.
79. USFDA 2007. US Food and Drug Administration. Sales of Supplements Containing Ephedrine Alkaloids (*Ephedra*) Prohibited (http://www.fda.gov/oc/initiatives/ephedra/)
80. Ved D.K. 2008. Demand and supply of medicinal plants. *Medplant- ENVIS Newsletter on Medicinal Plants.* 1(1): 2.
81. Von R.S.and Lipp F. 1982. New Plant Sources for Drugs and Foods from the New York Botanical Garden Herbarium. Cambridge and London: Harvard University Press.
82. Vukovich M.D., Schoorman R., Heilman C., Jacob P., Benowitz N.L. 2005. Caffeine-herbal ephedra combination increases resting energy expenditure, heart rate and blood pressure. *Clin. Exp. Pharmacol. Physiol.* 32(1-2): 47-53.
83. Walter K.S. and Gillett H.J. 1998. IUCN red list of threatened plants. World Conservation Monitoring Centre. IUCN, Gland, Switzerland.
84. Wu Z., Kong X., Zhang T., Ye J., Fang Z., Yang X. 2014. Pseudoephedrine /ephedrine shows potent anti-inflammatory activity against TNF-á-mediated acute liver failure induced by lipopolysaccharide/d-galactosamine, *European J. Pharmacol.* 724: 112-121.
85. Yamada I., Goto T., Takeuchi S., Ohshima S., Yoneyama K., Shibuya T., Kataoka E., Segawa D., Sato W., Dohmen T.,Anezaki Y. 2008. Mao (*Ephedra sinica* Stapf) protects against D-galactosamine and lipopolysaccharide-induced hepatic failure. *Cytokine.* 41(3): 293-301.
86. Yamasaki, K. and Fujita, K. 1979. Qualitative and quantitative analysis ofephedra alkaloids in Ephedra Herba by carbon-13 nuclear magnetic resonance. Chem. Pharm. Bull. 27: 43-47.
87. Zhao W., Deng A.J., Du G.H., Zhang J.L., Li Z.H., Qin H.L. 2009. Chemical constituents of the stems of *Ephedra sinica*. *J. Asian Nat. Prod. Res.* 11(2): 168-171.

9

Crataegus rhipidophylla

English: Hawthorn, Haw, Hedgethorn, May bush, May Day Flower, White Thorn

Dinesh Kumar and S.K. Vats

Crataegus rhipidophylla **Gand.**
(A) Herbarium sheet, **(B)** Branches with profuse flowering, **(C)** Single flower, **(D)** Dry leaves, **(E)** Seeds, **(F)** Seed with pericarp, **(G)** Seed without pericarp

INTRODUCTION

Crataegus species, or hawthorn, are shrubs or small trees that have been widely and globally used for treating cardiac problems (arrhythmia, angina, hypertension and gastrointestinal ailments. Fruits of the plant are also consumed as food and nutritional supplement. *Crataegus* species has several pharmacological properties attributable primarily to presence of flavonoids and anthocyanins. The major flavonoids reported are hyperoside, vitexin, and glycosylated derivatives of these compounds. *Crataegus* has been listed as the plant for future in terms of its utility and significance. Though the species can be easily grown, its commercial cultivation is not yet popular. This presentation highlights the medicinal significance of the species for its greater utilization.

A. BIOLOGICAL ASPECTS

• Systematic classification

Kingdom : Plantae

Division : Angiospermae

Class : Magnoliopsida

Order : Rosales

Family : Rosaceae

Genus : *Crataegus*

Species : *rhipidophylla*

• Synonyms

B. laevigata, C. *monogyna,* C. *oxycantha*

• Common names

English: Hawthorn, Haw, Hedgethorn, May bush, May Day Flower, White Thorn

German: Zweigriffeliger Weißdorn

Greek: *kratos*

Spanish: Azzarola

Hindi: *Vansanglior, Ban-sangli*

• Diversity within genus

Crataegus genus contain approximately 280 species such as *C. aestivalis, C. aronia var. aronia, C. azarolus, C. azarolus var. Aronia, C. azarolus var. azarolus, C. brettschneideri, C. cuneata, C. germanica, C. hupehensis, C. kansuensis, C. laevigata, C. macrocarpa, C. maximowiczii, C. microphylla, C. monogyna, C. opaca, C. oxycantha, C. pentagyna, C. pinnatifida, C. pinnatifida var. major, C. pinnatifida var. psilosa, C. pseudoheterophylla, C. rufula, C. scabrifolia, C. sanguine, C. rhipidophyllavar. rhipidophylla, C. xmacrocarpa,* etc.

• Geographical distribution

Crataegus rhipidophylla Gand. is found and grown in Europe, France, Mexico, North America, Australia, North Africa, Western Asia, Southeast Serbia, China and India.Its distribution in India is confined to temperate Himalayas, in Kashmir and Himachal Pradesh (The Wealth of India, 2004; Amy *et al.*, 2006; Verma *et al.*, 2007; WHO, 2003; Huang *et al.*, 2004; Kashyap *et al.*, 2012; Kumar *et al.*, 2012).

• Habitat and Ecology

It occurs and is grown in Europe, France, North America, North Africa, Western Asia. Within Asia, *C. rhipidophylla* is mostly located in Chinese and Indian temperate Himalayas, inhabiting areas across altitude of 1800-3000 meters (Amy *et al.*, 2006; The Wealth of India, 2004). British settlers introduced *C. oxycantha* into Tasmania and other parts of Australia in the 1800's. Later it spread into other parts such as in Victoria, Adelaide and New South Wales (WHO, 2003; Huang *et al.*, 2004; Kashyap *et al.*, 2012; Kumar *et al.*, 2012). Larvae of nectar-feeding insect Lepidopteran species feed on the flowers of Hawthorns and many species of birds and mammals use it as food and for shelter (Jolivet, 1998; Hemery and Simblet, 2014). During the winter months, birds like thrushes and waxwings eat fruits and disperse seeds through droppings.

• Taxonomic enumeration

The leaves of Crataegus are dark green in colour, and its margins vary from entire to serrate to deep lobed. These species produce dense clusters of white flowers which result in yellow to bright red colored fruits when unripe that turn black when ripe (Phipps *et al.*, 2003). The flowers have characteristic trimethylamine flavour (El-Sayed, 2011). *C. rhipidophylla* (*C. oxycantha*: Gr. Oxus = sharp, Akantha = a thorn) is well known among all other species of *Crataegus*. It is a spiny small to medium sized deciduous tree. The leaves are green glossy with toothed margins, flowers white in colour forming an umbrella shaped clusters, and fruits as bright shiny red berries. Both types of inflorescences, corymbs (flat-topped inflorescence) or umbels (globular) are reported in *C. oxycantha*. The flower usually has five petals, 5 to 18 stamens. The fruits of the species are famous as pomes, although the seeds and their bony endocarps are termed pyrenes. Each pome produces 1 to 5 pyrenes. Calyx of the fruit is persistent from the top. From the trunk or other branches arise small thorns, which are sharp-tipped, and are typically 1-3 cm long. The bark on trunk or stem is ash-gray (WHO, 2003; Huang *et al.*, 2004; Kashyap *et al.*, 2012).

• Phenological period

Crataegus is a small tree, deciduous in nature. The seeds germinate by autumn and grow slowly during the first year. After 2-3 years of growth, flowering takes place during spring. The flower bearing period of a tree ranges from 1 to 2 weeks under natural conditions. Flowering time of Crataegus is species-specific and is also dependent on environmental conditions. *C. rhipidophylla* is included in this group (Donmez, 2004). Fruit ripen by summer, and leaves senesce by autumn due to declining temperature. The species show fresh growth with advent of spring. The estimated age of the tree is about 70 years.

• Cytotypes

Chromosome counts in Crataegus are not often reported due to the difficulty of obtaining mitotic preparations. More often, cytology is studied from pollen mother cells. Chromosomes are difficult to be distinguished individually, though some measurements of size for Eurasian species are stated (Ptak, 1986). The identification of taxa has been considered difficult, especially for American species, and it is suggested that a voucher specimen is prepared for every recorded chromosome count.

C. REPRODUCTIVE BIOLOGY

Sexual and asexual gametophytes: In plants which arediploid, a single sporocyte (archespore, megaspore mother cell) undergoes normal female meiosis to produce 4 megaspores in each ovule, with the haploid chromosome number $n = 17$, and one of these spores develops through mitotic divisions into an 8-nucleate embryo sac. Female sporophytes experience various meiotic irregularities in triploids plants. Sexual gametophytes have been reported in tetraploids.

• Life Cycle

- Cotyledons: Two
    - **Pollination:** Crataegus is self-incompatible because it sets very few fruits on self-pollination (Clapham *et al.* 1989). *Crataegus* spp. are described as apomictic, with seeds developing without fertilization. Study on British hawthorn population showed that fruits were set in the absence of insects, indicating self-pollination or apomixes. In the Rosaceae subfamily Maloideae, apomixis is usually associated with polyploidy. The contrast between bagged flowers and flowers in the open-pollinated treatment was strong, with lower fruit-set when insects were excluded, suggesting insect pollination to be important. In the absence of pollinators, reduced fruit-set of hawthorn supports these studies (Jacobs *et al.*, 2009). The Crataegus flowers during April, May and sometime by early June. Fruit sets in open-pollinated and manually cross-pollinated flowers were double those for flowers bagged to exclude pollinators and those manually self-pollinated.
    - **Pollinators:** The major pollinators are the honey bees (*Apis mellifera*, 52% of pollinator visits), and others like members of *Apoidea* (12.3%) and *Diptera* (18.3%) (Jauregui *et al.*, 2014). Pollination is essential for the formation of viable seeds, even for egg cell developed by parthenogenesis.
    - **Seed development**: In gametophytic apomixes, it has been shown that seedlings produced by plants are very similar genetically to the parent plants. In pseudogamous apomixes, the mechanism involves some or the entire second half of double fertilization which is a characteristic of flowering plants. In this, fertilization of the polar nuclei produces the endosperm tissue which provides nutrition for the embryo. Seed formed from meiotically reduced embryo sacs (sexual seed) and some from aposporous initials (asexual seed), are referred to as facultative apomixis. No recent studies have been done on the extent of sexual and asexual seed production in *Crataegus* spp. Although it has not been directly observed in aposporous of Crataegu*s,* but on the basis of observations gathered from other plants, it was concluded that the egg cell may (or another cell in the embryo sac) develop by parthenogenesis (without fertilization) to give an embryo that also has unreduced chromosome number.

C. MEDICINAL ASPECTS

• Plant material used as crude drug

- **General appearance:** It is a small tree with bright green leaves with 3, 5 or 7 converging lobes, with secondary venation curved inwardly. The flowers are corymbs, consisting of 5 triangular sepals, 5 petals white in colour, and an androecium of 15–20 stamens inserted on the edge of a bi- or tricarpellate receptacle. The floral peduncles and sepals are glabrous and stamens have red anthers with 2–3 styles.
- **Organoleptic properties:** The fruits are astringent and slightly bitter-sweet in taste, having a faint, characteristic odour (Deutsches, 1998; Pharmacopée française, 1996; Pharmacopoea Helvetica, 1997; British Herbal Pharmacopoeia, 1996).

• Microscopic characteristics

The dorso-ventral leaf contains polygonal and straight-walled with striated cuticle cells on upper epidermis while lower epidermis cells are mostly sinuous. Leaves have anomocytic stomata with covering trichomes on both the epidermises but more numerous at the lower epidermis. It contains unicellular, tapering and long Trichomes, which are uniseriate with 2 cells and walls. Epidermis of floral pedicel and receptacle abundantly contains the covering trichomes similar to leaves. Leaves also contain cluster of prismatic crystals of calcium oxalates along the veins in cells. Calyx has anomocytic stomata with a striated cuticle on both inner and outer epidermis while epidermal cells of corolla are distinctly papillose. Anthers are fibrous with characteristic thickenings while pollens are elliptical to spherical with faintly granular exine and 3 germinal pores. The stem contains the epidermal cells, pericyclic fibres with lignified walls, parenchymatous cortex with prismatic and cluster of calcium oxalate crystals. Xylem is lignified, composed of scattered vessels, thick walled fibers whereas, parenchyma is separated by distinct medullary rays containing brown-coloured matter. The vessels are larger with bordered pits, and smaller elements with annular/spiral thickening. (Deutsches, 1998; British Herbal Pharmacopoeia, 1996).

• Powdered plant material

Various parts of the plant, such as leaves, berries, flowers and seeds constitute powdered material. Phytochemically, leaves, flowers and berries are similar in composition though the ratio of flavonoids and procyanidins may vary (Tassell *et al.,* 2010). It was reported that flowers of Crataegus contained higher levels of flavonoids, while leaves had the highest levels of oligomeric procyanidins

(Mills and Bone, 2000). Yang and Liu (2012) reported that the total content of phenolic compounds was higher in leaves and flowers compared to fruits, while procyanidins dominated in the fruits.

• Major Chemical constituent

D. rhipidophylla contains sugars and sugar alcohols, organic and phenolic acids, terpenes, monoamines, cholines, purine derivatives, polyphenols, flavonoids, lignins, napthoquinones, amygdalin, pectins, cratetegin and saponins.

• Substitutes and adulterants

C. rhipidophylla is generally substituted by plants such a*s C. sinaica, C. crus-galli, C. laevigata, C. phaenopyrum.* The common *a*dulterants are blackberry, cotoneaster.

• General identity tests

Macroscopic and microscopic examinations and thin-layer chromatography are the general identity test for *C. rhipidophylla* (European Pharmacopeia, 1999). Procyanidins of *C. rhipidophylla* can be tested using micro-chemical test (Pharmacopoeia Hungarica, 1986).

• Purity tests

- **Microbiological tests:** For *C. rhipidophylla,* the suggested tests for specific microorganisms and microbial contamination are as per WHO guidelines (WHO, 1998).
- **Foreign organic matter:** There should not be more than 8% lignified branches with a diameter greater than 2.5mm (European Pharmacopeia, 1999) and not more than 2% foreign matter (European Pharmacopeia, 1999, British Herbal Pharmacopoeia, 1996).
- **Total ash:** It should not be more than 10% (European Pharmacopeia, 1999).
- **Loss on drying:** Should not be more than 10% (European Pharmacopeia, 1999).
- **Pesticide residues:** The recommended maximum limit of aldrin and dieldrin is less than 0.05mg/kg (European Pharmacopeia, 1996). For other pesticides, refer to the European pharmacopoeia and the WHO guidelines on quality control methods for medicinal plants (WHO, 1998) and pesticide residues (WHO, 1997).

- **Other purity tests**: Chemical, ash value, both acid-insoluble and sulfated ash, water-soluble and alcohol-soluble extractive values, have not been studied for the species and are required to be established in accordance with national requirements.
- **Heavy metal determination**: For heavy metals analysis, the WHO guidelines on quality control methods for medicinal plants (WHO, 1998) are suggested for sample analysis.
- **Radioactive residues:** Similarly, for the analysis of radioactive isotopes, the WHO guidelines on quality control methods for medicinal plants (WHO, 1998) are suggested as above.
- **Chemical assays:** It should not contain less than 1.5% of flavonoids, calculated as hyperoside (European Pharmacopoeia, 1999), and not less than 0.6% of flavone C-glycosides, calculated as vitexin (Pharmacopoeia Helvetica, 1997). A high-performance liquid chromatography method is also available for the same (Rehwald, 1994).

• Medicinal usage

General use: *C. rhipidophylla* is widely used as a Cardio-protective.

Ethnobotanical uses: Reports on extensive use of hawthorn are highlighted in the ethno-botany of numerous First Nations. The berries as well as decoctions of the shoots, roots, and bark of different. *Crataegus* species are used to treat gastrointestinal ailments by native American tribes (Meskwaki, Blackfoot, Ojibwa, Potawatomi, Okanagon, Okanagan-Colville, Iroquois, and Cherokee tribes) (Moerman, 1998, 2009; Turner *et al.*, 1990). The Cherokee and Thompson tribes also used decoctions of the bark for heart ailments. Arnason *et al.* (1981) provided detailed reviews on *Crataegus* species-based food by eastern Canadians. Fruits of *C. douglasii* are cooked as sauce and more than 10 other spp. are eaten by native peoples of North America (Hellson and Gadd, 1974; Turner, 1978). *Crataegus* species are traditionally used in Mexico for the treatment of respiratory problems such as coughs, flu, bronchitis, asthma and diabetes (Arrieta *et al.*, 2010; Rigelsky and Sweet, 2002). In Arab traditional medicinal system, the decoction of leaves and unripe fruits of *C. aronia* are used to treat cardiovascular diseases, cancer, diabetes and sexual weakness (Miller, 1998aand b; Ju, 2005). Hawthorn thorns are reportedly used in dermatological and rheumatism problems and also as witchcraft medicine (Moerman, 1998, 2009). The aerial parts (fruits, leaves, and flowers) of hawthorn were used in Europe for the management of cardiovascular problems owing to its cardiotonic, anti-atherosclerotic and antispasmodic effects (Chang *et al.*, 2002). In China, the fruits of hawthorn have been utilized to improve digestion and blood circulation, and to treat diarrhea, abdominal pain, hyperlipidemia as well as hypertension (Barceloux, 2008; Chang *et al.*, 2002; Hobbs and Foster, 1990). In Europe and China, hawthorn fruits are also used to make commercial products such as wine, jam, and candy (Chang *et al.*, 2002; Kumar *et al.*, 2012). In Taiwanese Traditional system of medicine, dried fruits of *C. pinnatifida* are used as medicine and for preparation of soft drink (Kao *et al.*, 2007).

C. rhipidophylla is an officially declared cardio-protective plant in homeopathic system of medicine (Krzeminski *et al.*, 1993; Peschel *et al.*, 2008). Traditionally, it has been used as a cardiac tonic (Miller, 1998a and b). Popular Chinese drink *shan zha* contain *C. rhipidophylla* as an active therapeutic principle that has been used for lowering blood lipid levels in humans and rats (Chen *et al.*, 1995). In Celtic community in Europe, the tree represented the Goddess. The flowering of hawthorn indicated the beginning of ancient Celtic festival of Beltane and as a result the tree is famous as May-flower. A well-known physician, named late Dr. Green, of Ennis, County Clare of Ireland, attained the reputation for treating various ailments of heart. In 1894, his daughter revealed after his death that the mystery of his famous cure was a tincture of the ripe berries of hawthorn. The plant had significance in the oriental medicine as well, where the fruits were considered to have slightly warming qualities. The plant is also used in the treatment of digestive ailments, dyspnea, kidney stones (Ju, 2005; Kashyap *et al.*, 2012). The fruits and flowers constitute a drug which is official in many pharmacopoeias including the Homeopathic Pharmacopoeia of India. The drug from flowers has cardiotonic, hypotensive, diuretic, anti-spasmodic and nervine-sedative properties (Hawthorn Berry info and Products, 2006; http://www.bodybuilding.com/store/haw.html).

D. CHEMISTRY ASPECT

• Chemical composition

The genus *Crategus* is rich in polyphenols and terpenes. *C. rhipidophylla* species is reported to contain cholines, purines, monoamines, sugar alcohol, organic acid and terpenes. The details of chemical constituents isolated and identified in *Crataegus* genus and *C. rhipidophylla* are depicted in Fig. 3 and Table 1. Hyperoside is the essential molecule in Crataegus and is used as a chemical marker for the quality control as described in French Pharmacopoeia.

Rutin

Quercetin glucoside

Quercetin

Lutoeolin

Hyperoside

Naringenin

Catechin

Epicatechin

Orientin

Isoorentin

Vitexin

Isovitexin

Kaempferol glucoside

Chemical structure of important active constituents of *Crataegus rhipidophylla*

Table 1. Phytochemicals of *Crataegus* genus and *C. oxycantha*

Sr. No.	Major class of compounds	Compounds present in*Crataegus*	Compounds reported in *C. rhipidophylla*
1.	Sugars and sugar alcohols	Glucose, sucrose, fructose, xylose, sorbitol, and myo-inositol (Urbonaviciute *et al.*, 2006; Edwards *et al.*, 2012)	Sarbitol (Patri and Silano, 1994)
2.	Organic and phenolic acids	Malic, citric, succinic, ascorbic, tartaric, quinic, protocatechuic, 3- and 4-hydroxybenzoic, salicylic, and syringic acids, coumaric acid, caffeic acid, sinapic acid (Dewick, 2009), chlorogenic acid, ferulic acid,	Ascorbic acid (Kashyap *et al.*, 2012), chlorogenic and caffeic acids (Verma *et al.*, 2007)
3.	Terpenes	Oleanolic acid, Ursolic acid, Euscapic (Park*et al.*, 1994), Corosolic acid (Ahn*et al.*, 1998; Park *et al.*, 1994), Cuneataol (Ikeda *et al.*, 1999), Linarionosides A and B, 3β-glucopyranosyloxy-b-ionone, Icariside B6, Pisumionoside, (3*S*,5*R*,6*R*,7*E*,9*R*)-3,6-epoxy-7-megastigmen-5,9-diol-9-*O*-β-D-glucopyranoside, (6*S*,7*E*,9*R*)- roseoside (6*S*,7*E*,9 *R*-vomifoliol -9-*O*-β-D-glucopyranoside), and (6*R*,9*R*)-3-oxo-a-ionol-9-*O*-β-D-glucopyranoside.(5*Z*)-6-[5-(2-hydroxypropan-2-yl)-2-methyltetrahydrofuran-2-yl]-3-methylhexa-1, 5-dien-3-ol (35), (5*Z*)-6-[5-(2-*O*-β-D-glucopyranosyl-propan-2-yl)-2-methyl tetrahydrofuran-2-yl]-3-methylhexa-1,5-dien-3-ol, 5-ethenyl-2-[2- *O*-β-D-glucopyranosyl-(1''- 6')-β-D-glucopyranosyl-propan-2-yl]-5-methyltetrahydrofuran-2-ol, 4-[4β-O-β-D-xylopyranosyl-(1''-6')-β-D-glucopyranosyl-2,6,6-trimethyl -1-cyclohexen-1-yl]- butan-2-one, (Z)-3-hexenyl *O*-β-D-glucopyranosyl-(1''- 6')-β-D-glucopyranoside, (Z)-3-hexenyl *O*-β-D-xylopyranosyl-(1'- 6') -β-D-glucopyranoside, (Z)-3-hexenyl *O*-β-Drhamnopyranosyl-(1''-6') -β-D-glucopyranoside, (3*R*,5*S*, 6*S*,7*E*,9*S*)-megastiman-7-ene-3,5,6,9 -tetrol, (3*R*,5*S*,6*S*,7*E*,9*S*)-megastigman-7-ene-3,5,6,9-tetrol 9-*O*-β-D-glucopyranoside, (6*S*,7*E*,9*R*)-6,9-dihydroxy-4,7-megastigmadien-3-one 9-*O*-[β-D-xylopyranosyl-(1''- 6')-β-D-glucopyranoside, Linarionoside C, and (3*S*,9*R*)-3,9-dihydroxy-megastigman-5-ene 3-*O*-primeveroside (Song *et al.*, 2011; Edwards *et al.*, 2012)	Oleanolic acid, ursolic acid and Crategolic acid (Patri and Silano, 1994; Verma *et al.*, 2007)

4	Monoamines	Noradrenaline, adrenaline, dopamine and L-DOPA, Isobutylamine, O-methoxy phenylethylamine, Phenylethylamine, Tyramine (Edwards *et al.*, 2012)	Noradrenaline, adrenaline, dopamine and L-DOPA, Isobutylamine, O-methoxy phenylethylamine, Phenylethylamine, Tyramine (Patri and Silano, O-methoxy phenylethylamine, Phenylethylamine, Tyramine (Patri and Silano, 1994; Verma *et al.*, 2007)
5	Cholines	Choline and acetylcholine (Edwards *et al.*, 2012)	Choline and acetylcholine (Verma *et al.*, 2007)
6	Purine derivatives	Adenosine, Adenine, Guanine (Edwards *et al.*, 2012)	Adenosine, Adenine, Guanine (Verma *et al.*, 2007)
7	Flavonoids (Dewick, 2009).	Quercetin-3-O-galactoside, vitexin-2''-O-glucoside, vitexin-2'' -O-rhamnoside, vitexin-4''-O-glucoside, vitexin-4''-O-rhamnoside, saponaretin, saponaretin rhamnoside (isovitexin rhamnoside; (Nikolov, 1975), acetylvitexin-2''-O-rhamnoside, cratenacinand desacetylcratenacin (Batyuk, 1966; Batyuk *et al.*, 1972, 1966). naringenin, apigenin, quercitrin, isoquercitrin, hesperetin, luteolin, luteolin-7-O-glucoside, orientin, isoorientin, orientin-2''-O-rhamnoside, (-)-epicatechin, (+)-catechin, leucocyanidin, taxifolin, pinnatifinosides A, B, C and D (Zhang and Xu, 2001), pinnatifins C, D (Zhang *et al.*, 2001a), and I, pinnatifidin (Bykov *et al.*, 1972), vicenins 1, 2 and 3 (Nikolov *et al.*, 1981; Nikolov *et al.*, 1982; Nikolov and Vodenicharov, 1975), schaftoside, isoschaftoside, neoschaftoside and	Quercetin, isoquercitrin, hyperoside, procyanidin B_2, procyanidin B_5, procyanidin C_1, Vitamin P (citrin bioflavonoids), rutin, vitexin, (Ju, 2005), or vitexin-22 2 -O-α-L-rhamnoside, and catechin/ epicatechin (Verma *et al.*, 2007).

		isoneoschaftoside (Nikolov *et al.*, 1982; Nikolov and Vodenicharov, 1975), crateside (Nikolov *et al.*, 1973a; Nikolov *et al.*, 1982), Glogoside, (Nikolov *et al.*, 1973b). procyanidin B_2, procyanidin B_5, procyanidin C_1, Vitamin P (citrin bioflavonoids) (Edwards *et al.*, 2012)	
8	Lignins (Lignan glycosides)	monolignols (Dewick, 2009), verbascoside, ((-)-2α-*O*- (β-D-glucopyranosyl)-lyoniresinol, tortoside A, erythro-1-(4-*O*-β-D-glucopyranosyl-3-methoxyphenyl)-2-[4-(3-hydroxypropyl) -2,6-dimethoxyphenoxy]-1,3-propanediol, (7S,8R)-urolignoside, (7S,8R)-5-methoxydihydrodehydrodiconiferyl alcohol-4-O-β-D-glucopyranoside and acernikol-4''-*O*-β-D-glucopyranoside) (Gao *et al.* 2010; Edwards *et al.*, 2012)	
9	Napthoquinones	Crataequinone A and crataequinone B (Min *et al.*, 2004).	
10	Miscellaneous	Amygdalin, Pectins, Cratetegin, saponins (Verma *et al.*, 2007)	Amygdalin, Pectins, Cratetegin, saponins (Verma *et al.*, 2007)

• Chemical markers

Hyperoside, Quercetin, Procyanidins, Cratetegin, Amygdalin, Oleanolic acid, Ursolic acid and Crategolic acid

E. PHARMACOLOGICAL ASPECTS

Bioactivity

- **Anti-cancer activity:** Procyanidins, known to exhibit anti-carcinogenic activity, have been reported to be abundant in *C. rhipidophylla*. A study was conducted to evaluate chemo-preventive potential of Crataegus bark extract using THLE-2 and HepG2 cells. The results clearly suggest chemo-preventive properties of *C. rhipidophylla* that may be due to the presence of procyanidins types of constituents (Krajka-Kuzniak *et al.*, 2014).
- **Anti-bacterial activity:** *C. rhipidophylla ssp monogyna* exhibited excellent antimicrobial activity. *C. rhipidophylla* ssp. *monogyna* was evaluated in vitro against four bacterial test species known to cause humans infections. Berries and sheet of *C. rhipidophylla* ssp. *monogyna* (flavonols and anthocyanins) were more effective than their extracted flavonols. The flavonols only inhibited growth of *Pseudomonas aeruginosa* but failed to inhibit *S. aureus (*Benmalek *et al.*, 2013*)*. The ethanol extract of hawthorn berries also possess moderate bactericidal activity, especially against Gram-positive bacteria (Tadic *et al.*, 2008).
- **Anti-inflammatory/ anti-arthritic activity:** Alcoholic extract of *C. rhipidophylla* berries showed anti-inflammatory potential and decreased isoproterenol-induced inflammation (Vijayan *et al.*, 2012). Aqueous fraction of hawthorn fruit showed anti-inflammatory effects which could be attributed to down-regulation of COX-2, TNF-á, IL-1â, and IL-6 expression in LPS-stimulated RAW 264.7 cells (Li and Wang, 2011). Oral administration of ethanolic extract of hawthorn berries caused anti-inflammatory effect in a dose dependent manner in carrageenan-induced rat paw edema model. In this model, the extract given in the highest tested dose (200 mg/kg) showed 72.4% activity (Tadic *et al.*, 2008).It was observed that hawthorn exert anti-inflammatory effect by preventing synthesis and release of inflammatory mediators (Al-Abdaly, 2011). In another study, the anti-inflammatory mechanism of triterpene fraction of Crataegus was very closely related to the inhibition of peritoneal leukocyte infiltration and weak inhibition of phospholipase A2 (PLA2) (Ahumada *et al.*, 1997).

- **Gastro-protective/anti-ulcer effect:** Gastro-protective activity of *C. rhipidophylla* was evaluated by ethanol induced acute stress ulcer model in rats with reference drug as a ranitidine. The extract produced dose dependent gastro-protective activity with efficacy comparable to that of the reference drug (Tadic *et al.*, 2008; Kashyap *et al.*, 2012).
- **Antioxidant activity:** The methanol-water (50/50, v/v %) extract of fruits showed high antioxidant activity with DPPH assay (Kostic *et al.*, 2012). DPPH radical-scavenging capacity of the extract was concentration-dependent, with EC50 value of 52.04 μg/ml. Flavonol component of the extracts presented high antioxidant potential as compared to that of anthocyanins and standard antioxidants *viz* ascorbic acid and quercetin (Benmalek*et al.*, 2013). Antioxidant activity of hawthorn fractions was also tested in vitro with linoleic acid oxidation by AAPH-generated radicals. *C. rhipidophylla* extracts, in ABTS assay, indicated antioxidative activity (Silva *et al.*, 2000). It has also been reported that flavonoids and oligomeric procyanidins from hawthorn possess antioxidant properties (Sokol-Letowska *et al.*, 2007).
- **Anti-obesity activity:** For many decades, hawthorn has received much attention due to its potential to decrease plasma cholesterol and triacylglycerol concentrations. Several studies have shown the therapeutic value of Hawthorn in lowering lipid levels. Tincture of Crataegus (alcoholic extract of the berries of *C. oxycantha*) significantly decreases the deposited lipids in liver and aorta of hyperlipidemic dieted rats (Akila and Devaraj, 2008). Analytical reports of plasma lipoprotein profile have revealed that tincture reduced the levels of cholesterol, triglycerides (TG) and phospholipids in the low density lipoprotein (LDL) and very low density lipoprotein (VLDL) fractions in hyperlipidemic rats. Agarose gel electrophoretic pattern of plasma lipoproteins also indicated that tincture lowers the levels of the atherogenic beta-lipoproteins in hyperlipidemic rats (Shanthi *et al.*, 1994). Many herbs can decrease low-density lipoprotein oxidation. One study was conducted where, effect of Hawthorn fruit was found to be effective in lowering the blood lipid levels (Xu, 2004). Hawthorn fruit significantly decreased the ratio between low-density lipoprotein cholesterol (LDL-C) and serum cholesterol (TC): (LDL-C)/TC, TG levels (Xu *et al.*, 2009). Zhang *et al.* (2002), reported that hawthorn significantly lowered down the plasma non-HDL (VLDL + LDL) cholesterol concentrations and could reduce the hepatic cholesterol ester content in animal models (Zhang *et al.*, 2002). For the reduction of blood cholesterol, a 6:1 preparation of hawthorn is recommended which is about 2 g of concentrated extract of hawthorn in one-fourth teaspoon of water, juice

or other liquid, prescribed three times daily. Per dose, Hawthorn is standardized to 2% vitexin, 1.8% vitexin-4'-rhamnoside, and/or 20 % procyanidins. (Kamhi and Zampieron, 2005). Lin (2011) investigated the effect of hawthorn on cholesterol metabolism in hamsters and in human Caco-2 cells. The inhibitory activity was positively associated with oleanolic acid and ursolic acid contents in the extracts. The dichloromethane extract of hawthorn decreased the plasma non- HDL-C by 8% without changing HDL-C. Hawthorn extract lower down the plasma cholesterol concentration in rats and rabbits(Zhang *et al.*, 2002; Lin *et al.*, 2011). Crataegus tincture was administered to hyperlipidemic diet fed rats and it prevented the elevation of lipid levels in plasma. A significant decrease in lipid deposits in liver and aorta was noticed. The tincture of hawthorn produces remarkable reduction in the levels of cholesterol, triglycerides and phospholipids in LDL and VLDL hyperlipidemic rats. Histological examination of HLD fed rats also revealed the severe fatty vacuolation and degeneration of liver. The tincture of hawthorn had an ameliorating effect on these changes (Shanthi *et al.*, 1994). In another case, methanolic extract of *C. rhipidophylla* significantly decreased the levels of serum cholesterol, triglycerides, blood glucose, leukocytes count and platelets (Kanyonga *et al.*, 2011).

- **Anti-anxiety activity:** Combination of *C. oxycantha, Eschscholzia californica* and magnesium was found more effective than the placebo in reducing anxiety of 264 generalised anxiety disorder individuals (Hanus *et al.*, 2004). Moreover, the phytotherapeutic product-CPV, which is a combination of dry extract of *C. oxycantha, Passiflora incarnata* and *Valeriana officinalis* had shown the central effects as an anxiolytic in animal models (Min *et al.*, 2000; Tabach *et al.*, 2009).

- **Renoprotective activity:** A 70 % ethanolic extract of *C. rhipidophylla* fruit was evaluated for reno-protecive potential using H_2O_2 Rat model. The physiological functions of kidney in male rats were checked when it was treated with 1% hydrogen peroxide. The treatment showed no clear pathological lesions and the finding was documented for the deleterious effect of H_2O_2, which revealed its reno-protective potential (Al-Abdali, 2014).

- **Immunomodulatory activity:** Hawthorn extract may enhance the function of immune system. Chemical constituents of hawthorn may encourage the production of white blood cells, which attack and destroy disease causing viruses, bacteria and fungi. Immuno-modulatory effect of hawthorn extract was also assessed in experimental stroke model in which alcoholic extract of *C. rhipidophylla* showed immune-modulatory potential by alleviating pro-inflammatory immune responses associated with I/R-induced injury, boosting IL-10 levels, and increasing Foxp3-positive T (regs) in the brain (Elango and Devaraj, 2010).

- **Hepato-protective:** *C. rhipidophylla* is a rich source of procyanidins which have been reported to confer hepatoprotective activity. A study was conducted to evaluate hepato-protective potential of Crataegus bark extract using HepG2 cells. The results clearly suggest that the Nrf2/ARE pathway may play a major role in the regulation of procyanidin-mediated antioxidant as well as detoxification of hepatocytes. Hepato-protective properties of *C. rhipidophylla* may be due to the presence of procyanidins types of constituents (Krajka-Kuzniak *et al.*, 2014).

- **Beneficial role in Cardio and respiratory problems**

  - Crataegus is widely used for treating cardiovascular problems (Blesken, 1992). Hawthorn improves cardiac circulation by stimulating or depressing its activity. It dilates coronary blood supply and stabilizes the contractility of the heart muscle.It is safe for long-term therapy or in failing heart, and has enormous potential for cardiac arrhythmias. Hawthorn berries reduce blood levels of pyruvic and lactic acid, normalize prolonged systole and prevent ECG changes due to hypoxia. It is a valuable herb for a strong and healthy heart because of its potent antioxidant, anti-inflammatory and lipid-lowering properties (Lakshmi *et al.*, 2012). It is used for treating congestive heart failure (CHF) and other cardiac disorders by producing a number of potentials viz. anti-hypertensive, positive inotropic and anti-cardiac remodeling, anti-platelet aggregation, vaso-dilating, endothelial protective effect. Crataegus also produces cardio-protective effects by reducing smooth muscle cell migration and proliferation and by reducing the deterioration of contractile function and infarct size of heart. It showed potential for tackling arrhythmia, myocardial infarction and other types of problems associated with heart.

  - **Congestive cardiac failure:** Hwang *et al.* (2008) assessed the effects of hawthorn on remodeling and functioning of the left ventricle of heart in pressure overload induced cardiac hypertrophy model of rats for one month. Sprague male rats (300 g) were used for sham operation or aortic constriction in 4 weeks study. Rats were treated for 3 weeks after surgery with hawthorn extract (WS1442; 1.3, 13, 130 mg kg (-1) day (-1); AC-L, AC-M, AC-H) or vehicle (SH-V, AC-V). The cardiac functions such as systolic and diastolic function were monitored using echocardiograph at baseline and 4 weeks after dose. It was observed that hawthorn modifies left ventricular remodeling and counteracts myocardial dysfunction in the early pressure overload-induced cardiac hypertrophy (Hwang *et al.*, 2008). Degenring *et al.* (2003) also conducted a clinical trial in the patients of congestive heart failure NYHA (New York Heart Association Functional Classification) class II with double blind placebo controlled, *C. rhipidophylla* and *C. monogyna.* The standardized extract of fresh berries was used in patients with cardiac failure NYHA class II. A total of 143 patients were treated with orally administered three times dose of extract or placebo for eight weeks. The results revealed that fresh Crataegus berries have significant potential to improve heart condition under long term therapy (Degenring *et al.*, 2003; Kumar *et al.*, 2012). The potency of special hawthorn extract (WS 1442) was evaluated in NYHA stage II cardiac insufficiency patients in yet another case, involving 136 patients treated with WS 1442 or placebo for 8 weeks. An improvement was observed in the performance of the heart of WS 1442-treated group, while the condition of placebo group was observed to worsen progressively (Weikl *et al.*, 1996). Thus, results of clinical investigations confirm those of previous studies showing that Crataegus-Special extract WS 1442 is an effective and low-risk phytotherapy for the NYHA II cardiac insufficiency patients (Weikl *et al.*, 1996). In 30 NYHA II cardiac insufficiency patients, a placebo-controlled randomized double-blind study was carried out to check the efficacy of Crataegus special extract for 8 weeks. The extract was administered in the form of capsule, twice in a day. The treatment group with test drug showed a statistically significant advantage over placebo in terms of alteration in pressure-x-rate product (at a load of 50 W) and in improving heart rate. The systolic and diastolic blood pressure was mildly reduced in both groups (Leuchtgens *et al.*, 1993).

- **Anti-hypertensive activity:** Hypertension is a major problem worldwide and causes severe complications (Chobanian *et al.*, 2003). It affects 1 billion people globally, and is expected to rise to 1.6 billion by 2025 (Kearney *et al.*, 2005; Wang *et al.*, 2013). With the popularity and prevalence of traditional and folk medicine, there has been growing interest in traditional herbal medicine for patients with hypertension around the globe. Since the middle ages, *C. rhipidophylla* has been used as an anti-hypertensive agent. In 1880, Green (a non-registered practitioner in Ireland) introduced this drug as medicine in Britain, and followed by others including homoeopathic school (Assmann, 1930). Jennings (1906) promoted the use of Crataegus tincture as an adjuvant as well as digitalis substitute for the treatment of cardiac disorders. Reilly (1910) recommended it for the cardiac failure with arteriosclerosis and high blood pressure in America. In France, Huchard (1903), Leclerc (1912), and Renon (1915) recommended the use of five minims of the hawthorn tincture, twice a day for moderating blood pressure in cases of hypertension within reasonable bounds (James, 1939). A number of clinical trials have described modest blood pressure reduction with use of hawthorn. Ten cases of hypertension were reviewed and the result of treatment with one drachm of Crataegus tincture three times daily showed uniform reduction of the systolic and diastolic blood pressure (James, 1939). A randomised controlled trial was also carried out to check hawthorn efficacy for hypertension in type 2 diabetic patients for 16 weeks. Results showed that hawthorn group has greater reductions than the placebo group in diastolic blood pressure. Hawthorn also showed hypotensive effect in diabetic patients (Belz *et al.*, 2002). A pilot study for investigating hypotensive effect of hawthorn extract and magnesium dietary supplements individually and in combination, compared with a placebo, was performed. In this study, hawthorn extract produced a decline in systolic and diastolic blood pressure and achieved promising reduction in the resting diastolic blood pressure at week 10 (Walker *et al.*, 2006). Similarly, camphor-Crataegus berry combination (CCC) was evaluated on 24 orthostatic dysregulation patients. The CCC drops reduce orthostatic blood pressure, especially affecting diastolic blood pressure after 1 minute of orthostasis in all dosages as compared to placebo. Eighty drops dose was found statistically significant on diastolic blood pressure (Walker *et al.*, 2002). Clinical investigations explored the effects of Crataegus and its various preparations in hypertension but also suggested some contradictory results. A trail was conducted to show the relationship between hawthorn extract dose and brachial artery flow mediated dilation (FMD). 1000 mg, 1500 mg, and 2500 mg randomized dose of extract and placebo was assigned to each participant. In this study, results showed that there was no relation of a dose-response effect for FMD or any other outcomes, such as absolute change in brachial artery diameter and blood pressure (Asher *et al.*, 2012).

- **Positive inotropic effect:** It was reported that WS 1442 extract of *C. rhipidophylla*, and its ethyl acetate fraction, increased the force of contraction in left ventricular papillary muscle strips via cAMP-independent mechanism, probably by inhibition of the sodium pump. Hawthorn likely acts on the Na^+/K^+-ATPase and enhanced the efficiency of calcium transport in cardiomyocytes (Wang *et al.*, 2013).

- **Anticardiac remodeling effect:** Hwang *et al.* (2008) studied the effect of Crataegus on remodeling and function of the left ventricle (LV) after 1 month of pressure overload-induced cardiac hypertrophy. It was reported that the treatment could modify left ventricular remodeling and counteracts myocardial dysfunction in early stage of the induced stress.

- **Antiplatelet aggregation effect:** Activated platelets have significant role in pathological development of arterial disorders (strokes and acute coronary syndromes), which are initiated by plaque disruption and subsequent platelet-thrombus formation. Crataegus extract had effective anti-platelet activity at 100, 200, and 500 mg/kg, probably because of increase in bleeding time, decrease in platelet aggregation assessed by PFA-100 as well as reduction in serum levels of thromboxane B_2 (Mehta, 2002; Abdullah *et al.*, 2012; Wang *et al.*, 2013).

- **Vasodilating effect:** The effectiveness of hawthorn to lower blood pressure had been linked to nitric oxide (NO)-mediated vasodilation, but recent evidence suggested that it is likely to be mediated via an NO-independent mechanism (Asher *et al.*, 2012). Wang *et al.* (2013), has reviewed the possible mechanisms causing vasodilation. NO and endothelium derived hyperpolarizing factor (EDHF) are the two factors credited to control vascular homeostasis (Palmer *et al.*, 1988; Busse *et al.*, 2002). Miller (1998a) suggested that both coronary circulation and peripheral vasculature may not be influenced by hawthorn, but it acts by inhibiting angiotensin-converting enzyme (ACE).

- **Endothelial protective effect:** Formation of edema in cardiac endothelium causes serious diseases like atherosclerosis, sepsis, or heart failure (Wang *et al.*, 2013). WS 1442 extract of Crataegus effectively protects against endothelial barrier dysfunction and also prevents the deleterious hyperpermeability-associated rise, by interfering with sarcoplasmic/endoplasmic reticulum Ca^{2+} ATPase and inositol 1,4,5-trisphosphate pathway without inducing store-operated calcium entry (Elisabeth *et al.*, 2012). The research work suggested that sustained intake of Crataegus also prevents aging-related endothelial dysfunction by decreasing prostanoid-mediated contractile responses (Idris-Khodja *et al.*, 2012).

- **Reduction of smooth muscle cell migration and proliferation properties:** There are limited studies on migration and proliferation effects of herbal drugs. However, hawthorn exhibits some cadio-protective effects because of decrease in smooth muscle cell migration and proliferation properties (Wang *et al.*, 2013). WS 1442 extract reduces the vascular smooth muscle cell migration by 38% when proliferation was by 44%. It inhibited vascular smooth muscle cell DNA synthesis induced by platelet-derived growth factor, blocked recombinant human platelet-derived growth factor receptor (PDGFR)-â kinase activity and decreased PDGFR-â activation and extracellular signal-regulated kinase activation in platelet-derived growth factor (Furst *et al.*, 2010).

- **Protection in ischemia/reperfusion injury:** *C. rhipidophylla* have been reported in animal models to have cardio-protective role in case of ischemia/reperfusion. Ischemia and reperfusion causes multiple injuries in microcirculation (Han *et al.*, 2008). WS 1442 extract significantly reduced the deterioration of contractile function and infarct size of myocardium in rat (Veveris *et al.*, 2004). The extract has also shown antiarrhythmic activity by reducing average prevalence of malignant arrhythmias and ventricular tachycardia (Al Makdessi *et al.*, 1999).
- **Antiarrhythmic effect:** Alcoholic extract of *C. rhipidophylla* helps to prevent digoxin induced arrhythmia in Wistar rats and produce san anti-arrhythmic effect. This may be an alternative medication for arrhythmias digoxin induced toxicity in humans (Alp *et al.*, 2014). Anti-arrhythmic effect was also estimated using a cultured cardiomyocyte assay method. The ethanol extract of the species was compared with other known cardioactive drugs such as ouabain, epinephrine, and propranolol for its anti-arrhythmic potential. *C. rhipidophylla* extract have been shown to have potential to induce rhythmicity in quiescent cardiomyocytes and treating arrhythmias with no chronotropic effect, and without *â*-adrenergic receptor blockade (Long *et al.*, 2006). Crataegus pretreatment significantly reduced the prevalence of malignant arrhythmias (Al Makdessi *et al.*, 1999). Anti-arrhythmic mechanism of extract remains elusive, however. Crataegus extract mainly results in significant reduction in total number of ventricular ectopic beats (ventricular tachycardia) (Garjani *et al.*, 2000).
- **Myocardial infarction:** *C. rhipidophylla* possess positive inotropic effect of amines such as phenethylamine, *O*-methoxyphenethylamine and tyramine. These amines were reported to be responsible for the increase of intracellular calcium, which helps in prolongation of effect and inotropic activity (Wagner and Grevel, 1982; Kocyildiz *et al.*, 2006; Kumar *et al.*, 2012). In another study, alcoholic extract was found to maintain antioxidant status in mitochondria and decrease the Kreb's cycle enzymes induced by isoproterenol in rat heart. It was also suggested to prevent the mitochondrial lipid peroxidative damage (Jayalakshmi *et al.*, 2006; Kumar *et al.*, 2012).

Clinical Studies

- **Role in management of specific ailments:** *C. rhipidophylla* is beneficial for lowering the blood pressure, increasing the effectiveness of the heart's pumping action, strengthening the heart muscle, slowing the heart beat, dilating coronary arteries, preventing heart disease, heart attack and stroke etc. It also helps in healing from heart surgery, supporting immune system and increasing longevity. The German Commission E is a scientific body which determines the effectiveness of herbal medicines. They recommended tea or tincture of hawthorn for cardiac insufficiency of stages I and II as per NYHA classification. It helps to reduce the feelings of pressure and tightness in the cardiac region (Lakshmi *et al.*, 2012).
- **Toxicity related information:** Hawthorn does not produce any serious side effects, hence is considered to be a safer drug and is recommended for long term use (Lakshmi *et al.*, 2012). Some side effects related to hawthorn extracts are reported in the literature but these are rare or mild, such as mild rash, headache, sweating, dizziness, sleepiness, agitation, gastrointestinal complaints (Kumar *et al.*, 2012).

• Precautions and warning

Despite its wide use by many groups across the world, consumption of Hawthorn for medical purpose, such as for treatment of cardiovascular diseases need caution. Avoid the use Crataegus or any of its components where history of hypersensitivity or allergic reaction is known. The contraindication has also been suggested for the children under the age of 12 years (Weikl *et al.*, 1996; Ammon and Handel, 1981b). Moreover, the drug is contra indicated in pregnancy, breast feeding, infants and children (Lakshmi *et al.*, 2012).

• Dosage

Dosages of Crataegus are dependent upon concentration of extracts. Therapeutic dose of standardized extract (1.8 % of vitexin-4- rhamnoside) is 100-250 mg, three times a day. A dose of standardized extract (18% oligomeric procyanidins) is in the range of 250-500 mg/day (Monograph, 2010).

• Toxicology

The LD_{50} of Crataegus extract is 25 mg/kg, which showed low toxicity (Ammon and Handel, 1981a). Commission E monograph described that mice and rats showed the safety of standardized extract at doses up to at 3 g/kg body weight (Blumenthal *et al.*, 1988). The studies in rats using excessive dosing of flower extract of hawthorn (600 mg/kg/day; flavonoids) over 30 days showed unremarkable side effects. In humans, the acute oral toxicity of hawthorn was 6 g/kg.

• **Formulations**

The formulation are suggested by Heather and Michael (2009):

- Tincture - 5-120 drops (4-6mL),t.d.s
- Fluid extract - 10-60 drops,t.d.s
- Infusion - 3-5g (flowers and berries)
- Freeze dried - 1-1.5g/day
- Standardized extract (1.8% vitexin-4'-rhamnoside or 10% procyanadin)-100-250mg, t.d.s
- Solid extract - 1/4 tsp, t.d.s/q.d.s

F. THREAT AND CONSERVATION ASPECT

Population status

Threats

• **Habitat loss/degradation/fragmentation**

The loss of habitat for any species it local or regional scale is the most serious threat for its status. Forest felling or fires, developmental activities in forests, and land/ landuse changes are some of the human activities those have increased in recent times and considered as the primary cause of risk for over 80% endangered plant species. No systematic studies have been carried out to ascertain threat status of *C. rhipidophylla*. Since the species inhabits temperate or upper temperate regions in Himalayas, it is likely to face less threat from changes in landuse pattern, which are more prominent at lower elevations. Also, the useful part of the plant are berries which are collected seasonally, the species apparently face little risk from destructive harvesting. However, on the basis of its natural population being sparse or rare, conservation for species has been stressed for. According to a regional study, the species has been listed as Critically Endangered as per IUCN Red List criteria (Khan, 2012).

• **Over-exploitation**

In Himalayas, there is no definite information regarding the changing status of *C. rhipidophylla* in its natural habitat. However, studies suggest growing demand of the plant parts (leaves, flowers and berries) to agencies catering to the demand of various indigenous schools of medicines. Natural plantations of the species in the wild therefore are constantly under the extraction pressure and a dwindling status of its population may quite be likely. According to a report from Finland, changes in forest management practice and overgrowing of meadows and other open habitats pose main threats to hawthorn status in wild (Rassi *et al.*, 2010).

• Introduction and spread of exotic/alien species

C. rhipidophylla was introduced as a hedge plant in some parts of Australia, including Tasmania, in the 1800's. Being an aggressive and tenacious in its persistence, the species is stated to be difficult to eradicate (Kashyap *et al.*, 2012).

• Pollution and diseases

No specific pest/diseases have been reported for *C. rhipidophylla*

• Climate change

Plant species distributed in temperate and alpine environment are more vulnerable to climate change related impacts. Under the anticipated rise in atmospheric temperature, *Crataegus* species, like many others, is likely to move upwards and inhabit newer areas around or above the current tree line limits. Its populations at the lower elevation range may face altered phenology and be subjected to renewed biotic and abiotic stress interaction. Plant pollinator interaction is also expected to change. There is little information regarding how these parameters would affect *C. rhipidophylla* in its natural habitat under climate change scenario and calls for such studies to be taken up.

Conservation strategies

There are currently no large scale efforts to generate the source material through organized cultivation though some pioneering studies have been taken up. Vegetative propagation through stem cuttings, from plantlets or suckers from wild and through seed is known. Information regarding seeds, which requires stratification and require one to two years to germinate, can be found with companies advertising on net.

• *Ex-situ* conservation

Ex situ cultivation is an important conservation measure for species sought by herbal industry, for the purpose of both, generation of desired quantity as well as authentic raw material for crude drug. The available genetic diversity from its threatened habitat can be conserved and suitably exploited for better yield of secondary metabolites, under protected conditions. Many medicinal plant species have been taken up under organized *ex situ* cultivation at national and international level, but there are no reports available for *C. rhipidophylla*.

• ***In situ* conservation**

No reports are available for *in situ* cultivation of *C. rhipidophylla* or such efforts for its conservation.

G. CULTIVATION ASPECTS

- **Climate:** Preferably humid and sub humid climate in temperate areas with an annual rainfall of more than 600mm. Cooler areas are suited for its cultivation.
- **Soil:** *C. rhipidophylla* prefers sandy clay-loam soil for the cultivation.
- **Nursery raising and planting**:
- **Seed:** For better results, ripened seeds should be sown during the autumn. Generally, seeds take more than a year to germinate, though some may germinate during the next spring. Stored seed can be very slow and erratic to germinate. For achieving better germination, the seeds should be stratified for 3 months at temperature of 15°C and then again for 4°C for another 3 months. Further, 18 months may still be required to germinate them. The time for germination can be reduced by the scarification of seed before the stratification. Fermentation of the seed with its own pulp for a few days may also speed up the process of germination. Another approach is to harvest the "green" seed and sow them immediately in a cold frame. If timed well, it can germinate during the next spring. For requirement of plants in smaller number, it is better to grow them in individual pots for the first year, and then transplant them in nursery beds or in field directly in late spring. For requirement in larger quantities, it is better to sow them directly outdoors in a seedbed, but with protection from mice and other seed-eating creatures.

• **Vegetative propagation:**

Cultivation through seeds is cumbersome, owing to seed dormancy and poor viability. Even after scarification and stratification of seeds, the success rate of plants raised through seeds remains less than 20%. Stem cutting, of diameter smaller than a pencil, therefore provides better way to generate plants. These cuttings may be planted at a distance of 10 cm apart in sandy soil, under protected nursery/ polyhouse conditions. Indole Butyric Acid (IBA) at 4000 ppm, for promoting callus, is recommended. Rooted cuttings should be hardened before taking to field. About 2,500 plants are required for plantation over 1 hectare area. After hardening, the plants can be transferred to field and planted at a space of 2m apart.

- **Manure:** Application of Farm Yard Manure (FYM) in the field, before and after is better for growth. A basal dose of 15 t/ha during July-August and about 20-22 t/ha of FYM at the time of flowering during March-April is recommended.
- **Irrigation:** *C. rhipidophylla* can tolerate medium to high drought and its water requirements for irrigation are modest.
- **Harvesting:** The fruits are ready for harvest, usually after 5-6 years of growth in the field, and generally harvested during August to October. A higher yield of up to 600 kg fruits per hectare area can be obtained after nine years of planting. Fruits should be shade dried and stored under cool and dry conditions.

H. COMMERCIAL ASPECTS

Trade data

No authentic data is available on global production or demand of *C. rhipidophylla*. However, information for some European countries is available. Hawthorn is listed as one of the top six medicinal plants in Bulgaria, and about 400 tonnes of leaves, fruits and flowers are collected annually from the wild (Evstatieva and Hardalova 2004; Kathe *et al.,* 2003). It figures among top ten and top eighteen most collected medicinal plants in Albania and Bosnia and Herzegovina respectively, mostly for trade in the international market. Approximately, 28 tonnes of the collected flowers and leaves traded by Romania fetch around $52,000 in the international market (Kathe *et al.,* 2003).

- **Major users:** Several countries in Europe, America, Asia are demand centers for hawthorn.

I. FUTURE DIRECTIONS FOR RESEARCH / VALUE ADDITION

C. rhipidophylla is one of the many medicinal plant species, the use of which is cautioned in the absence of scientific validation and research data. Studies carried out over a long period of time have supported the use of *C. rhipidophylla* to treat several conditions associated with cardiovascular disorders, such as hypertension, ischemic heart disease, and congestive heart problem. A recent study (Saeedi *et al.*, 2018) showing therapeutic and protective potential of *C. rhipidophylla* for treating non-alcoholic fatty liver disease in rats opens future scope of assessments for clinical studies. Its potential role in treating inflammatory conditions offers opportunities as well as invites new research initiatives to treat series of associate ailments. At the same time, organized and sustained research efforts are also required for establishing plantations of viable size to meet the growing demand of crude drug material by the pharmaceutical industry.

J. PATENTS

- There are over 30 patents are granted or applied for *Crataegus rhipidophylla*. The following section provide information on 10 patents, which have granted status. For details please referto*(https://patents.google.com/patent/US5976568A/en?q=crataegus&q=rhipidophylla&q=granted&q= patent&oq =crataegus+oxycantha).
- Riley, P. A. Modular system of dietary supplement compositions for optimizing health benefits and methods. Application number - US5976568A. 1999-11-02.US5976568A.
- William E. Hahn, Charles A. Arnold. Liberating intracellular matter from biological material. Application number -US6405948B1. 2002-06-18. US6405948B1.
- Yasushi Sakai, Yoshiharu Yokoo. Process for producing a plant extract containing plant powder. Application number -US20040161524A1. 2009-04-21. US7521079B2
- Giovanni Pauletti, Donald Harrison, Kishor kumar Desai. Method for augmentation of intraepithelial and systemic exposure of therapeutic agents having substrate activity for cytochrome P450 enzymes and membrane efflux systems following vaginal and oral cavity administration.Application number -US20070036834A1. 2012-05-15. US8178123B2.
- Takehisa Iwama, Keigo Matsumoto, Takayuki Imoto, Nobuhide Miyachi, Masahiro Goto. Cosmetic, external skin preparation, and medical instrument. Application number -US20120258059A1. 2015-04-07. US8999300B2.
- Keigo Matsumoto, Takayuki Imoto, Takehisa Iwama. Production method of cosmetic, preparation method of gel for cosmetics, and method of reducing use amount of polymer thickener blended in cosmetic raw materials. Application number -US20140113976A1. 2017-02-14. US9566218B2.
- Masahiro Gotoh, Takayuki Imoto, Tsubasa Kashino, Takehisa Iwama, Nobuhide Miyachi. Gel sheet comprising lipidic peptide type gelling agent and polymeric compound. Application number -EP2638921A1. 2018-04-25. EP2638921B1.
- Seiki Tamura, Tomohiro Iimura, Tatsuo Souda, Haruhiko Furukawa. Cosmetic for hair containing co-modified organopolysiloxane. Application number -US8715626B2. 2014-05-06. US8715626B2.
- Takayuki Imoto. Thickened composition containing lipid peptide-type compound. Application number -US20170042783A1. 2018-09-11. US10071044B2.

- Hogne Hallaraker,Jan Remmereit, Alvin Berger. Lipid compositions with high DHA content. Application number -US9458409B2. 2016-10-04. US9458409B2.

AKNOWLEDGEMENT

Authors are grateful to the Director of the institute for providing necessary facilities. Thanks are due to CSIR, New Delhi for financial support under the network project BSc 0109and to SERB-DST, New Delhi for the Fast Track Scheme for Young Scientist (SB/YS/LS-81/2014).

LITERATURES CITED

1. Abdullah SS., Hesham S., Fahaid AH. 2012. Effect of hawthorn (*Crataegus aronia* syn. *Azarolus* (L)) on platelet function in albino Wistar rats. *Thrombosis Research*. 130(1): 75-80.
2. Ahn KS., Hahm MS., Park EJ, Lee HK, Kim IH. 1998. Corosolic acid isolated from the fruit of *Crataegus pinnatifida* var. *psilosa* is a protein kinase C inhibitor as well as a cytotoxic agent. *Planta Med*. 64: 468-470.
3. Ahumada C., Saenz T., Garcia D, De La Puerta R, Fernandez A, Martinez E. 1997. The effects of a triterpene fraction isolated from *Crataegus monogyna* Jacq. on different acute inflammation models in rats and mice. Leucocyte migration and phospholipase A2 inhibition. *JPP*, 49(3): 329-331.
4. Akila M., Devaraj H., 2008. Synergistic effect of tincture of Crataegus and *Mangifera indica* L. extract on hyperlipidemic and antioxidant status in atherogenic rats. *Vasc Pharmacol*, 49(4-6): 173-177.
5. Al Makdessi S., Sweidan H., Dietz K., Jacob R. 1999. Protective effect of *Crataegus oxycantha* against reperfusion arrhythmias after global no-flow ischemia in the rat heart. *Basic Res. Cardiol*. 94(2): 71-77.
6. Al-Abdali AIO., 2014. Study effect of crude alcoholic extract of Hawthorn (*Crataegus Oxycantha*) and crude polyphenolic compounds from blackolive (*Olea Europae*) fruits on some physiological parameters of kidney of male rats treated with hydrogen peroxide. *Iraqi J. Cancer Med. Genet*.7(2): 113-120.
7. Al-Abdaly AIO., 2011. Effect of flavonoids extracted from Hawthorn (*Crataegus oxycantha*) on some hematological parameters of female mice. *J. Vet. Sci*. 4(2): 145-151.
8. Alp H., Soner BC., Baysal T., Sahin AS. 2015. Protective effects of Hawthorn (*Crataegus oxycantha*) extract against digoxin-induced arrhythmias in rats. *Anatol J Cardiol*. , 15 (0), 000-000. DOI:10.5152/akd.2014:5869
9. Ammon HPT., Handel M., 1981a., *Crataegus*, toxicology and pharmacology, Part 1: *toxicity Planta Med*, 43: 105-120.
10. Ammon HPT., Handel M., 1981b *Crataegus*, toxicology and pharmacology, Part III: *Planta Med.*, 43: 313-322.
11. Amy B., Deanna M., Christy W., Hawthorn. 2006 URL: http//www.geocities.com/chadrx/hawth.html.
12. Arnason T., Hebda RJ., Johns T. 1981. Use of plants for food and medicine by native peoples of eastern Canada. *Can. J. Bot*., 59: 2189-2325.
13. Arrieta J., Siles-Barrios D., Garcia-Sanchez J., Reyes-Trejo B., Sanchez-Mendoza ME. 2010. Relaxant effect of the extracts of *Crataegus mexicana* on guinea pig tracheal smooth muscle. Pharmacog J., 2: 40-46.

14. Asher GN., Viera AJ., Weaver MA., Dominik R., Caughey M., Hinderliter AL. 2012. Effect of hawthorn standardized extract on flow mediated dilation in prehypertensive and mildly hypertensive adults: a randomized, controlled cross-over trial. *BMC Complement Altern Med.*:12-26.
15. Barceloux DG., Hawthorn (Crataegus Species). 2008. In: Barceloux, D.G. (Ed.), Medical toxicology of natural substances: foods, fungi, medicinal herbs, plants, and venomous animals. Wiley Interscience: 510-513.
16. Batyuk VS., Chernobrovaya NV., Kolesnikov DG. 1972. Flavonoids of Crataegus-the structure of cratenacin. *Chem. Nat. Compd.*, 5: 199-200.
17. Batyuk VS., Chernobrovaya NV., Prokopenko AP. 1966Cratenacin-a new flavone glycoside from *Crataegus curvisepala. Chem. Nat. Compd.*, 2: 90-93.
18. Belz GG., Butzer R., Gaus W., Loew D. 2002. Camphor-Crataegus berry extract combination dose-dependently reduces tilt induced fall in blood pressure in orthostatic hypotension. *Phytomedicine*, 9(7): 581-588.
19. Benmalek Y., Yahia OA,Belkebi., A. Fardeau M. 2013. Anti-microbial and anti-oxidant activities of *Illicium verum, Crataegus oxycantha ssp monogyna* and *Allium cepa* red and white varieties.*Bioengineered.*, 4(4): 244-248.
20. Blesken R., 1992. Crataegus in cardiology. Fortschr Med, 15: 290-292.
21. Blumenthal M., Busse W., Goldberg A. 1988. The complete Germen commission E Monographs, Boston MA:American Botanical Council: 142-144.
22. British Herbal Pharmacopoeia 1972. London, British Herbal Medicine Association, 1996.
23. Busse R, Edwards G, Félétou M, Fleming I, Vanhoutte PM, Weston AH. 2002. EDHF: bringing the concepts together. *Trends Pharmacol Scis.* 23(8):374–380.
24. Bykov VI., Glyzin VI., Ban'Kovskii AI., 1972.Pinnatifidin-a new flavonol glycoside from *Crataegus pinnatipida*. Khimiya Prirodnykh Soedinenii, 8:715-718.
25. Chang Q., Zuo Z, Harrison F., Chow MS. 2002. Hawthorn. *J Clin Pharmacol*, 42,:605-612.
26. Chen JD., Wu YZ., Tao ZL., Chen ZM., Liu XP., 1995. Hawthorn (shan zha) drink and its lowering effect on blood lipid levels in humans and rats. World Rev Nutr Diet, 77: 147-154.
27. Chobanian AV., Bakris GL., Black HR., Cushman WC., Green LA., Izzo JL Jr., Jones DW., Materson BJ., Oparil S., Wright JT., Jr, Roccella EJ. 2003. Seventh report of the joint national committee on prevention, detection, evaluation, and treatment of high blood pressure, Hypertension, 42(6):1206-1252.
28. Clapham A.R., Tutin T.G., Moore D.M. 1989. Flora of the British Isles. Cambridge: Cambridge University Press, Cambridge.
29. *Crataegus oxycantha*: The Wealth of India. 2004First supplement series. NISCAIR, CSIR, New Delhi, Vol 2,
30. Degenring FH, Suter A., Weber M., Saller R.Bioforce AG, Roggwil. A 2003. Randomised double blind placebo controlled clinical trial of a standardised extract of fresh Crataegus berries (Crataegisan) in the treatment of patients with congestive heart failure NYHA II. Phytomedicine,, 10(5): 363-369.
31. Deutsches 1998., Arzneibuch., Stuttgart, Deutscher Apotheker Verlag,.
32. Dewick PM., 2009. Medicinal Natural Products. A Biosynthetic Approach. John Wiley andSons Ltd., Chichester, SXW, UK,.
33. DonmezA.A. 2004. The Genus Crataegus L. (Rosaceae) with special reference to hybridisation and biodiversity in Turkey. *Turk J Bot.* 28: 29-37.
34. Edwards J.E., Brown P.N., Talent N., Dickinson T.A., Shipley P.R. 2012.A review of the chemistry of the genus Crataegus. *Phytochem.* 79: 5-26.

35. Elango C., Devaraj S.N.2010. Immunomodulatory effect of Hawthorn extract in an experimental stroke model. *J Neuroinflamm*. 7: 97.
36. Elisabeth A.W., Roland M., Alexander I.B. 2012. The vascular barrier-protecting hawthorn extract WS 1442 raises endothelial calcium levels by inhibition of SERCA and activation of the IP3 pathway. *J Mol Cell Cardiol.*53(4): 567-577.
37. Kamhi, E. Zampieron, E. 2005. Managing Cholesterol Naturally. Available at http://www.naturalnurse .com/cholesterol.htm
38. El-Sayed A.M. 2011. The Pherobase: Database of Insect Pheromones and Semiochemicals. Available at: http://www.pherobase.com.
39. European Pharmacopoeia. 1996. Strasbourg Cedex: Council of Europe.
40. European pharmacopoeia, ThirdEdition, Suppl. 2000. Council of Europe, Strasbourg, 1999.ISBN 10: 9287138818 / ISBN 13: 9789287138811
41. Evstatieva L., HardalovaR.2004. Conservation and sustainable use of medicinal plants in Bulgaria.Medicinal Plant Conservation: Medicinal Plant Specialist Group 9/10: 24-28.
42. Furst R., Zirrgiebel U., Totzke F., Zahler S., Vollmar A.M., Koch E. 2010. The Crataegus extract WS 1442 inhibits balloon catheter-induced intimal hyperplasia in the rat carotid artery by directly influencing PDGFR-â. *Atherosclerosis* ,211(2): 409-417.
43. Gao P.Y., Li L.Z., Peng Y., Li F.F., Niu C., Huang X.X., Ming M., Song S.J. 2010 Monoterpene and lignan glycosides in the leaves of *Crataegus pinnatifida.Biochem. Syst. Ecol*. 38: 988-992.
44. Garjani A., Nazemiyeh H., Maleki N., Valizadeh H. 2000. Effects of extracts from flowering tops of *Crataegus meyeri* A. Pojark. on ischaemic arrhythmias in anaesthetized rats. *Phytother Res*. 14(6): 428-431.
45. Hanus M., Lafon J., Mathieu M.2004. Double-blind, randomised, placebo-controlled study to evaluate the efficacy and safety of a fixed combination containing two plant extracts (*Crataegus oxycantha* and *Eschscholtzia californica*) and magnesium in mild-to-moderate anxiety disorders. *Curr Med Res Opin.*20: 63-71.
46. Hawthorne Berry info and Products-great for the heart, 2006. http://www.bodybuilding.com/store/haw.html.
47. Hellson J.C., Gadd M.1974. Ethnobotany of the Blackfoot Indians. National Museum of Man Mercury Series.*Canadian Ethnology Service Paper*. 19:1-138.ISBN: 9781772821819
48. Hemery G., Simblet S. 2014. The New Sylva: A Discourse of Forest and Orchard Trees for the Twenty-First Century; Chapter III: HAWTHORN. A and C Black. 131-134.Bloomsbury Publishing ISBN: 9781408835449.
49. Huang Y, Chen ZY, Ho WKK (2004) Crataegus (Howthorn). In: Lester P, Choon NO, Barry H, Herbal and traditional medicine, biomolecular and clinical aspects. China 21 CRC Press.
50. Hwang H.S., Bleske B.E., Ghannam M.M., Converso K., Russell M.W., Hunter J.C., Boluyt M.O. 2008. Effects of hawthorn on cardiac remodeling and left ventricular dysfunction after 1 month of pressure overload-induced cardiac hypertrophy in rats. *Cardiovasc Drugs Ther*. 22(1): 19-28.
51. Hwang H.S., Bleske B.E., Ghannam M.M., Converso K., Russell M.W., Hunter J.C., Boluyt M.O. 2008. Effects of hawthorn on cardiac remodeling and left ventricular dysfunction after 1 month of pressure overload-induced cardiac hypertrophy in rats.*Cardiovasc Drugs Ther*. 22(1): 19-28.
52. Idris-Khodja N., Auger C., Koch E., Schini-Kerth V.B. 2012. Crataegus special extract WS 1442 prevents aging-related endothelial dysfunction. *Phytomedicine* 19(8-9): 699-706.
53. Ikeda T., Ogawa Y., Nohara T. 1999. A new triterpenoid from *Crataegus cuneata*. *Chem. Pharm. Bull*. 47:1487-1488.
54. James D.P.G. 1939. *Crataegus oxycantha* in hypertension. *The B M J.* 951-953.

55. Jayalakshmi R., Thirupurasundari C.J., Niranjali D.S. 2006. Pretreatment with alcoholic extract of *Crataegus oxycantha* (AEC) activates mitochondrial protection during isoproterenol-induced myocardial infarction in rats. *Mol Cell Biochem*. 292: 59-67.
56. Jacobs J.H., Clark S.J., Denholm I., Goulson D., Stoate C., Osborne J.L. 2009. Pollination biology of fruit-bearing hedgerow plants and the role of flower-visiting insects in fruit-set. *Ann. Bot*. 104(7): 1397-1404.
57. Jauregui A., Montalti D., Segura L.N. 2014.First record of *Crataegus monogyna* Jacq. (Rosales: Rosaceae) in Buenos Aires province, Argentina. 10(5): 1167-1169.
58. Jolivet P. 1998.Interrelationship Between Insects and Plants; Chapter-3: An early twentieth century classification of food plant selection among phytophagous insects CRC Press L.L.48-50.
59. Ju L.Y. 2005.*Crataegus oxycantha* (aubepine) in the use as herb medicine in France. *Zhongguo Zhong Yao Za Zhi*. 30(8): 634-640.
60. Kanyonga M.P., Faouzi M.Y.A., Zellou A., Essassi M., Cherrah Y. 2011. Effect of methanolic extract of *Crataegus oxycantha* on blood haemostasis in rats. *J Chem Pharm Res*. 3(3): 713-717.
61. Kao E.S., Wang C.J., Lin W.L., Chu C.Y., Tseng T.H. 2007. Effects of polyphenols derived from fruit of *Crataegus pinnatifida* on cell transformation, dermal edema and skin tumor formation by phorbol ester application. *Food Chem Toxicol*. 45:1795-1804.
62. Kashyap C.P., Arya V., Thakur N. 2012. Ethnomedicinal and phytopharmacological potential of *Crataegus oxycantha* Linn.- A review. *Asian Pac J Trop Biomed*.S1194-S1199.
63. Kathe W., Honnef S., Heym A. 2003. Medicinal and Aromatic Plants in Albania, Bosnia-Herzegovina, Bulgaria, Croatia and Romania. BfN Skripten 91.
64. Kearney PM., Whelton M., Reynolds K., Muntner P., Whelton PK., He J. 2005. Global burden of hypertension: analysis of worldwide data. 36 (9455): 217-223.
65. Khan, S.M. 2012.Plant communities and vegetation ecosystem services in the naran valley, western Himalaya. Ph. D. thesis, Department of Biology, University of Leicester England, United Kingdom.
66. Kocyildiz Z.C., Birman H., Olgac V., Akgun-Dar K., Meliko G., Mericli A.H. 2006. *Crataegus tanacetifolia* leaf extract prevents L-NAME-induced hypertension in rats: a morphological study. *Phytother Res*. 20: 6-70.
67. Kostic D.A., Velickovic J.M., Mitic S.S., Mitic M.N., Randelovic S.S. 2012. Phenolic content, antioxidant and antimicrobial activities of *Crataegus Oxycantha* L (Rosaceae) fruit extract from outheast Serbia. *Trop J Pharm Res*. 11(1): 117-124.
68. Krajka-Kuzniak V., Paluszczak J., Oszmianski J., Baer- Dubowska W. 2014. Hawthorn (*Crataegus oxycantha* L.) bark extract regulates antioxidant response element (ARE)-mediated enzyme expression via Nrf2 pathway activation in normal hepatocyte cell line. *Phytother Res*.28(4): 593-602.
69. Krzeminski T., Chatterjee S.S. 1993. Ischemia and early reperfusion induced arrhythmias: beneficial effects of an extract of *Crataegus oxycantha* L. Pharmacy and *Pharmacol Letters*. 3: 45-48.
70. Kumar D., Arya V., Bhat Z.A., Khan N.A., Prasad D.N. 2012. The genus *Crataegus*: chemical and pharmacological perspectives. *Revista Brasileira de Farmacognosia*. 22(5): 1187-1200.
71. Lakshmi T., Geetha R.V., Roy A. 2012. *Crataegus oxycantha* Linn. commonly known as Hawthorn-A Scientific Review. *Int.J.PharmTech Res*. 4(1): 458-465.
72. Leuchtgens H. 1993. Crataegus special extract WS 1442 in NYHA II heart failure. A placebo controlled randomized double-blind study. *Fortschr Med*. 111(20-21): 352-354.
73. Li C., Wang M.H. 2011. Anti-inflammatory effect of the water fraction from hawthorn fruit on LPS-stimulated RAW 264.7 cells.*Nutr Res Pract*. 5(2): 101-106.

74. Lin Y., Vermeer M.A., Trautwein E.A. 2011. Triterpenic acids present in Hawthorn lower plasma cholesterol by inhibiting intestinal ACAT activity in hamsters.*Evid Based Complement Alternat Med.* 201(1): 1-9.
75. Liu Y., Wang M.W. 2008. Botanical drugs: challenges and opportunities: contribution to Linnaeus Memorial Symposium. *Life Sci.* 82: 445-449.
76. Long S.R., Carey R.A., Crofoot K.M., Proteau P.J., Filtz T.M. 2006. Effect of hawthorn (Crataegus oxycantha) crude extract and chromatographic fractions on multiple activities in a cultured cardiomyocyte assay. *Phytomedicine.* 13: 643-650.
77. Mehta S.R. 2002. Appropriate antiplatelet and antithrombotic therapy in patients with acute coronary syndromes: recent updates to the ACC/AHA guidelines. *J Invasive Cardiol.* 14(1-2): 27-34.
78. Miller A.L. 1998b. Botanical influences on cardiovascular disease. *Altern Med Rev.* 3(6): 422-431.
79. Miller L.G. 1998a. Herbal medicinals: selected clinical considerations focusing on known or potential drug-herb interactions.*Arch. Intern. Med.* 158(20): 2200-2211.
80. Mills S., Bone K. 2000.London: Churchill Livingstone; *Principles and Practice of Phytotherapy.*
81. Min B.S., Kim Y.H., Lee S.M., Jung H.J., Lee J.S., Na M.K., Lee C.O., Lee J.P., Bae K. 2000. Cytotoxic triterpenes from *Crataegus pinnatifida. Arch Pharm Res.* 23: 155-158.
82. Min B.S., Huong H.T.T., Kim J.H., Jun H.J., Na M.K., Nam N.H., Lee H.K., Bae K.H., Kang S.S. 2004. Furo-1,2-naphthoquinones from *Crataegus pinnatifida* with ICAM-1 expression inhibition activity. *Planta Med.* 70: 1166-1169.
83. Moerman D.E. 1998. Native American Ethnobotany. Timber Press, Portland, OR.
84. Moerman D.E. 2009. Native American Medicinal Plants. Timber Press, Portland, OR.
85. Nikolov N., Batyuk V.S., Ivanov V. 1973a. Crateside-a new flavonol glycoside from *Crataegus monogyna* and *Crataegus pentagyna. Khim. Prir. Soedin.* 9: 153-156.
86. Nikolov N.T. 1975. New flavone C biosides from *Crataegus monogyna* and *Crataegus pentagyna. Khim. Prir. Soedin.* 3: 422-423.
87. Nikolov N.T., Dellamonica G., Chopin J. 1981. Di-C-Glycosylflavones from *Crataegus monogyna. Phytochemistry.* 20: 2780-2781.
88. Nikolov N.T., Litvinenko V.I., Kovalev I.P. 1973b. Glogoside-a new flavonoid from *Crataegus pentagyna. Khim. Prir. Soedin.* 9: 148-151.
89. Nikolov N.T., Seligmann O., Wagner H., Horowitz R.M., Gentili B. 1982. New flavonoid-glycosides from *Crataegus monogyna* and *Crataegus pentagyna. Planta Med.* 44: 50-53.
90. Nikolov N.T., Vodenicharov R.I. 1975. Di-*C*-glycosides from *Crataegus monogyna. Chem. Nat. Compd.* 11: 436-437.
91. Palmer R.M.J., Ashton D.S., Moncada S. 1988. Vascular endothelial cells synthesize nitric oxide from L-arginine. *Nature.* 333(6174):664–666.
92. Park S.W., Yook C.S., Lee H.K. 1994.Chemical components from the fruits of *Crataegus pinnatifida* var psilosa. *Kor. J. Pharmacog.* 25: 328-335.
93. Ptak K. 1986. Cyto-embryological investigations on the Polish representatives of the genus Crataegus L. I. Chromosome numbers; embryology of diploid and tetraploid species. *Acta Biologica Cracoviensia.* Xxviii: 107-122.
94. Patri G., Silano V. 1994. Plant preparations used as ingredients of cosmetic products, edition 1, Council of Europe, Belgium. 69.
95. Peschel W., Bohr C., Plescher A. 2008. Variability of total flavonoids in Crataegus-factor evaluation for the monitored production of industrial starting material. *Fitoterapia.* 79: 6-20.
96. Anonymous, 1986. Pharmacopoeia Hungarica, 7th ed. Budapest, Hungarian Pharmacopoeia Commission, Medicina Konyvkiado.

97. Anonymous, 1996. Pharmacopée française. Paris, Adrapharm.
98. Anonymous, 1997. Pharmacopoea helvetica, 8th ed. Berne, Département fédéral de l'intérieur.
99. Phipps J.B., O'Kennon R.J., Lance R.W. 2003. Hawthorns and Medlars. Timber Press, Inc., Portland, OR.
100. Rassi P., Hyvärinen E., Juslén A., Mannerkoski, I. (eds). 2010. The 2010 Red List of Finnish Species. pp. 685. Ympäristöministeriö and Suomen ympäristökeskus [Ministry of the Environment and Finnish Environment Institute], Helsinki.
101. Rehwald A., Meier B., Sticher O. 1994. Qualitative and quantitative reversed-phase highperformance liquid chromatography of flavonoids in Crataegus leaves and flowers.*J Chromatogr A*. 677: 25-33.
102. Rigelsky J.M., Sweet B.V. 2002. Hawthorn: pharmacology and therapeutic uses. *Am. J. Health Syst. Pharm*. 59: 417-422.
103. Saeedi G., Jeivad F., Goharbari M., Gheshlaghi G.H., Sabzevari O. 2018. More clinical studies are needed to establish the effects of hawthorn, especially in healthy humans. Drug Res (Stuttg).doi: 10.1055/a-0579-7532.
104. Shanthi S., Parasakthy K., Deepalakshmi P.D. 1994. Hypolipidaemic activity of tincture of Crataegus in rats. *Indian J Biochem Biophys*. 31(2): 143-146.
105. Shanthi S., Parasakthy K., Deepalakshmi P.D., Devaraj S.N. 1994. Hypolipidemic activity of tincture of Crataegus in rats. *Indian J Biochem Biophys*.31(2): 143-146.
106. Silva A.P., Silva M.L., Filomena M. 2000. Antioxidants in medicinal plant extracts. A Research Study of the antioxidant capacity of *Crataegus, Hamamelis and Hydrastis*. *Phytother Res*. 14: 612-616.
107. Sokol-Letowska A., Oszmianski J., Wojdylo A. 2007. Antioxidant activity of phenolic compounds of Hawthorn, pine, skullcap. *Food Chem*. 103(3): 853-859.
108. Song S.J., Li L.Z., Gao P.Y., Peng Y., Yang J.Y., Wu C.F 2011. Terpenoids and hexenes from the leaves of *Crataegus pinnatifida*. *Food Chem*. 129: 933-939.
109. Tabach R., Mattei R., Carlini E.L. 2009. Pharmacological evaluation of a phytotherapeutic product - CPV (dry extract of *Crataegus oxycantha* L., *Passiflora incarnata* L. and *Valeriana officinalis* L.) in laboratory animals. *Rev Bras Farmacogn*. 19: 255-260.
110. Tadic V.M., Dobric S., Markovic G.M., Ethordevic S.M., Arsic I.A., Menkovic N.R., Stevic T. 2008. Anti-inflammatory, gastroprotective, free-radical-scavenging, and antimicrobial activities of Hawthorn berries ethanol extract. *J Agric Food Chem*. 56 (17): 7700-7709.
111. Tassell M. C., Kingston R., Gilroy D., Lehane M., Furey A. 2010. Hawthorn (*Crataegus* spp.) in the treatment of cardiovascular disease. *Pharmacognosy Reviews*. 4(7): 32–41. http://doi.org/10.4103/0973-7847.65324.
112. Turner N.J. 1978. Food Plants of the British Columbia Indians. Part 2. Interior Peoples, British Columbia Provincial Musuem Handbook No. 36, Victoria.
113. Turner N.J., Thompson L.C., Thompson M.T., York A.Z. 1990. Thompson Ethnobotany: Knowledge and Usage of Plants by the Thompson Indians of British Columbia. The Royal British Columbia Museum, Victoria, BC.
114. Urbonaviciute A., Jakstas V., Kornygova O., Janulis V., Maruska A. 2006. Capillary electrophoretic analysis of flavonoids in single-styled hawthorn (*Crataegus monogyna* Jacq.) ethanolic extracts. *J. Chromatogr. A*.1112: 339-344.
115. Verma S.K., Jain V., Verma D., Khamesra R. 2007. *Crataegus oxycantha*-a cardioprotective herb. *Journal of herbal medicine and toxicology*. 1(1): 65-71.
116. Veveris M., Koch E., Chatterjee S.S. 2004. Crataegus special extract WS 1442 improves cardiac function and reduces infarct size in a rat model of prolonged coronary ischemia and reperfusion. *Life Sciences*. 74(15): 1945-1955.

117. Vijayan N.A., Thiruchenduran M., Devaraj S.N. 2012. Anti-inflammatory and anti-apoptotic effects of *Crataegus oxycantha* on isoproterenol-induced myocardial damage. *Mol Cell Biochem*. 367(1-2): 1-8.
118. Wagner H., Grevel J. 1982. Cardioactive drugs IV. *Planta Med.* 6: 98-101.
119. Walker A.F., Marakis G., Simpson E., Hope J.L., Robinson P.A., Hassanein M., Simpson H.C. 2006. Hypotensive effects of hawthorn for patients with diabetes taking prescription drugs: a randomised controlled trial," *Br J Gen Pract*. 56(527): 437-443.
120. Walker A.F., Marakis G., Morris A.P., Robinson P.A. 2002. Promising hypotensive effect of hawthorn extract: a randomized double-blind pilot study of mild, essential hypertension.*Phytother Res*. 16(1): 48-54.
121. Wang J., Xiong X., Feng B. 2013. Effect of Crataegus usage in cardiovascular disease prevention: An evidence-based approach.*Evid Based Complement Alternat Med.* 1- 16.
122. Weikl A., Assmus K.D., Neukum-Schmidt A., Schmitz J., Zapfe G., Noh H.S., Siegrist J. 1996. Crataegus Special Extract WS 1442. Assessment of objective effectiveness in patients with heart failure (NYHA II) *Fortschr Med.* 114(24): 291-296.
123. Anonymous, 1997. WHO, Guidelines for predicting dietary intake of pesticide residues, 2nd rev. ed. Geneva. (document WHO/FSF/FOS/97.7).
124. Anonymous, 1998. WHO, Quality control methods for medicinal plant materials. Geneva.
125. Anonymous, 2003. WHO monographs on medicinal plants newly independent states. 91-93.
126. Xu H., Xu H.E., Ryan D. 2009. A study of the comparative effects of hawthorn fruit compound and simvastatin on lowering blood lipid levels.*Am. J. Chin. Med.* 37(5): 903-908.
127. Xu H.E. 2004. The Progress of resource, environment and health in China, Scope China 3, Peking University Medical Press, Beijing, China.
128. Yang B., Liu P. 2012. Composition and health effects of phenolic compounds in hawthorn (*Crataegus* spp.) of different origins. *J. Sci. Food Agric.* 92: 1578–1590.
129. Zhang Z., Ho WKK., Huang Y., Anthony E.J., Lam L.W., Chen Z.Y. 2002, Hawthorn fruit is hypolipidemic in rabbits fed a high cholesterol diet. *J. of Nutrition*. 132(1): 5-10.
130. Zhang P.C., Xu S.X. 2001. Chemical constituents from the leaves of *Crataegus pinnatifida* Bge. var. *major* NEBr. *Yaoxue Xuebao*. 36: 754-757.
131. Zhang P.C., Zhou Y.J., Xu S.X. 2001a. Two novel flavonoid glycosides from *Crataegus pinnatifida* Bge. var. *major* NEBr. J. *Asian Nat. Prod. Res*. 3: 77-82.

10

Ginkgo biloba

Hindi : *Balkuwari*

Manu Sharma

***Ginkgo biloba* L.**

(A) Herbarium sheet, **(B)** Green leaves (source of crude drug), **(C)** Seeds
(D) Nursery raising in beds, **(E)** Saplings ready for transplantation
(F) *Ginkgo biloba* plantation, **(G)** Winter senescence

INTRODUCTION

The Ginkgo tree (*Ginkgo biloba*) is the only living representative of the order Ginkgoales, a group of gymnosperms of the family Ginkgoaceae. Ginkgo tree can live for as long as 1000 years and has been present for over 250 million years ("Living Fossil"). The name Ginkgo is believed to have come from the Chinese word *sankyo* or *yin-kuo*, meaning "hill apricot" or "silver apricot". The species name "*biloba*," meaning two lobes, refers to the unique two-lobed, fan-like leaves. Ginkgo has been used as both food and medicine for thousands of years. It is the only tree to have survived atomic blast in Hiroshima (Singh *et al.*, 2008).Seeds of Ginkgo were recorded in the early Chinese records on herbals as a medicine.

G. biloba extract (GBE), is the extracts of Ginkgo leaves, and is available as film-coated tablets, liquid formulations or inject-able in markets of Europe and America.

Standardized extract of the leaves of the *G. biloba*, labeled EGb761, is one of the popular herbal supplements. GBE has been used in traditional Chinese medicine for centuries for treating circulatory disorders, asthma, tinnitus, vertigo, and cognitive problems. Currently, GBE is among the most commonly taken phyto-medicines globally, and is approved in Germany for treatment of cerebral insufficiency (memory loss associated with Alzheimer, vascular or multi-infarct dementia) and conditions such as, tinnitus (ringing in the ears), vertigo and intermittent claudication that results in poor circulation (Mohanta, 2012).

A. BIOLOGICAL ASPECT

Systematic classification

Earlier, genus *Ginkgo* was placed in the class *Coniferosida* because it was thought to be more related to conifers than to other gymnosperm, even though the two groups appear to have evolved independently. Ginkgo is more like a conifer than a deciduous broadleaf tree, yet it has a unique position. Recent studies have suggested a much closer relationship to the cycads than to conifers. Research also suggests that green algae (Coccomyxa) live in association with Ginkgo tissues. This type of association is not known on any other tree though it occurs in the animal kingdom.

More recently, Ginkgo has been suggested as the sole living link between the lower and higher plants, *ie.* between ferns and conifers(Malischewsky, 2014).

Synonyms

Salisburia adiantifolia, Salisburia macrophylla and *Pterophylla* sp. (Isah, 2015)

Common names

Worldwide

Anny's dwarf, Autumn gold, Maidenhair tree, Silver apricot, Ginkgo, Kew tree, Ginkyo, Yinshing, Fossil tree (Puttalingamma, 2015).

In different Indian languages

Hindi	:	*Balkuwari.*
Sanskrit	:	*Vrikshamla, Kankuushta.*
Kashmiri	:	*Aziz* tree.
Marathi	:	*Acai*
Tamil	:	*Kumbakonam*

Diversity within genus

The prevalence of Genus *Ginkgo* goes back to Jurassic period, approximately 170 million years ago. Molecular evidence suggests that some stands in China are of natural origin. Genetic diversity and differentiation of Chinese *G.biloba* is higher than that of Korean and North American populations, which descended from China (Mohanta, 2012).It is widely believed that the recent revival of *G. biloba* was aided by Buddhist monks who venerated the tree and cultivated it in their temple grounds.

The distribution of various species of Ginkgo in the world is given in Table 1.

Table 1: Distribution of various species of Ginkgo in the world

1.	*Ginkgo biloba,*	China
2.	*Ginkgo yimaensis,*	China
3.	*Ginkgohuolihensis*	China
4.	*Ginkgo adiantoides*	Scotland
5.	*Ginkgo gardneri*	Scotland
6.	*Ginkgo apodes*	China
7.	*Ginkgo cranei*	North Dakota
8.	*Ginkgo digitata*	Greenland
9.	*Ginkgo dissecta*	Canada
10.	*Ginkgobaiera*	Canada
11.	*Ginkgo huttonii*	England
12.	*Ginkgo beckii*	North America
13.	*Ginkgo sibirica*	Oregon

Geographical distribution

Ginkgo is native to China and is scattered in broad leaved mixed-mesophytic forests (elevation up to 1,100 m) which are located on the border of the Yangtze River valley and on the hill country (Tang *et al.,* 2011).

This attractive tree has been widely planted as an ornamental in many parts of the world, and in temple gardens in Japan and China. Fossils records show that there were at least 16 genera of the family Ginkgoaceae contributing a significant part of the vegetation of the world. Now, it is represented by a single species *G. biloba* (Beek and Montoro, 2009).

For centuries, Gingko was believed to have gone extinct in the wild, till it was reported that reasonably good wild population exists in areas like Zhejiang, Guizhou, Sichuan and Guangxi provinces of China. The species is grown in at least two small areas in Zhejiang and Tianmushan in China, and cultivated in North America and Europe for over 200-300 years(Gong *et al.,* 2008).

Habitat and ecology

In a study in Tianmu Mountain of China (Hsieh and Duhai, 2004), the Ginkgo population comprised of about 244 individuals of mean diameter of 45 m at breast height (DBH) and a mean height of 18.4 m. Most of this population was growing on degraded sites, along stream beds, on rocky slopes and on the edges of soil beds. Ginkgo seedlings are rare on Tianmu Mountains and are found inpart of forests which were highly disturbed. Low genetic diversity was foundamong theseplants population. About 40% of the Tianmu Ginkgo had more than one trunk and DBH of more than 10 cm. Most of these secondary trunks have originated from lignotubers located at or just below ground level. Secondary trunk formation was most conspicuous in plantswhich were damaged bysevere stress. The sprouting ability of Ginkgo is an important trait that supports the species to perpetuate on eroded mountains and retain morphological stability (Hsieh and Duhai, 2004).

Taxonomic enumeration

Linnaeus (1771) was the first to describe *G. biloba*. Some other names like *S.adiantifolia* Smith and *Pterophyllus salisburiensis* Nelson are not considered to be its valid synonyms as per the rules of priority. The English name "Maidenhair tree" is based on the resemblance of its foliage to "Maidenhair fern" (*Adiantum* sp.). It is known by other names in Japan ("Ginkyo") and France ("L'arbre aux Quarante ecus" and "Noyer Du Japon"). Earlier,species of Ginkgo were included in family Taxaceae, which also included Podocarpaceae and Cepholotaxaceae. Later, in 1895, it was found that Ginkgo possesses multicoated spermatozoid.This formed the basis of Engler's classification of a particular family and class of Ginkgoopsida which dates back to the lower Jurassic (Chase, 2007). The group of these plants reached its highest position in upper Jurassic, but severely declined by Cretaceous period, reaching its critical status by the end of Cretaceous era. By modern times, only a single species has survived only in one continent, owing to natural or cultural reasons. Thus, *G. biloba* is the prime botanical example of a "living fossil". This term was earlier used by Darwin for the King Crab *(Limulus*ssp.*).* The systematic classification of *G. biloba* is as follows

Scientific classification

Division: Spermatophyta

1. **Sub-division**: Coniferophytina

   i) Class: Ginkgoopsida

   ii) Class: Pinopsida (with three sub-classes).

2. **Sub-division:** Cycadophytina

   i) Class: Lyginopteridopsida (Seed fern)

   ii) Class: Cycadopsida (Cycadaceae a. o.)

   iii) Class: Pinopsida (Only known in fossil form)

   iv) Class: Gnetopsida (Ephedra, Welwitschia, Gnetum)

3. **Sub-division:** Magnoliophytina (Angiospermae).

Binomial name: *G. biloba*

Phenological period

Ginkgo is a dioecious species, with separate male and female individuals occurring in a roughly 1:1 ratio. Ginkgo shows a long juvenile period, typically not reaching sexual maturity until 20 to 30 years of age. Male and female sex organs are produced on short shoots, in the axils of bud scales and leaves. The male catkins emerge before the leaves and fall off immediately after shedding their pollen. Pollination takes place by wind from early April in areas with mild winters to late May in areas with severe winters.

Ginkgo ovules are 2-3 mm long and are produced in pairs at the ends of 1-1.5 cm long stalks. When the ovule is receptive, it secretes a small droplet of mucilaginous fluid from its micropyle, which functions to capture airborne pollen. After this, the pollen enters into the pollen chamber. Once inside the ovule, the male gametophyte starts a four-month long development period that results in production of a pair of multi-flagellated spermatozoids. One of these fertilizes a waiting egg cell while the ovules are still on the tree. Depending on the date of pollination, this union can occur anytime between late-August to late-September(Echenard *et al.*, 2008).

B. REPRODUCTIVE BIOLOGY

Reproductive phenology

It is hypothesized that during late Palaeozoic, ginkgophytes (Ginkgophyta) were a distinctive lineage of gymnosperms. This group radiated in Middle or Late Triassic but got poorly diversified during the Permian. At present, *G. biloba* is the only species that represents Ginkgophytes. There are four groups of living gymnosperms (conifers, cycads, gnetophytes, and *Ginkgo*), none of which is directly related to each other. However, all lack flowers and fruits of angiosperms. The name gymnosperm combines the Greek root *gymnos*, or "naked," with sperma, or "seed" *i.e.* they are naked seeded plants (Tredici, 2007).

G. biloba being dioecious, possess male and female organs on different trees, and have distinct sex chromosomes XX and XY. Male leaves are divided into two different lobes, while female leaves have no differentiation. The fleshy outer coverings of the seeds of female Ginkgo plants exude the foul smell of rancid butter caused by butyric and isobutyric acids. Due to this, male plants, vegetatively propagated from shoots, are preferred for cultivation. Four distinct ontogenic stages start with germination and end by developing into fully differentiated leaves(Alessandro *et al.,* 2015). Male and female sex organs are produced on short shoots, in the axils of bud scales and leaves. The male catkins emerge before the leaves and fall off immediately after shedding their pollen.To study the complete reproductive phenology of the plant, it is important to know about the female and male part of the plants (Xie *et al.*, 2013).

Female plant

The female tree is responsible for generating seeds when the tree reaches maturity in 20-40 years. Seeds are packed in pairs inside a fleshy outer covering and are also referred to as fruits. Female trees can be easily distinguished frommale trees because when their fruits fall, they produce butyric acid [$CH_3(CH_2)_2COOH$] on decay. This outer covering changes colour at different phases: it is green in early spring, then changes to yellow and finally into brownish orange colour at the time of its fall (Bauer, 2013).

Male plant

Male plants of *G. biloba* are much less messy or smelly, and are preferred for planting compared to female plants which produce foul smelling seeds. On shoots of male plants, pollen is produced in small balls or cones. The pollen is then carried to the female plant by wind, where its goal is to land on a pollination droplet. The sperm are carried into the plant by this droplet, which then swims to the ovule and fertilize it. Male and female plants usually develop their reproductive organs at the same time. Female *Ginkgo* trees typically produce seeds first. About four months later, pollens are produced by the male plant and fertilize the female plant.

Life cycle of *G. biloba*

Reproductive cycle of plant from pollination to production of the seeds takes about 14 months. Pollination and gametophytic phase occurs within first year, and in the following spring the embryo develops. Pollen is produced by strobilus, which is a loose, pendulous, catkin like structure with a main axis. Numerous haploid microspores are produced following meiosis in the cells of microsporangia (present at the tip of appendages attached on strobilus). Cell division occurs in the microspores and results in formation of five-celled pollen grain. Arising on the axils of bud scales and foliage leaves of the spur branch are the ovuliferous structures, which may consist of 2-3 erect ovules. A seed coat and integument is present in the ovule surrounding the nucleus. Within nucellus of the ovule, meiosis takes place to give rise to four haploid megaspore cells. At this stage of development, pollen is released from the microsporangi as of the male tree. Wind carries pollen to the micropyle at the tip of integument. The pollen grain is then retracted into the pollen chamber of the nucellus, which results in development of multi-branched pollen tubes (Lin *et al.,* 2012).

A single megaspore undergoes three nuclear divisions upon enlargement by meiosis in the ovule. A wall is synthesized when about 8,000 haploid nuclei are produced resulting in formation of a cellular structure which then moves towards the micropylar end of the ovule. One egg is present in each ovule. Above the fertilization chamber, the male gametophyte is suspended in a cavity at the basal end. This male gametophyte divides into multiflagellated sperm. Then, within the fertilization chamber, contents of the pollen tube and sperm are released. Only one sperm,out of the approximately 1000 sperms released, fuses with the egg's nucleus in the archegonium (Wu *et al.* 2013). When male and female haploid (n) reproductive cells (seeds and pollen) come together, fertilization occurs. Then diploid (2n) zygote begins to grow shortly after fertilization. As many cells divide (by mitosis), the tree reaches maturity and produces its own haploid reproductive cells by the process of meiosis.

G. biloba is one of the most notable zoidogamous gymnosperms. Different studies on sexual reproduction in *G. biloba* have shown that archegonial chamber is full of liquid when spermatozoid is released from pollen tubes, through which the spermatozoids can swim. Pollen tubes and degenerated nucellar cells are responsible for generating this liquid (Galia *et al.*, 2012).

The life cycle of *G. biloba* is described as below.

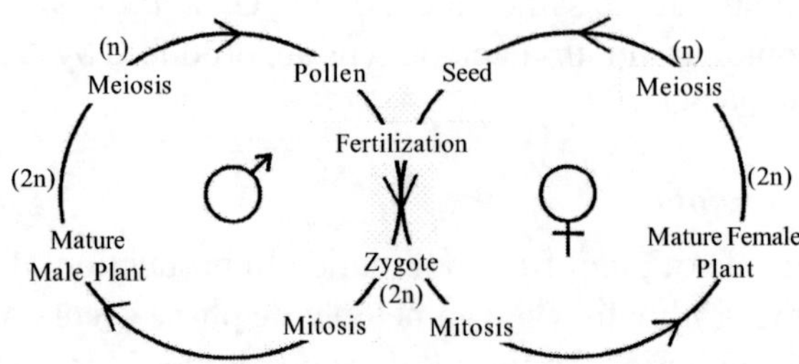

Reproductive life cycle of *Gingko biloba*.

The mature plants produce seeds or pollen after gaining maturity. Seeds are fertilized by pollen and *via* mitosis new tree starts to grow. The tree grows to maturity again and thus the cycle continuous.

Pollination and pollinators

Gymnosperms are mostly wind pollinated or anemophilous. Ginkgo, one of the most important among the oldest lineages of gymnosperms, is also wind pollinated and shows highly efficient and successful pollination. Earlier, wind pollination was considered as random process, but nowdays it is recognised as highly coordinated and synchronous. Wind pollination in case of Ginkgo occurs from the end of April to early May. A successful pollination banks onmechanisms that increase chances of pollination. Pollination drop (PD), which is secreted and extruded from the micropyle for capturing pollen and transporting it into the ovule, is one such mechanism (Jin, 2012 b). Pollination drops are produced by gymnosperms to capture airborne pollen grains and to trigger their germination. Thus, pollination drops act as landing site for pollens that scavenge them from air and transport the pollen grains to nuclear surface (Leslie, 2010).

G. biloba has non-saccate pollen and a more or less erect ovule during pollination. Its nuclear cells may undergo programmed cell death to form a pollen chamber where pollen grains germinate and develop pollen-tubes. Pollen chamber in the ovule is a structure with specific function. Between pollination and fertilization, there is a span of four months in which pollen grains undergo functional activities. Through successive degeneration of nucellar cells, a pollen chamber is developed in micropylar region.

Role of tentpole in reproductive process

In the reproductive process, a number of characteristics are important. These include structure of archegonium, pollen chamber, tentpole and haustorial pollen tubes. Out of these, tentpole is the most important feature in *G. biloba* reproduction. Tentpole is commonly situated between two or more archegonia of the female gametophyte and protrudes toward archegonial chamber. It has been found that tentpole helps in reproduction process by functioning as a secretory structurethat produces a liquid favourable for spermatozoid to swim *via* its flagella towards the archegonium just before fertilization.

In mid-May, when the cellular stage of female gametophyte is reached, differentiation of achregonia at micropylar end occurs. Some cells between the two archegonia differentiate to form tentpole. Approximately, at 55 day after pollination, early tentpole begins to protrude from central part of micropylar end of female gametophyte, forming a flat column like structure. Generally one female gametophyte has only one well developed tentpole (Jin, 2012a).

C. MEDICINALASPECTS

Plant material used as crude drug

Specific medicinal properties have been attributed to different parts of the plant, such as leaves, fruits, seeds etc. However, leaves are the most preferred source of crude drug. General appearanceof different plant-parts is as follows:

1. Leaves

- Leaf arrangement: alternate
- Leaf type: simple
- Leaf margin: lobed
- Leaf shape: fan-shaped
- Leaf venation: parallel; palmate
- Leaf type and persistence: deciduous
- Leaf blade length: 2 to 4 inches
- Leaf color: green
- Fall color: yellow
- Fall characteristic: showy

2. Flower

- Flower color: green
- Flower characteristics: pleasant fragrance; inconspicuous and not showy; spring flowering

3. Fruit

- Fruit shape: oval; round
- Fruit length: 1 to 3 inches
- Fruit covering: fleshy
- Fruit color: green; yellow
- Fruit characteristics: inconspicuous and not showy; fruit, twigs, or foliage

4. Trunk and Branches

- Trunk/bark/branches: showy trunk; branches droop as the tree grows, and require pruning; no thorns
- Pruning requirement: needs light pruning to develop a strong structure
- Breakage: resistant
- Twig color: brown; gray
- Twig thickness: medium; thick

5. Tree

- Height: 50 to 75 feet
- Spread: 50 to 60 feet
- Crown uniformity: irregular outline or silhouette
- Crown shape: round; pyramidal
- Crown density: open
- Growth rate: slow
- Texture: medium (Zhao *et al.,* 2010; Little, 2013)

Ginkgo nuts

Ginkgo nuts, present on female plants, are found within the seeds. Female plants bear paired ovules, which on fertilization become yellow, and Ginkgo nuts are present in their plum- like seeds, which are surrounded by fleshy outer covering. According to Ch'ng (2013) parameters of Ginkgo nuts are:

- Color: Silver
- Dimensions: 22.31-28.20 mm in length, 14.15-17.49 mm in width and 12.65-15.83 mm in diameter.
- Mass and volume: 2.583 g and 2.480 mm^3 respectively

Microscopic characteristics

Microscopic characteristics of different tissues from *G. biloba* have been discussed in detail by Wang *et al.* (2014a).

Microscopic structure of root

The root cross section reveals three major parts: *i.e.* epidermis cells, cortex and vascular cylinder. The epidermal cells are relatively closely packed and are bigger than other cells. The cortex parenchyma cells arealso somewhat larger and arranged with six laps.Endodermis cells of cortex areirregular in shape, narrow and slender in appearance. Vascular cylinder of primary xylem,composed of oval cells, has two xylem bundles. The primary phloem cells are much smaller and areof irregular rectangular shape.

Microscopic structure of leaf

Leaf epidermis and hypodermis are composed of monolayer cells of rectangular or square shapes, with mesophyll tissue well differentiated into spongy parenchyma and palisade tissues. The palisade tissue, composed of one or two layer of cells, has relatively large intercellular gap. The chloroplasts are present in spongy parenchyma as well as palisade tissue and are loosely arranged.

Microscopic structure of anther

About 60 anther cells of large dimension have been observed. The cells at the wall are thinner and smaller than those at other places.

Microscopic structure of stem

The pith is about half the diameter of cross section of stem. The number of pith ray is 30. That is apparently helpful for lateral transportation and storage of nutritional components. The secondary xylem cells are closely placed and arranged into 8 lines. The cross section of stem also shows that secondary phloem cells are relatively large and loosely arranged, while the epithelial cells are relatively small and closely arranged.

Microscopic structure of embryo

The longitudinal section of *G. biloba* embryo shows the presence of radicle, hypocotyl, cotyledon, and embryo in torpedo shaped arrangement. The cross section of embryo shows cells to be relatively small and closely arranged. The cells become larger and looser as one moves from inside to outer side.

Microscopic structureof callus

The cross section of callus displays the intercellular gap to be loosely arranged with the size of cells to be more consistent. Most cells appear elliptical.

Microscopic structure of ovule

The section of ovule reveals that its macrosporophyll is simplified as ovule collar and consists of one or two ovules. Ovule covering may vary in thickness, and embryo is placed in embryo sac.

Powder of *G. biloba*

The powder of *G. biloba* is greenish brown in color. Clusters of calcium oxalate have been found in abundance when observed under polarized light microscope, which are 5-15 to 100-120 µm in diameter, bright polychrome. Upper epidermis consists of sub-rectangular cells with wavy anticlinal walls. In lower epidermis sub-rectangular or sub-rounded cells with slightly wavy anticlinal walls, anisocytic stomata, with guard cells sometimes sunken with respect to adjacent epidermal cells are present. In parenchymatous cells,yellowish oil droplets are also found. Fibers present in the powder are 10-25 µm in diameter and mostly broken, long fusiform. Stone cells usually found in petioles, are sub-rectangular, sub-rounded or sub-square, 20-45 µm in diameter, bright yellowish-white in color. Tracheids are also present which are mainly bordered-pitted and reticulate, 7-30 µm in diameter and white in color.

MAJOR CHEMICAL CONSTITUENT

Extract from *G. biloba*, like in case of manymedicinal plants, is generally believed to have synergistic effects attributed to many of its constituents. The crude drug of Ginkgo is obtained from dried green leaves of the tree. It comprises flavonoids (including meletin, kaempferol and isorhamnetin) and lactones (including ginkgolides and bilobalide). GBE can remove free radicals, protect the endothelial cells of blood vessels, block platelet activating factors, and improve brain circulation. GBE has been widely used in the treatment of dementia, cognitive impairment, peripheral nerve problems, and vascular tinnitus. However, clinical studies about the efficacy of GBE in the treatment of dementia have been inconclusive. Some studies report beneficial effects on cognition and functioning, while others do not (Mullaicharam, 2013). Two main pharmacologically active groups of compounds are present in the Ginkgo leaf extract. They are the flavonoids and the terpenoids. Flavonoids, also called phenylbenzopyrones or phenylchromones, are a group of low molecular weight substances that are wide spread in the plant kingdom. Flavonoids present in the Ginkgo leaf extract are flavones, flavonols, tannins, biflavones (amentoflavone, bilobetol, 5-methoxybilobetol, ginkgetin, isoginkgetin and sciadopitysin), and associated glycosides of quercitin and kaempferol attached to 3-rhamnosides, 3-rutinosides, or *p*-coumaric esters. These compounds are known to act mainly as antioxidants/free radical scavengers, enzyme inhibitors, and cation chelators (Chan, 2007). In general, the bioavailability of flavonoids is relatively low due to its limited absorption and rapid elimination. Flavonoids in the glycosidic form are poorly absorbed in the intestine and can be absorbed directly only in the aglycone form. Once absorbed, flavonoids reach the liver where these are metabolized to conjugated derivatives. It is fairly well known that the biological activities of flavonoid metabolitesare not always the same as those of the parent compound. There are no conclusive studies to define the dose of GBE needed to achieve beneficial effects, although the recommended dose of standardized extract, GBE 761, is 40 to 60 mg, 3 to 4 times daily based on clinical trials (Bai, 2015). The different class of compounds, present in *G. biloba*, are described in Table 2.

Table 2: Classification of different classes of compounds present in *Gingko biloba*

Class	Major chemical constituents
Flavonoids	kaempferol, quercetin, isorhamnetin, rutin, luteolin, delphidenon, myricetin
Terpenoids	Diterpenes: ginkgolides A, B, C, J (M is found in the root) Sesquiterpene: bilobalide; Triterpenes: sterols
Biflavonoids	Sciadopitysin, ginkgetin, isoginkgetin, amentoflavone, bilobetin, 5'-methoxybilobetin
Organic acids	Benzoic acid derivatives (ginkgolic acid), N-containing acids
Polyphenols others	Di-trans-poly-cis-octadecaprenol, waxes, steroids, 2-hexenal, cardanols, sugars,catechins, proanthocyanidins, phenols, aliphatic acids, rhamnose

SUBSTITUTES AND ADULTERANTS

Ginkgo leaves contain more terpene lactones than flavonol glycosides. In order to enhance flavonol glycosides, adulteration is carried out with plants such as *Sophora japonica*, which contain flavonol glycosides (Demirezer, 2014).

GENERAL IDENTITY TESTS/ PURITY TESTS

Identification tests for *G. biloba* in the United States Pharmacopeia (USP) are based on high-performance thin-layer chromatography (HPTLC). Besides HPTLC, macroscopic and microscopic examinations are made to confirm botanical identity and composition. The solvent systems used, includes ethyl acetate, anhydrous formic acid, glacial acetic acid, and water (100:11:11:26). The spots are detected by using spraying reagents 5 mg/mL of 2-aminoethyl diphenylborinate in methanol and 50 mg/mL of polyethylene glycol 400 in alcohol. Moreover, identification for terpenelactones present in *G. biloba* are detected by using solvent system *i.e.* toluene, ethyl acetate, acetone and methanol (20:10:10:1.2) and acetic anhydride isused as spraying reagent.For the identification of flavonol glycosides, methanol, water, and phosphoric acid (100:100:1) is used as solvent system.The USP requires that a dry extract of *G. biloba* or dried leaves should be characterized by containing not less than 22% and not more than 27% of flavonoids calculated as flavonol glycosides via high-performance liquid chromatography (HPLC). Quantitative determination of Ginkgo extract have also been done by using combination methods including HPLC, gas chromatography (GC), nuclear magnetic resonance (NMR), and mass spectrometry (MS). Recent methods used for the quantization of the major components of Ginkgo include reversed-phase HPLC/ESI-MS (high-performance liquid chromatography/electrospray ionization-mass spectrometry), HPLC/MS-MS, GC-MS, and ^{1}H-NMR(Harnly, 2012).

Medicinal usage

General uses

1. Cerebral insufficiency: memory deficit, poor concentration,depression, and headache resulting from dementia syndromes (attention and memory loss that occur withAlzheimer's disease and multi-infarct dementia.
2. Vertigo and tinnitus (ringing in the ear) of vascular origin.
3. In Peripheral Arterial Occlusive Disease in Stage II according to Fontaine (intermittent claudication in a regimen of physical therapeutic measures, in particular walking exercise).
4. Sexual dysfunction associated with use of antidepressants, particularly selective serotonin reuptake inhibitors (SSRIs) use.
5. Control of acute altitude sickness symptoms and vascular reactivity to cold exposure.
6. Protective action in hypoxia.
7. Acute cochlear deafness (Mahadevan, 2008).

Ethnobotanical uses and information

Traditional use of *G. biloba* in China dates back to around 15th century to the middle of last century in some European countries such as Germany.Nuts of the Ginkgo are used in soups while fruit are cooked or fermented and used as delicacy for wedding parties. Nuts are also used to treat bladder inflammation and pulmonary disorders such as cough and asthma (Sadaf, 2016). It is believed that Ginkgo tree is resistant to insect, pests, fungal, viral and bacterial diseases. It is also believed that the species is resistant to environmental pollution and that prompted its use as a street tree. Studies have suggested that the tree has no trouble adapting to greenhouse effect. It is also grown for shade (Sierpina, 2003).

D. CHEMISTRY ASPECTS

Dating back to 1505 AD, *G. biloba* is the most valuable and ancient plant among medicinally important plants. As the discovery of pharmacologically important plants has intensifiedin last 40-50 years, *G. biloba* also attracted keen interest of researchers due to its important medicinal properties. Ginkgo leaf-extract EGb761 was developed in 1965 in Germany, but used commercially no earlier than in 1974. EGb761 consists of 24% flavonoids and 6% terpene lactones as major components (Drieu *et al.,* 2000).

Chemical composition

Various classes of secondary metabolites including terpenoids, polyphenols, allyl phenols, organic acids, carbohydrates, fatty acids, lipids, inorganic salts and amino acids have been isolated from *G. biloba*. The active principles are compounds with different chemical structures and on the basis of chemical structure these bioactive compounds comprise of two main groups, *i.e.* Flavonoids and Terpene lactones.

Flavonoids

These flavonoids are also known as phenylbenzopyrines or phenylchromones and are structurally related compounds lowin molecular weight. Presence of two aromatic rings and a heterocyclic ring containing one oxygen atom is the characteristic feature of this group. Flavonol glycosides, bioflavonoids, biflavones, proanthrocyanidins and isoflavonoids are major representative of this group. About twenty flavonol glycosides are generally present in the extract of *G. biloba*. In these flavonol glycosides, aglycone part consists of one or more hydroxyl groups, which are bound at 3 or 7 position with a carbohydrate moiety *i.e.* lycone part. Phenolic aglycones like quercetin, kaempferol or isorhametin are present in very low concentration, when alone. However, a large number of bioflavonoids are present in the form of dimmers, connected *via* C-O-C or C-C bond. These biflavone includes amentoflavone, 5-methoxybilobetol, bilobetol, isoginkgetin, ginkgetin, and sciadopitysin.

These compounds are poorly absorbed by intestine and are rapidly eliminated in their glycosylated form resulting in low bioavailability. Their absorption takes place only when presented as aglycones, and are metabolised into their conjugative derivatives in liver (Yoshitake *et al.,* 2010). The major compounds isolated in this class are described in Figure 2.

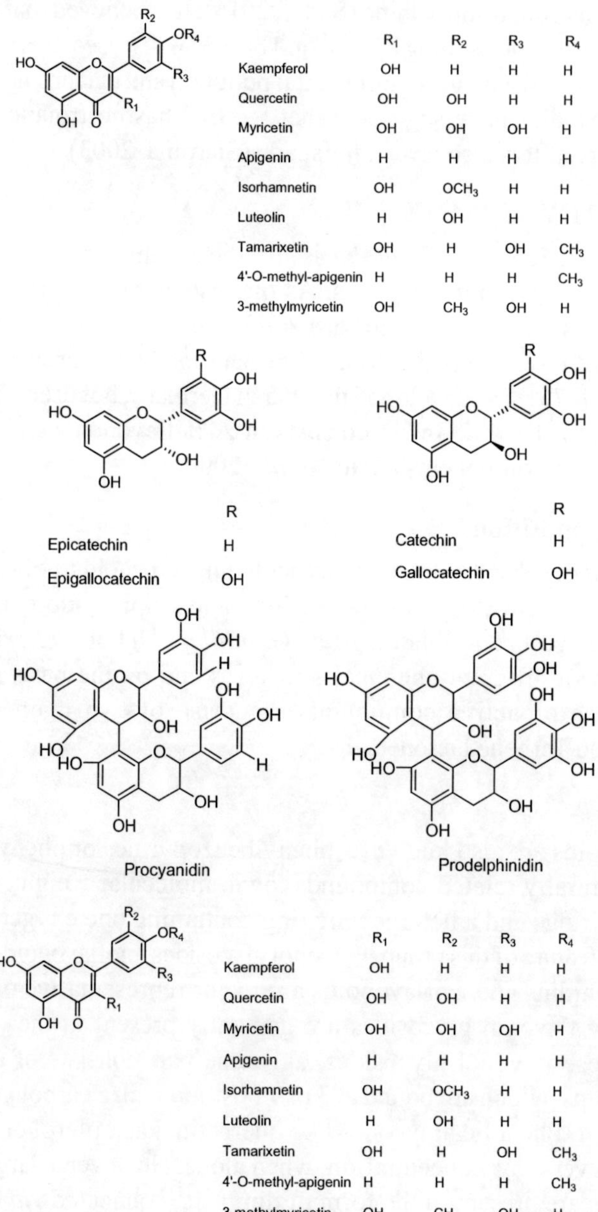

	R_1	R_2	R_3	R_4
Kaempferol	OH	H	H	H
Quercetin	OH	OH	H	H
Myricetin	OH	OH	OH	H
Apigenin	H	H	H	H
Isorhamnetin	OH	OCH_3	H	H
Luteolin	H	OH	H	H
Tamarixetin	OH	H	OH	CH_3
4'-O-methyl-apigenin	H	H	H	CH_3
3-methylmyricetin	OH	CH_3	OH	H

	R
Epicatechin	H
Epigallocatechin	OH

	R
Catechin	H
Gallocatechin	OH

	R_1	R_2	R_3	R_4
Kaempferol	OH	H	H	H
Quercetin	OH	OH	H	H
Myricetin	OH	OH	OH	H
Apigenin	H	H	H	H
Isorhamnetin	OH	OCH_3	H	H
Luteolin	H	OH	H	H
Tamarixetin	OH	H	OH	CH_3
4'-O-methyl-apigenin	H	H	H	CH_3
3-methylmyricetin	OH	CH_3	OH	H

Fig. 2: Major flavonoids, anthocyanidin and biflavones isolated from *Ginkgo biloba*

Terpene lactones: They are present as 20-carbon diterpene lactone derivatives (known as ginkgolides) and 15-carbon sesquiterpene (known as bilobalide). Presence of a tetra-butyl group in their structure is the characteristic feature of diterpenes. A rigid carbon skeleton consisting of 6 fused 5-membered carbocyclic forms a ginkgolide, *i.e.* a spiro[4.4]nonane carbocyclin ring, three lactones and a tetrahydrofuran ring. Diterpenes ginkgolides A, B, C, J, M, K and L were isolated from root bark. Two new diterpenoids were isolated from leaves *i.e.* ginkgolides P and Q.

On the other hand, only 5-rings are present in biobalide, resulting in the formation of more flexible structure, that closely relate to ginkgolides. They are different from ginkgolides as tetrahydrofuran ring is absent. Bilobide and bilobol are the two main constituents of this group (Ding *et al.,* 2004). Chemical structures of various terpenoids isolated from *G. biloba* are shown in Figure 3.

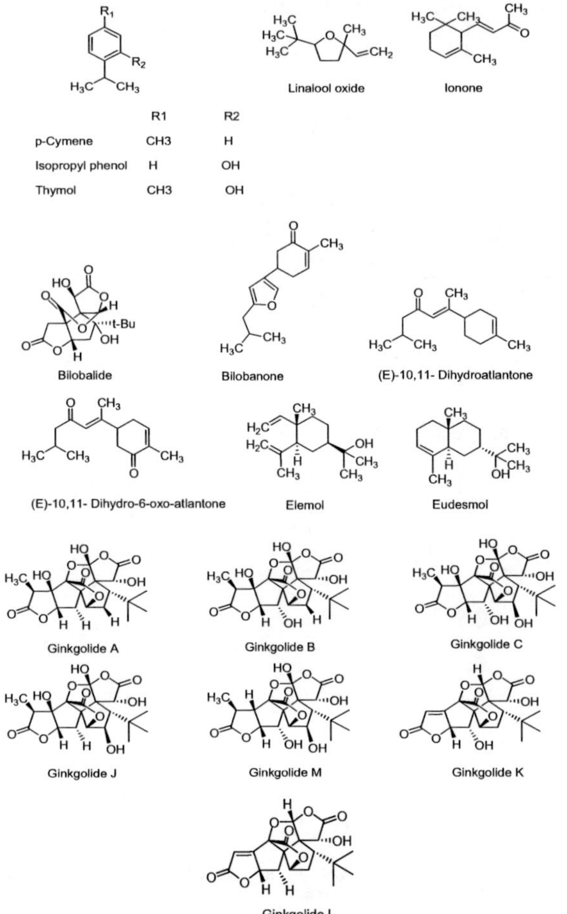

Fig. 3: Chemical structures of terpenoids isolated from *Ginkgo biloba*

Other than flavonoids and terpene lactones, *G.biloba* also contains steroids, phytosterols (Figure 4), cartenoids (Figure 5), polyprenols (Figure 6), cardanols and cardols (Figure 7), anacardic acid and resorcylic acid (Figure 8) and organic acids (Figure 9) (Beek and Montoro, 2009; Yao, 2013; Nuhu, 2014).

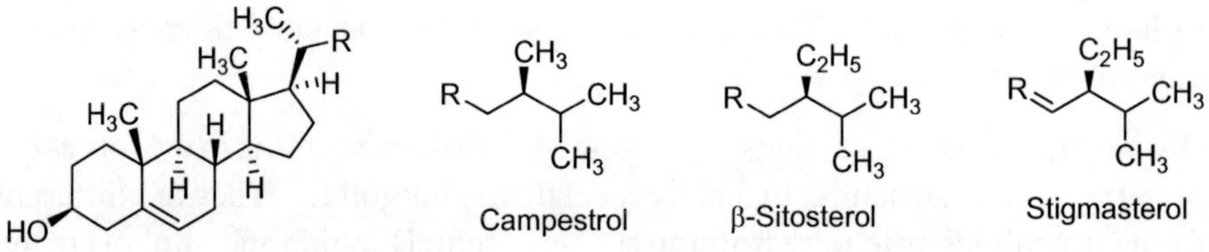

Campestrol β-Sitosterol Stigmasterol

Fig. 4: Chemical structures of steroids and phytosterols isolated from *Ginkgo biloba*

α-Carotene γ-Carotene Lutein Zeaxanthin

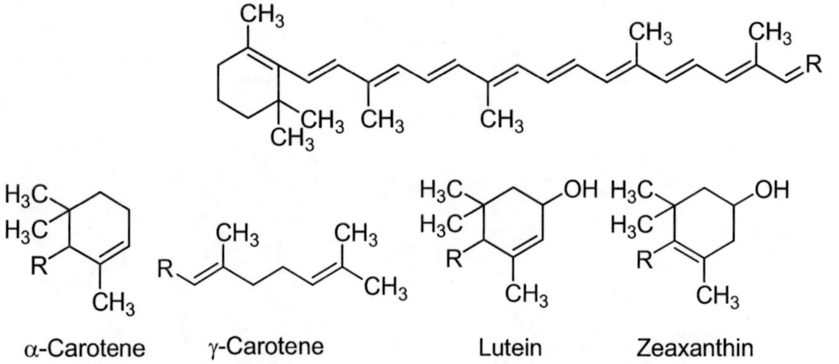

Fig. 5: Chemical structures of carotenoids isolated from *Ginkgo biloba*

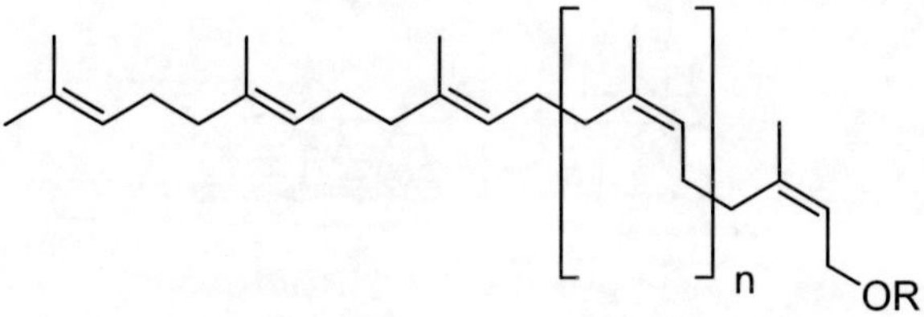

	R	
Ginkgo polyprenols	H	n = 10-20
Ginkgo polyprenols acetates	Ac	n = 10-20 (n =14 is most abundant)

Fig. 6: Chemical structures of polyprenols isolated from *Ginkgo biloba*

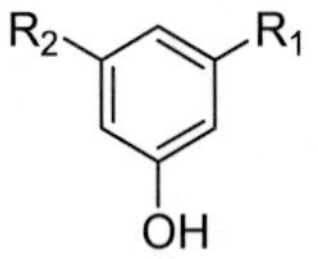

R2 / R1 / OH (structure)

	R_1	R_2
3-Tridecylphenol	$(CH_2)_{12}CH_3$	H
3-Tetradecylphenol	$(CH_2)_{13}CH_3$	H
3-Pentadecylphenol	$(CH_2)_{14}CH_3$	H
3-Heptadecylphenol	$(CH_2)_{16}CH_3$	H
Ginkgol	$(CH_2)_7(CH)_2(CH_2)_5CH_3$	H
5-Tridecylresorcinol	$(CH_2)_{12}CH_3$	OH
5-Tetradecylresorcinol	$(CH_2)_{13}CH_3$	OH
5-Pentadecylresorcinol	$(CH_2)_{14}CH_3$	OH
5-Heptadecylresorcinol	$(CH_2)_{16}CH_3$	OH

Fig. 7: Chemical structures of cardanols and cardolsisolated from *Gingko biloba*

R2 / R1 / COOH / OH (structure)

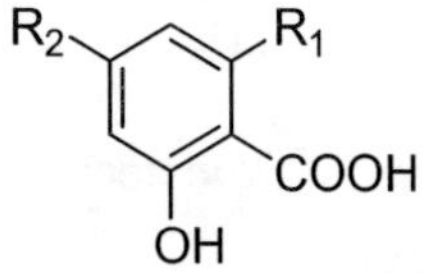

	R_1	R_2
6-Tridecylsalicylic acid	$(CH_2)_{12}CH_3$	H
6-Tetradecylsalicylic acid	$(CH_2)_{13}CH_3$	H
6-Pentadecylsalicylic acid	$(CH_2)_{14}CH_3$	H
6-[8-Pentadecenyl]-salicylic acid	$(CH_2)_7(CH)_2(CH_2)_5CH_3$	H
6-Hexadecylsalicylic acid	$(CH_2)_{16}CH_3$	H
6-[9,12-Heptadecadienyl]-salicylic acid	$(CH_2)_8(CH)_2CH_2(CH)_2(CH_2)_3CH_3$	H
6-[8-Heptadecenyl]-salicylic acid	$(CH_2)_7(CH)_2(CH_2)_7CH_3$	H
6-[8-Pentadecenyl]-resorcylic acid	$(CH_2)_7(CH)_2(CH_2)_5CH_3$	OH
6-Tridecylresorcylic acid	$(CH_2)_{16}CH_3$	OH

Fig. 8: Chemical structures of anacardic acid and resorcylic acid isolated from *Gingko biloba*

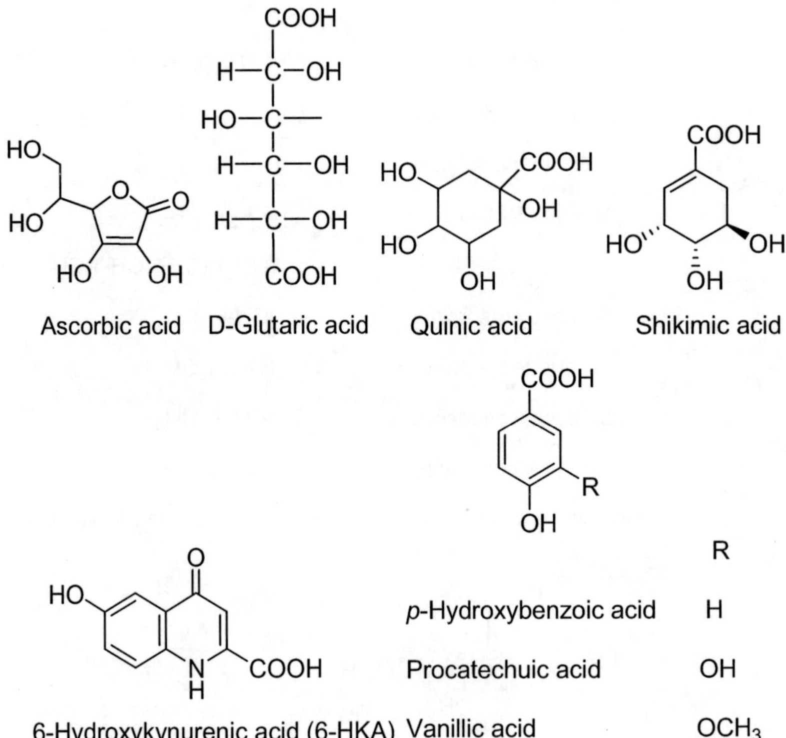

Fig. 9: Chemical structures of organic acids isolated from *Ginkgo biloba*

G. biloba also containssome other compounds like sucrose, pinitol, sequoyitol, polysaccharides. *G. biloba* has some long chain hydrocarbons and lipidslike 1-hexacosanol, 2-hexenal, *n*-nonacosane, *n*-non- acosan-10-*ol* (ginnol), *n*-non acosan-10-one (ginnon), *n*-octacosan-1-*ol*, linolenic acid, linoleic acid, 14-methylhexadecanoic acid, 5,9,12-octadecatrienoic acid, oleic acid, palmitic acid, eicosanedicarboxylic acid, phellonic acid. *G. biloba* also has some inorganic salts/ complexes like selenium, copper, zinc, chromium, iron, manganese, calcium oxalate and phosphoric acid.

Apart from these compounds some miscellaneous organic compounds are also present in *G. biloba* (Figure 10).

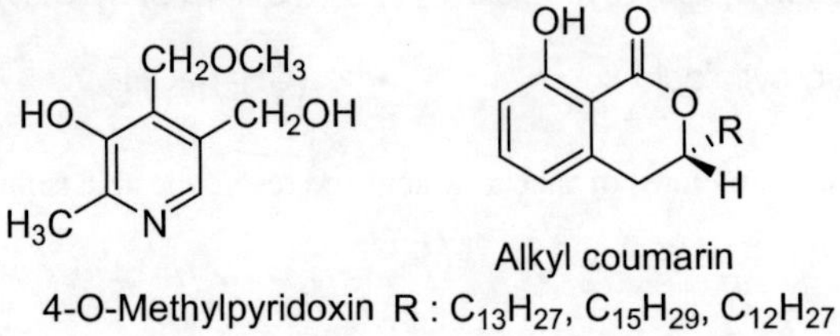

Alkyl coumarin

4-O-Methylpyridoxin R : $C_{13}H_{27}$, $C_{15}H_{29}$, $C_{12}H_{27}$

Fig. 10: Chemical structures of miscellaneous compounds isolated from *Ginkgo biloba*

Chemical markers

For quality assessment of *G. biloba,* mainly ten compounds are selected as chemical markers. These markers are selected from two classes of phyto-constituents occurring in the plant leaves *i.e.* terpene trilactones and the ubiquitous flavonoids.

The first group *i.e.* terpene trilactones includes platelet-activating factor inhibitors such as ginkgolides A, B, C and J as well as neuroprotective agent bilobalide. A wide variety of flavonols and flavonol glycosides are present in second group, including quercetin, kaempferol, isorhamnetin, isoquercetin (quercetin-3-*O*-*β*-D-glucopyranoside) and rutin [quercetin-3-*O*-α-L-rhamnopyranosyl-(1-6)-*β*-D-glucopyranoside]. These chemical markers are examined by ^{1}HNMR in DMSO-d_6(Napolitano, 2012).

E. PHARMACOLOGICAL ASPECTS

BIOACTIVITY

Neurological Activity

Gavrilova(2014) studied the effects of GBE 761® on neuropsychiatric symptoms (NPS) and cognition in patients with mild cognitive impairment (MCI). In this study, 240 mg GBE 761 daily, or a placebo, were given for a period of 24 weeks to one hundred and sixty patients with MCI. In this sample,at least 6 of the 12-items scored neuropsychiatric symptoms. The state sub-score of the State-Trait Anxiety Inventory and the Geriatric Depression Scale were used to access the effects on NPS. The result showed that GBE 761 improved NPS and cognitive performance in patients with MCI. The drug was found safe and well tolerated (Gavrilova, 2014).

Zhang (2012) discussed the clinical effects of Western medicine therapy assisted by *G. biloba* tablet (GBT) on patients with vascular cognitive impairment of none dementia (VCIND). A total of 80 patients with VCIND were divided into two groups randomly: Conventional treatment group (control group) and combined treatment group. Conventional treatment group was given conventional treatment with anti-platelet aggregation. In this group, 75mg aspirin was given three times a day for 3months. While in combined treatment group, 19.2mg GBT was given three times a day for 3months together with conventional treatment (anti-platelet aggregation drugs). Montreal cognitive assessment (MoCA) and transcranial Doppler (TCD) were used to observe changes of cognitive ability and cerebral blood flow in VCIND patients before and after treatment in both groups. Then the clinical data were analyzed so as to compare the efficacy in two groups. After 3 month-treatment in combined treatment group, the scores of executive ability, attention, abstract, delayed memory, orientation in the MoCA were significantly increased compared with those before treatment and those in control group after treatment. Besides, blood flow velocity of anterior cerebral artery increased significantly than that before treatment and that in control group after treatment.

Christoph *et al.* (2015) showed that GBE 761, in combination with cholinesterase inhibitors,to be effective in the treatment ofpatients.In addition to neuroprotective effects, GBE 761 wasreported to elevate certain neurotransmitters such as dopamine, noradrenaline, and acetylcholine in the brain. The study further analysed the effect of GBE 761, donepezil and the combination of both drugs on the central cholinergic system in aged rats. GBE 761 (100mg/kg/day), donepezil (1.5 mg/kg/day), the combination of both drugs were given to 24 month old rats for 14 days. To calculate the extracellular concentrations of acetylcholine (ACh), choline, glucose and lactatemicro-dialysis were carried out in hippocampus of rat. To measure the activities of acetylcholinesterase (AChE), choline acetyltransferase (ChAT) and high affinity choline uptake (HACU), brain homogenates were prepared. The result showed that GBE 761 alone had no effect, while donepezil and the combination of donepezil and GBE 761 increased basal ACh levels by 2-3folds. Significant reductions of AChE and HACU were also recorded. Moreover, treatment with GBE 761 decreased extracellular choline release and showed slight increase in the ChAT activity (2010). Shan-Tan *et al.*(2015) studied neurological activity *i.e.* dementia and showed that GBE 761 at 240 mg/day was able to stabilize neuropsychiatric symptoms after 22–26 weeks of treatment.

Klin *et al.* (2009) evaluated the effect of standardized extract of G. *biloba* leaves (EGb 761) on learning, memory and exploratory behavior. Water maze and hole-board tests were used to estimate the learning, memory and exploratory behavior parameters. GBE 761 was given at different doses (50, 100 and 150 mg/kg b.w. per day to rats) for three months. After completion of the behavioral experiment, neurotransmitters concentration was estimated in selected brain regions. The result suggested that the increased level of 5-hydroxytryptamine (5-HT) in the hippocampus and 5-HIAA (5-HT metabolite) in the prefrontal cortex correlated with the retention of spatial memory. The result also suggested that long-term administration of *G. biloba* extract can improve spatial memory.

In another study, Mdzinarishvili *et al.* (2012) showed anti-edema and anti-ischemic effects on administration of EGb 761 in brain. Neuro-protective effects of EGb761 in experimental stroke while monitoring of brain metabolism were estimated by micro dialysis.Oxygen glucose deprivation in brain slices was used *in vitro* and middle cerebral artery occlusion (MCAO) *in vivo* to induce ischemia in mouse brain. The result displayed that *in vitro* EGb761 reduced ischemia-induced cell swelling in hippocampal slices by 60%. *In vivo* administration of EGb761 (300 mg/kg) was shown to reduced cell degeneration and edema formation by 35-50%.

The effect of ginkgolide B on brain metabolism and tissue oxygenation in severe hemorrhagic stroke was studied by Chi *et al.* (2015). The aim of study was to show the effect of ginkgolide B in osmotherapy on brain metabolism and tissue oxygenation.The parameters *i.e.* multimodality monitoring including intracranial pressure (ICP), cerebral perfusion pressure (CPP), partial pressure of brain tissue oxygen, lactate/pyruvate ratio (LPR) and micro dialysis were used to study the effect of ginkgolide B osmotherapy. The results demonstrated that peak lactate/pyruvate ratio was recorded at the time of initiation of osmotherapy with an 18% decrease after 2h on administration of gingkolide B. Also, the brain glucose remained unaffected.

Koh *et al.* (2012) investigate deffect of GBE 761 in ischemic brain injury to show that it protects neuronal cells from ischemic brain injury *via* a number of neuroprotective mechanisms. Hippocalcin is a calcium sensor protein that regulates intracellular calcium concentrations and apoptotic cell death. In this study, GBE 761 regulated hippocalcin expression in cerebral ischemia. Male rats, used to screen this activity, were treated with vehicle or GBE 761 (100mg/kg) prior to middle cerebral artery occlusion (MCAO), and cerebral cortex tissues were collected 24 h after MCAO. The results suggested that the maintenance of hippocalcin during cerebral ischemia contributes to the neuro-protective role of GBE 761.

On similar lines, Snitz *et al.* (2009) reported *G. biloba* to prevent cognitive decline in older adults. Ginkgo Evaluation of Memory (GEM) study was done in which a randomized, double-blind, placebo-controlled clinical trial of 3069 community-dwelling participants aged 72 to 96 years was taken in 6 academic medical centers in the United States. The twice-daily dose of 120 mg extract of *G. biloba* (n=1545) was given to participants. The results revealed that the use of *G. biloba*, 120 mg twice daily, did not result in less cognitive decline in older adults with normal cognition or with mild cognitive impairment.

Hepato-protective

Elatrash and Haliem (2015) carried out a study with monosodium glutamate (MSG) to check the hepato-protective activity of *G. biloba*. In this study, MSG was administrated to rats at dose of 1.5 mg/kg body weight for four weeks. In liver, the activities of ALT and AST significantly increased in the serum on MSG administration while serum total protein and albumin significantly decreased. *G. biloba* (80 mg/kg body weight) co-treatment with MSG, significantly reduced

ALT and AST in the serum and was found to be comparable with control. Also, total protein and albumin were significantly increased in the serum, while serum urea, serum creatinine and uric acid were significantly decreased after administration of *G. biloba*.

In a recent experiment, liver fibrosis index, Masson's trichrome staining and liver function index was studied to demonstrate that *G. biloba* extract (GBE) mitigated fibrosis and improved function of the liver. The study also showed that GBE reduced liver fibrosis not only by inhibiting p38 MAPK and NF-êBp65 degradation but also by inhibiting hepatocyte apoptosis via down-regulation of Bax, up-regulation of Bcl-2, and subsequent inhibition of caspase-3 activation (Wang *et al.*, 2015).

EGb761 extract was shown to have a protective effect against renal ischemia-reperfusion injury in rats such that malondyaldehyde (MDA) levels were significantly high and superoxide dismutase (SOD) enzyme activity was significantly lower. Furthermore, microscopic examination showed that lesser tubular necrosis was observed on administration of GBE (Akdere *et al.*, 2014).

Antioxidant Activity

Koczka *et al.* (2015) studied total phenolic content and antioxidant capacity (FRAP method) of GBE made from leaves of male and female trees were determined and compared. Water and aqueous-ethanolic (water/ethanol 80/20,v/v) extracts were prepared by varying the time of infusing, boiling and steeping in order to determine the effect of the extraction method on the above parameters. Antioxidant activity and phenolic content of Ginkgo leaf extracts correlated well with significant correlation coefficients.Total phenolic content was measured by using the method of Singleton and Rossi and was expressed as milligram of gallic acid equivalent per gram of dry mass.To determine the antioxidant capacity, the FRAP assay (ferric reducing ability of plasma) was used. The result showed that total phenolic contents of GBE exhibited significant differences, depending on the sex of the tree. More phenolics were measured from leaves collected from male trees than from female trees, and in almost every extraction method.The extent of these differences was up to 10 % for decoctions and 6 to 30 % for infusions. Similarly, for phenolic contents, extracts of leaves collected from male trees had higher antioxidant capacity than that for female trees.

Cisowska (2010) evaluated the antioxidant activities of GBE by investigating the effect of ethanol extracts from green and yellow leaves of *G. biloba* (0.05%) on the rate of lipid oxidation in meat dumplings stored under freezing conditions. The results revealed that the added ethanol extracts from *G. biloba* leaves limited the rate of oxidation of the fats in the product. The study showed that ethanol extracts from *G. biloba* leaves may be used as an antioxidant to increase the stability of meat dumplings.

Anti-Inflammatory Activity and Anti-Arthritic Activity

Anti-inflammatory activity of GBE, studied in rats after ischemia-reperfusion injury, indicated its potential to alleviate inflammatory reactions by promoting the activity of IL-4, an anti-inflammatory cytokine, and inhibit the activity of IL-6, an inflammatory cytokine (Bao, 2008).

In yet another study, Chen *et al.* (2013) observed anti-inflammatory effect of *G. biloba*. They determined whether EGb761 could inhibit lipopolysaccharide (LPS)- and IL-1b-induced inflammatory responses in human articular chondrocytes and apply the chondroprotection in OA (osteoarthritis) rats. They found that LPS markedly induced the production of PGE2 and NO, and the protein expressions of COX-2 and iNOS in human chondrocytes. LPS was also seen to up-regulate the expressions of toll-like receptor-4 (TLR4), its downstream signal TNF receptor-associated factor 6 (TRAF6), and nuclear factor (NF)-kB signaling. These LPS-induced inflammatory responses were efficaciously reversed by EGb761 and its active components quercetin and kampferol. The similar results were observed by using IL-1b as an *in vitro* model to mimic an inflammatory response. In an OA rat model, PGE2 and NO levels in blood, the histological alterations, and COX-2 and nitrotyrosine expressions in cartilages were markedly increased, which were effectively reversed by EGb761. Their results suggested that EGb761 exerts the anti-inflammatory effects on human articular chondrocytes and OA rats.

Ho *et al.* (2013) studied the effect of GBE against osteoarthritis (OA), a common joint disorder with varying degrees of inflammation. Ideally, the anti-OA drug should have immune-modulatory effects but limited or no toxicity. In this study, the anti-inflammatory effects of GBE in interleukin-1 (IL-1)-stimulated human chondrocytes and chondrocytes (prepared from cartilage specimens taken from patients with osteoarthritis who had received total hip or total knee replacement) were examined. The concentrations of chemokines and the degree of cell migration were determined by ELISA and chemotaxis assays, respectively. The activation of inducible nitric oxide synthase (iNOS), mitogen-activated protein kinases (MAPKs), activator protein-1 (AP-1), and nuclear factor-kappaB (NF-kB) was determined by immune-blotting, immune-histochemistry, andelectrophoretic mobility shift assay. The result showed that GBE inhibited IL-1-induced production of chemokines, which in turn resulted in attenuation of THP-1 cell migration toward GBE-treated cell culture medium. GBE also suppressed IL-1- stimulated iNOS expression and release of nitric oxide (NO).

The EGb-mediated suppression of the iNOS-NO pathway correlated with activator protein-1 (AP-1) but not nuclear factor-kappaB (NF-kB) DNA-binding activity. GBE inhibited only c-Jun N-terminal kinase (JNK) andunexpectedly,GBE selectively caused degradation of c-Jun protein. Further investigation revealed that EGb-mediated c-Jun degradation was preceded by ubiquitination of c-Jun and could be prevented by the proteosome inhibitor MG-132. The results imply that GBE protects against chondrocyte degeneration by inhibiting JNK activation and inducing ubiquitination-dependent c-Jun degradation. The results suggested that GBE is a potential therapeutic agent for the treatment of OA.

Zhou *et al.* (2014) carried out work on *G. biloba* leaf extract (P-PGBL) and screened its anti-inflammatory activity. Theirresults showed that P-PGBL treatment inhibited the LPS-induced protein and mRNA expression of Toll-like receptor 4 (TLR4).However, P-PGBL treatment down regulatedthe nuclear factor-êB (NF-êB) p65 protein expression in LPS-stimulated RAW264.7 macrophage cells. Simultaneously, the levels of pro-inflammatory cytokines *i.e.* IL-1â and IL-6 were suppressed by P-PGBL. This showed that P-PGBL exerts an anti-inflammatory effect by inhibiting TLR4/NF-êB signaling pathway.

Anti-Cancer Activity

Dias *et al.*(2013) showed effects of G. *biloba* on chemically-induced mammary tumors in rats receiving tamoxifen (TAM).The study investigated the effects of GBE in female Sprague–Dawley (SD) rats bearing chemically-induced mammary tumors and receiving TAM.Animals bearing mammary tumors (e"1 cm in diameter) were divided into four groups: TAM [10 mg/kg, intragastrically (i.g.)], TAM plus GBE [50 and 100 mg/kg, intraperitoneally (i.p.)] or an untreated control group. After 4 weeks, the therapeutic efficacy of the different treatments was evaluated by measuring the tumor volume and the proportions of each tumor that were alive, necrotic or degenerative. In addition, labeling indexes (LI%) were calculated for cell proliferation (PCNA LI%) and apoptosis (cleaved caspase-3 LI%), expression of estrogen receptor-alpha (ER-á) and p63 biomarkers. The result displayed that co-treatment with 100 mg/kg GBE presented a slightly beneficial effect on the therapeutic efficacy of TAM in female SD rats bearing mammary tumor.

Wu *et al.* (2011) isolated, purified and tested the *in vitro* antitumor activities of polysaccharides from *G. biloba* sarcotesta with remarkable results. Mesallamy *et al.* (2011) showed chemopreventive role of *G. biloba* against hepato-carcinogenesis induced by nitrosodiethylamine (NDEA) that may be executed through its antioxidant, antiangiogenic and antigenotoxic activities as demonstrated by decreased levels of alanine aminotransferase (ALT), aspartate aminotransferase (AST), and gamma glutamyltransferase (GGT), and decreased comet assay parameters and improved liver architecture. Similarly, Feng *et al.* (2009) also observed anti-cancer effect of G. *biloba*. In this study,GBE inhibited proliferation in sarcoma and leukemia cell lines by increasing the levels of free radical scavenging enzymes, and this activity was enhanced by increasing the concentration of aglycone portion of the constituent flavonoid glycosides.

Ma *et al.* (2010) studied action of Ginkgolic acid (GA), which suppressed the development of pancreatic cancerby inhibiting pathways driving lipogenesis. The results indicated that GA inhibited the *denovo* lipogenesis of cancer cells through inducing activation of AMP activated protein kinase (AMPK) signaling and down regulated the expression of key enzymes *i.e.* acetyl-CoA carboxylase [ACC], fatty acid synthase [FASN]) involved in lipogenesis.

Liou *et al.*(2015) isolated Ginkgolide C from *G. biloba* leaves and screened it for antiadipogenic effectin 3T3-L1 adipocytes. In this study, cell supernatant was collected to assay glycerol release, and cells were lysed to measure protein and gene expression related to adipogenesis and lipolysis by western blot and real-timePCR, respectively.Ginkgolide C suppressed lipid accumulation in differentiated adipocytes. It also decreased adipogenesis-related transcription factor expression, including peroxisome proliferator-activated receptor and CCAAT/enhancer bindingprotein. Furthermore, ginkgolide C enhanced adipose triglyceride lipase and hormone-sensitive lipase production for lipolysis and increased phosphorylation of AMP-activated protein kinase (AMPK), resulting in decreased activity of acetyl-CoA carboxylase for fatty acid synthesis. Moreover, ginkgolide C also improved activation of sirtuin 1 and phosphorylation of AMPK in differentiated 3T3-L1 cells. The results suggested that ginkgolide C is an effective flavone for increasing lipolysis and inhibiting adipogenesis in adipocytes through the activated AMPK pathway.

Vardy *et al.* (2013) investigated herb-drug interactions with *G. biloba* in women receiving hormonal treatment for early breast cancer. The study suggested that co-administration of *G. biloba* does not significantly affect the pharmacokinetics of tamoxifen, anastrozole or letrozole. There was no difference in the toxicity profile of hormone therapy with *G. biloba* use in women with early stage breast cancer.

Zhou *et al.* (2012) studied the influence of GBE on proliferation, apoptosis of ACC-2 cell and Survivin gene expression in adenoid cystic carcinoma of lacrimal gland. The study showed that GBE had inhibitory effect on the proliferation of ACC-2 cell. Furthermore, flow cytometer test indicated that GBE can gradually increase ACC-2 cell in G_0-G_1 stage and decrease it in G_2-M and S stage.

Hoenerhoff *et al.* (2013) showed that hepato-cellular carcinomas in B6C3F1 mice treated with GBE for two years differ from spontaneous liver tumors in cancer gene mutations and genomic pathways.This study provided a molecular context for the genetic changes associated with hepato-carcinogenesis in GBE exposed mice.

Jeon *et al.* (2015)isolated ginkgetin and screened its anti-cancer activity for prostate cancer through inhibition of signal transducer and activator of transcription 3 activity. The results revealed that Ginkgetin inhibited STAT3 localization to the nucleus by dephosphorylating STAT3 at Try705, which may leadto the inhibition of expression of STAT3 target genes *i.e.* cell survival-related genes (cycling D1 and Survivin) and anti-apoptotic proteins (Bcl-2 and Bcl-xL). Therefore, ginkgetin inhibited the growth of STAT3-activated tumor cells.

Tsai *et al.* (2014) evaluatedanti-cancer activity of GBE for small cell lung cancer. The extract decreased the migration of non-small cancer cell by down-regulating metastasis associated factor heat-shock protein 27 (HSP27). The results suggested that EGb761 inhibited HSP27 expression and migratory ability of A549/H441 cells. Furthermore, EGb761 also decreased the migration ability of A549/H441 by inhibiting HSP27 through AKT and p38 MAPK pathways activation.

Saker *et al.*(2011) investigated anti-tumor activity of EGb761 against topsin induced ovarian cancer. In this study, ovarian cancer induced by topsin which may lead to histopathological changes *i.e.* number of ovarian follicles decreased and most of them degenerated.Topsin also significantly decreased the levels of both LH (luteinizing hormone) and FSH (follicle-stimulating hormone) and increased estradiol. Furthermore, histo-chemical results showedan increase in carbohydrate content and decrease in total protein in the ovarian tissue. Treatment with EGb 761, resulted decreased toxicity and oxidative effect of topsin in the ovary.

Antiviral Activity

Lu *et al.* (2012) showed that ginkgolic acid inhibited HIV protease activity and HIV infection *in vitro.* In this study, purified ginkgolic acid and recombinant HIV-1 HXB2 KIIA protease were used for the HIV protease activity assay. The result suggested that ginkgolic acid (31.2 ìg/ml) inhibited HIV protease activity by 60%, compared with the negative control, and the effect was concentration-dependent. In addition, ginkgolic acid treatment (50 and 100 ìg/ml) effectively inhibited the HIV infection at day 7 in a concentration-dependent manner. Ginkgolic acid, at a concentration of upto 150 ìg/ml, displayed very limited cytotoxicity.

Wang *et al.*(2015) screened the antiviral activity of a nano-emulsion of polyprenols from *G.biloba* leaves (GBP) against Influenza AH_3N_2 and Hepatitis B virus *in vitro.*The study showed that GBP is non-toxic to normal cells, hepatitis B virus DNA, hepatitis B virus antigen and HepG2215. Furthermore, GBP at a high concentration of 100 µg/mL displayed 70% virucidal activity and a 74.9% protection rate on MDCK (Madin-Daby Canine Kidney) cells infected with H_3N_2 virus. The GBP nano-emulsion was very stable and non-toxic and showed very strong antiviral activity against influenza A H_3N_2 and hepatitis B virus *in vitro*. The mechanisms involved in lengthening the incubation time and increasing the drug concentration. GBP has promising potential as an antiviral drug.

Beneficial Role in Respiratory /Cardio Problems

Bernatoniene (2010) screened the effect of GBE on mitochondrial oxidative phosphorylation in normal and ischemic rat heart.GBE in 70% ethanol (1:5) was administered orally to rats under normal and ischemic conditions. Mitochondrial respiration rates were determined by oxygraph.The results suggested that GBE reduced generation of free radicals in mitochondria and diminished the degree of respiration stimulation by exogenous cytochrome c. Thus, these results indicated that GBE exerts cardio protective effects by reducing ischemia-caused impairment of the functions of heart mitochondria.

Li *et al.*(2015) showed effect of GBE on experimental cardiac remodeling. The aim of the study was to investigate the ameliorated effects of GBE on experimental cardiac remodeling in rats induced by acute cardiac infarction The study was focused on myocardial type I collagen, transforming growth factor beta 1 (TGF-β1), matrix metalloproteinase 2 (MMP-2) and matrix metalloproteinase 9 (MMP-9). The results demonstrated that GBE inhibited experimental rat myocardial remodeling after acute myocardial infarction through reduced transcription of TGF-β1, MMP-2 and MMP-9 genes and by the decreased expression of type I collagen, MMP-2 and MMP-9 proteins in myocardial cells.

Zhao *et al.* (2013) isolated Ginkgolide from *G. biloba* extract and evaluated its anti-arrhythmic activity by targeting hERG and ICa-L channel. The study investigated the effects of G. *biloba* extract (GBE) and ginkgolide (GLD) on human ether-a-go-go-related gene (hERG)-encoded K^+ channels and mechanisms involved in the hERG-HEK293 cell line by determining GBE- and GLD-induced changes in action potential duration (APD), L-type calcium current, and the intracellular calcium concentration in guinea-pig ventricular myocytes. The results demonstrated that low concentration of GBE (0.005 mg/ml) increased hERG currents, but the high concentration of GBE (0.05 to 0.25 mg/ml) reduced hERG currents. Furthermore, GLD decreased hERG currents in a concentration-dependent manner (0.005 to 0.25 mg/ml). This showed that both GBE and GLD altered kinetics of the hERG channel.GLD did not shift the activation or the inactivation curve, but only decreased the time constant of inactivation.GBE accelerated the activation of hERG channels without changing the inactivation curve, but decreased the time constant of inactivation. Both GBE and GLD decreased the APD, inhibited the current, and decreased the calcium concentration in isolated guinea-pig ventricular myocytes. The results indicated that GBE and GLD can prevent ischemic arrhythmias and also had an antiarrhythmic effect by inhibition ofcurrentscarried out (Zhao *et al.*, 2013).

Wei *et al.* (2013) investigated the role of GBE in cardiovascular disease *i.e.,* atherosclerosis through down-regulating the expression of connexin 43 in rabbits.The effect of GBE was studied on atherosclerotic lesion development in rabbits maintainedon a high-fat diet. The plasma lipid measurement showed that GBE inhibited high-fat diet induced increased serum triglyceride (TG)by 59.1% (0.9 ±0.2 4 mmol/l vs. 2.2±0.4 mmol/l), total cholesterol (TC) by 18.2% (31.1 ±1.4 mmol/l vs. 38.0 ±0.4 mmol/l), and low density lipoprotein cholesterol (LDL-C) by 15% (28.9±1.3 mmol/l vs. 34.0±1.0 mmol/l) at 12 weeks ($p < 0.01$). Moreover, enface Sudan IV-positive lesion area of the aorta in the GBE group (51.7 ±3.1%) was significantly lowered compared with that in the HF (high fat) group (88.2 ±2.2%; $p< 0.01$) and mean atherosclerotic lesion area of the GBE group was reduced by 53.2% compared with the HF group ($p <0.01$). Furthermore, GBE suppressed high-fat diet induced up regulation of connexin 43 (Cx43) in rabbits analyzed by immune histochemistry and western blot. Thus, the study suggested that GBE prevented atherosclerosis progress by modulating plasma lipid as well as by suppressing atherosclerotic lesion development, and declined expression of Cx43 protein.

Protective effect of EGB761 against age-associated diastolic dysfunction in cardiomyocytes of d-galactose-induced rats was studied by Jing *et al.* (2012). The results indicated *in vitro* stimulation with D-galactose induced AGEs (advanced glycation end products) production. Administration of EGB761 decreased the number of cells positive for SA-â-gal (Senescence-associated â-gal). Furthermore, decreased diastolic [Ca^{2+}] and increased reuptake of Ca^{2+} stores in the SR (Sarcoplasmic reticulum) was observed. In addition, the level of p- Ser16-PLN(Phospholamban protein) as well as SERCA (Sarco/ endoplasmic reticulum calcium adenosine triphosphate)was significantly increased. The study indicated that EGb761 improved SERCA2a function by increase the amount of Ser16 sites PLN phosphorylation as well as increased formation of AGEs products on SERCA2a.

Ran (2014) reported EGb 761 to cause reduction in myocardial ischemia reperfusion injury. The study showed that on administration of EGb761, serum cTnI (Cardiac troponin I) concentrations in the E (post conditioning) group during reperfusion decreased significantly, and concluded that *G. biloba* extract had myocardial protective effects by reducing the generation of oxygen-free radicals and increasing the antioxidant capacity of the myocardial cells.

Clinical studies

EGb761 is the standardized leaf extract of *G. biloba*, launched by Dr. Willmar Schwabe (GmbH & Co) in 1976. A number of preclinical and clinical research studies carried out in recent years suggest that GBE has many positive effects. These include scavenging radicals, anti oxidation, anti-tumour and protective effects in central nervous system and therapeutic effects for cerebral and peripheral vascular diseases. It has been presumed that these therapeutic effects are due to synergistic effect of two distinct groups of active compounds present in the extract *i.e.* flavonoids (quercetin, kaempferol and isorhamnetin) and terpenoid lactones (bilobalide, ginkgolides A, B and C). However, oral bioavailability of flavonoids is relatively low due to their poor solubility (Nicolai *et al.,* 2009).

Role in the management of specific ailments

1. Neurological disorder

EGb761 can treat a variety of neurological disorders including Alzheimer's disease and age related dementia (Weinmann *et al.*, 2010; Wang *et al.*, 2014b). The antioxidant activity of EGb761 may play a substantial role in neuroprotection by decreasing bax/bcl-2 ratios, reversal of ischemia-induced reductions of cyclooxygenase III, mRNA in hippocampal CA1 neurons, inhibition of nitric oxide synthesis, scavenging of free radicals and attenuation of lipid peroxidation (Tan *et al.*, 2014). Another study (Gauthier and Schlaefke, 2015) showed that when Ginkgo extract and anti-platelet aggregation drug is given to vascular cognitive impairment of none dementia patients (VCIND) for 3 months, it helps in improving cognitive functions (attention, deferred memory) of VCIND patients. The cerebral blood flow in temporal and frontal lobe increases and this has a close relation with cognitive function.

Interference of GBE with pathogenic mechanism of Alzheimer's disease (AD) and pure vascular dementia (VaD) has been reported. GBE helps in restoration of mitochondrial function, thus provides energy to neurons, improves hippocampal neurogenesis, inhibits aggregation and decreases toxicity of Aâ protein, and blood viscosity and enhances microperfusion (Ihl *et al.*, 2012). Additionally, electroencephalographic (EEG) studies showed that *G. biloba* has vigilance-enhancing and cognitively activating effects on performance and brain electrical activity(Wang *et al.,* 2014c).

Since year 2000, along with anti-cholinesterase inhibitors and memantine, GBE (120-240mg) is listed among the group of anti-dementia drugs. The standardized extract is given in 2-3 daily doses (liquid/solid dosage) for dementia or AD. However, it was reported that 240 mg of EGb761 had no effect in older patients for the prevention of dementia onset (Banachewski *et al.*, 2011).AD is the most common neurodegenerative disease associated with progressive cognitive and memory loss. Extracellular deposition of amyloid â peptide (Aâ) in senile plaques, intracellular neuro-friability tangles, cholinergic deficit and neuronal loss are major molecular hallmarks of this disease. As EGb761 is demonstrated to possess different profiles of action, it has the potential to treat or prevent AD (Herrschaft *et al.,* 2012). Plausible mechanism of action of extract against AD includes antioxidant and anti-apoptotic properties, as well as potential inhibition of caspase-3 activation and amyloid- â-aggregation (Feucht and Patel, 2011).

The German Institute for Quality and Efficacy in Health Care (IQWIG) and the European Medicines Agency (EMA) reccommended that treatment with GBE 761 in a dose of 240mg/d shows a beneficial effect on individuals suffering from mild to moderate or moderate to severe AD (Kyriakis and Avruch, 2012).On the other hand, some clinical trials were not able to confirm the efficacy of GBE 761 (160-240 mg/day for 24weeks) to ameliorate cognitive function in mild to moderate AD patients.

Brain ischemia leads to cessation of ATP generation in the affected areas resulting in generalized depolarization, release of excitatory transmitters in the selectivity vulnerable neuronsand opening of the voltage-dependant and glutamate-regulated calcium channels. Ginkgo extracts increases blood flow, improve peripheral arterial insufficiency, reduce blood viscosity and thrombocyte aggregation induced by ischemia (Kim and Lee, 2013).

MPP (1-methyl-4-phenyl-pyridine) is an active metabolite of the MPTP (1-methyl-4-phenyl-1,2,3,6-tetrahydropyridine) toxin that induces Parkinsonism by causing dopaminergic cell death throughinhibition of mitochondrial complex. Studies have shown that neuroprotective effect was rendered by extract against this toxin; also the extract regulates the homeostasis of copper in the brain of animal. It also attenuates neuro-degeneration in the nigrostriatal pathway and inhibits oxidative stress (Schwarzkopf *et al.*, 2013).

2. Cardiovascular diseases

Myocardial ischemia-reperfusion (IR) can result in damage to the myocardium followed by blood restoration after a critical period of blood occlusion. Ginkgo extracts are protective against myocardial ischemic injury. A possible role can be scavenging of free radicals and anti-lipid peroxidation, resulting in dilated coronary blood vessels, ameliorated microcirculation and inhibited cardiac myocytes apoptosis(Szulc and Radwan, 2012).

3. Diabetes

Studies have shown that patients suffering from type-II diabetes might benefit from ingesting *G. biloba* extract through improved platelet function, altered platelet-vessel wall interactions and reduced malonidialdehyde levels in platelets. *In vitro* and *in vivo* studies on extract have indicated that it reduces total superoxide dismutase (SOD) activity after adjusting the expressions of cytokines in patients with diabetic nephropathy. In other studies, Ginkgo extract inhibited high glucose induced IL-6 release and ICAM-1 accumulation, leading to reduction in endothelial adhesiveness to monocytes (Vupputuri *et al.*, 2011).

4. Cancer

During recent years, it has been found that Ginkgo extract has anti-neoplastic activity, having lethal effect on gastric cancer, liver cancer, intestinal cancer and breast cancer, though the precise mechanism of anti-tumour activity is unclear (Adler and Fosket, 1999).

5. Inflammation

Heat, redness, swelling and pains are some of the ways by which a living tissue expresses its reaction to irritation, injury or infection. The extracts of Ginkgo alleviate these symptoms and thus show positive results. Among several mechanisms, interleukin (IL) regulation plays a key role in anti-inflammatory functions. Suppression of monocyte activation *via* inhibition of nitric oxide and tristetrapolin-mediated toll-like receptor 4 expression is another way of regulation of inflammation by extract. In the oral chronic inflammatory disease periodontitis, *G. biloba* plays an anti-inflammatory role by up-regulating heme-oxygenase 1 (HO-1) and Nrf-2 levels in the nucleus (Liu *et al.*, 2014).

Toxicity Studies

Recently, a National Toxicology Program (NTP) study reported the results of two genotoxic assays (*in vitro* and *in vivo*). In a bacterial gene mutation assay (Ames assay), GBE was mutagenic in *Salmonella typhimurium* strains, TA98 and TA100, and the *Escherichia coli* strain, Wp2 *uvr*A (pKM101) with and without S9 mix. The peripheral blood micronucleus assay in male and female mice administered GBE lasted for 13 weeks. Furthermore, the NTP 2-year study of GBE reported significant increases in the incidence of hepatocellular adenoma/carcinoma and hepatoblastomas in B6C3F1 mice of both sexes at the lowest dose tested (200 mg/kg bw/day). The effect showed dose dependency, indicating clear evidence of hepato-carcinogenicity in mice. Therefore, clarification of *in vivo* genotoxicity of GBE is a critical issue for human cancer risk assessment (Rider *et al.*, 2014).

Ginkgotoxin (4'-*O*-methylpyridoxine also known as MPN) is a neurotoxin that occurs naturally in *G. biloba*. Studies show that accidental ingestion or deliberate use of *G. biloba* results in overdose of ginkgotoxin. This causes epileptic convulsions, vomiting, unconsciousness and irritability and even death. First case of toxicity by ginkgotoxin was observed in 1881and many other cases have been reported thereafter (Jang *et al.*, 2015).

MPN is present in the concentration of 170-400 μg g^{-1} in the raw seeds of *G. biloba.* Roasted seeds of Gingko are commonly eaten in Japan, Korea and China for its nutritive value. In Japan, these seeds are known as 'Gin-nan'. But the MPN is heat resistant, so cooking or other heat treatments only partially inactivate it. MPN is known to compete with vitamin B_6, which is a cofactor of glutamate decarboxylase. Consequently, MPN indirectly inhibits the activity of this enzyme resulting in decrease of ã-amino-butyric acid (GABA) levels in brain, which results in convulsions (Fiehe *et al.*, 2000).

Toxicity of *G. biloba* leaf extract

Commercially, *G. biloba* extracts may be available as full extracts, crude extracts or simple extracts and sold in the form of tablets or capsules. Standardized extract EGb761®, according to Dr. Willmar Schwabe Pharmaceuticals, consists of 5-7% terpenene lactones, 22-24% flavonoids, less than 5ppm ginkgolic acid and 5-10% organic acids and other constituents. However, the composition of extract may vary owing to variability in the multi-step extraction and concentration process. For example, recent studies show that high amounts of free flavonol aglycones, especially quercetin and kaempferol were found in some extracts. Furthermore, some products contain 200 times more ginkgolic acids compared to others. These variations are the cause ofmajor differences between the pharmacological and toxicity profiles of the extract (Chan *et al.*, 2007).

Formulations

According to the Commission E Monographs, 120–240 mg of standardized extract in liquid or solid pharmaceutical formfor oral intake is suggested in two or three daily doses for dementia syndromes (primary degenerative dementia, vascular dementia, or mixed forms of both). Ginkgo ingredients are available commercially in several forms, including in teas, liquids, colas, capsules, extracts, tablets, sprays, etc. Pharmacokinetic testing of Ginkgo in capsule, drop, and tablet forms has been evaluated, including testing for IV administration. When administered orally during fasting, bioavailability was reported to be high; food did not change the AUC quantitatively but increased the time to maximum plasma concentration (Ude *et al.*, 2013). Doses of 120–160 mg of native dry extract is recommended in two or three daily doses to improve pain-free walking in case of peripheral arterial occlusive disease, vertigo and tinnitus of vascular origin. These doses correspond to an estimate of 50 fresh Ginkgo leaves to yield one standard dose of the extract. Dried extracts of leaves in the form of tablets, standardized to contain 24% flavone glycosides and 6% terpenes, are also available commercially. Commercial extracts of the Ginkgo leaves are enriched water-acetone or water-ethanol extracts of the Ginkgo leaves, and are standardized on their flavonoid content or their terpene trilactone content. In Japan, *G. biloba* extract is also sold as a health food product while in Germany it is sold as an over-the-counter (OTC) preparation.

Some of the branded products of *G. biloba* are:

Bio-Biloba ®: Concentrated Ginkgo leaf liquid. Each millilitre contains 14.5 mg flavone glycosides and 2.8 mg terpene lactones. EGb761 from Dr. Willmar Schwabe Pharmaceuticals is the standardized mono-extract of dried leaves from Ginkgo (50:1) consists of 24% Ginkgo flavonol glycosides and 6% terpene lactones, of which 2.9% is bilobalide and 3.1% is the ginkgolides A, B and C. It contains less than 5 ppm ginkgolic acids and conforms to the German Commission E requirements for GBE.

Geriaforce®: Each tablet contains an ethanolic extract of fresh leaves 1:4, with a content of flavonol glycosides of 0.20 mg/ml and ginkgolides 0.34 mg/ml.

Ginkgold®: The standardized mono-extract (EGb-761) of dried leaves from Ginkgo (50:1) consists of 24% Ginkgo flavonol glycosides and 6% terpene lactones, of which 2.9% is bilobalide and 3.1% is the ginkgolides A, B, and C. Contains less than 5 ppm ginkgolic acids and conforms to the German Commission E requirements for GBE.

Kaveri®: Prepared from the standardized mono-extract (LI-1370) of dried leaves from Ginkgo (50:1) containing at least 25% flavonol glycosides and 6% terpene lactones. Contains less than 5 ppm ginkgolic acids and conforms to the German Commission E requirements for GBE

Rokan®: The standardized mono-extract (EGb-761) of dried leaves from Ginkgo (50:1) consists of 24% Ginkgo flavonol glycosides and 6% terpene lactones, of which 2.9% is bilobalide and 3.1% is the ginkgolides A, B, and C. Contains less than 5 ppm ginkgolic acids and conforms to the German Commission E requirements for GBE.

Tanakan®: The standardized mono-extract (EGb-761) of dried leaves from Ginkgo (50:1) consists of 24% Ginkgo flavonol glycosides and 6% terpene lactones, of which 2.9% is bilobalide and 3.1% is the ginkgolides A, B, and C. Contains less than 5 ppm ginkgolic acids and conforms to the German Commission E requirements for GBE.

Tebonin®: The standardized mono-extract (EGb761) of dried leaves from Ginkgo (50:1) consists of 24% Ginkgo flavonol glycosides and 6% terpene lactones, of which 2.9% is bilobalide and 3.1% is the ginkgolides A, B, and C. Contains less than 5 ppm ginkgolic acids and conforms to the German Commission E requirements for GBE.

F. THREAT AND CONSERVATION ASPECT

Population status

Natural wild population of the species is confined to the Zhejiang province of China. The species can also be found in some other parts,such as Guizhou, Sichuan and Guangxi Provinces but these populations are not up to sufficient numbers (Royer *et al.*, 2003). Presently, it is considered a monotypic genus, the sole survivor of the ancient family Ginkgoaceae, and also referred to as the living fossil. *G. biloba* is endemic to China, where the persistence of natural populations has been in question on the grounds that no unequivocally natural stands remain today. Researchers have reported that *G. biloba* trees were found in wild or semi wild condition in Dahongshan of Hubei Province, Tianmushan of Zhejing Province, Wuchuan of Guizhou Province, Jinfoshan of Chongqing Municipality. The inference is based on historic ages claimed for old Ginkgotrees, Ginkgo distribution habitats, and the general species composition of those regions. However, there is lack of (1) reliable data on the floristic composition of Ginkgo forests, (2) age structure of Ginkgo populations, (3) credible scientific data for the claims for great ages of *G. biloba* trees (such as those "lived over 1000 years"), and (4) solid information on the presence or absence of human influence on *G. biloba* populations (Gong *et al.*, 2008).

To summarize, it is believed that *G. biloba* is native of China (specific provinces), Japan, Taiwan, Turkey, and exotic in countries like, Czechoslovakia (Former), Estonia, Germany, India, Russian Federation, South Africa, United States of America, Yugoslavia (Former).

Threats of its extinction

The status of *G. biloba* in wild has shrunk alarmingly, getting severely threatened.The fossil record of Ginkgo shows a decline in its diversity and distribution. The diminution of Ginkgo's range continued into the tertiary, and was particularly striking from the Oligocene (38 to 26 myr). The genus disappeared from polar areas, through the end of the Miocene (24 to 7 myr). These dramatic changes were most likely the result of the extensive cooling that occurred throughout the Northern hemisphere during these time periods. The genus Ginkgo disappeared from Europe by the end of the Pliocene (1.8 myr) as temperatures dropped down and the rainfall regimen gradually shifted from summer wet to summer dry. The only known Pleistocene (1.8 myr to present) occurrences of the genus Ginkgoare from South Western Japan (McCarney *et al.*, 2008).

Hardly any wild population of Ginkgo exists today. Only a single wild population of 244 individuals was located in Tianmushan in 1984, with no poor seedlings presence (Sadaf, 2016). Ginkgo is thought to be a "pioneer" species in its local environment. Fresh surveys have been encouraged in relatively recent decades. In 1989, only one hundred sixty seven (167) individuals were recorded in one location, and three additional wild populations were alsoreported in Longchou (Guangxi Province), Wuchuanshan (Guizhou Province) and Dahong (Hubei Province) (Royer *et al.*, 2003).

Conservation strategies

The International Union for Conservation of Nature (IUCN) listed the species under the category of "endangered" due to the rapid decline in the numbers of individual trees around the world (Fu and Jin, 1992). The existing population of *G. biloba* throughout the world is considered to be the cultivated ones. Among the suggested conservation strategies for Ginkgo plants isthe selection of high yielding individuals and developing efficient propagation techniques. Propagation is recommended through *in vitro* micro cuttings and possibly tissue culture techniques. The identification of right genotype at an early stage and its clonal propagation through *in vitro* techniques will play a great role in establishing productive populations and production of ginkgolides and bilobalide for developing authentic drug. Because of extremely slow growth and poor regeneration through seeds, vegetative propagation through stem cuttings is a practical way of augmenting productive *G. biloba* (Sohier, 2002). Conventional cuttings are valuable option to conserve the species.

Vegetative cuttings are rooted by using root growth-inducing agents such as indole-3-butyric acid (IBA) and alphanaphthalene acetic acid (NAA) (Purohit *et al.*, 2009). *In vitro* regeneration of the Ginkgo shoot is derived from embryo and cotyledon, immature zygotic embryos, apical and nodal bud explants, and rooting performed on a medium added with endosperm extract (Tommasi and Scaramuzzi, 2004). Somatic embryogenesis, or development whereby embryoids are produced from somatic cells, is obtained from microspore, haploid protoplast, mega gametophyte, and immature zygotic embryo explants. Despite experimental successes of shoot regeneration and somatic embryogenesis, complete plantlet regeneration and later establishing them in field is yet to be achieved.

China, France, and Germany are undertaking largescale propagation, and such initiatives are implemented through plantations for conserving the species' population and for harvesting herbal medicine (Schmid and Balz, 2003). In India, a few isolated populations exist in the Himalayas and other locations (Xiang and Xiang, 1999) and these could be source of mother plants.The seedlings developed by clonal regeneration have shown survival in nature but because of the dioecious nature of plant, propagation from seeds appears difficult (Peter, 2007). Low seed germination rates and recalcitrance, a long juvenile phase are some of the problems in propagating *G. biloba*. However, high environmental adaptability of the species and its unparalleled tolerance to environmental stress makes it a favourite for planting, throughout the temperate and subtropical world, for medicinal and ornamental purposes.

Cryopreservation offers an alternative to labour/cost-intensive *in vitro* culture.Samples stored in liquid nitrogen do not need periodic subcultures. Cryopreservation of 2 year old *G. biloba* callus, through desiccation method with optimal pre-culture on sucrose and ABA amended medium for 14 days was achieved (Popova *et al.,* 2009; Lu *et al*., 2009).

Recently a study was conducted (Kumar *et al.*, 2018) on rhizosphere microflora associated with ginkgo trees in order to augment propagation and conservation of the tree species underambient climatic conditions of IHR (Indian Himalayan Region). The occurrence of supportive endophytic microorganisms (bacteria and fungi) was the major findings of this study.

G. CULTIVATION ASPECTS

Climate

The species is capable of growing under varied conditions of climate, ranging from dry temperate to moderate hot climate. Generally, the species prefers full sunlight.Ginkgo seeds are shed in late summer under warmer conditions, but under cold temperature, these are shed late in the season. The colder temperature delays full embryo development until the ensuing spring.

Soil

It can grow in all soil types, from heavy clay, loam or light sandy soil but perform better in sandy soil andrequires a well drain conditions, with mean soil temperature 15 to 27 °C. The pH of soil should be maintained at 5-5.5.

Nursery raising and planting

The seeds should be sown in the nursery beds at a density of 300,000 to 400,000 plants/ha. After two years, the seedlings are transplanted to their final site in rows at a density of 25,000 plants/ha. Fields should be kept weed-free throughout the year. However, Ginkgo tree is extremely slow growing and its regeneration through seeds is very poor. Stimulation of adventitious root formation of stem cuttings treated with auxins and commercial rooting mixture containing auxins is used for its propagation. Roots stimulation can also be achieved by treatment with carbendazim. In China, dipping the base of cutting into ABT-1 (aminobenzotriazole) rooting powder for 1h gave the best results.

Irrigation

Adequate irrigation is necessary to obtain optimal yields if rainfall is insufficient during summer months. In Central Europe, temperatures are too low for an optimal production and late frost may damage sprouting leaves in springtime.

Harvesting

Depending on climatic conditions, green Ginkgo leaves are harvested in July (USA), August (China) or September/October (France). Since, manual weed control is carried out simultaneously with harvesting and drying, the need for manpower increases during this period on plantations in the US and France varies from 52 to about 150 persons. Leaves are harvested mechanically using modified cotton pickers. In China, leaves are hand-picked by thousands of small holders and their family members. Yields vary from 2 to 4 ton of dry leaves per ha depending on site and tree canopy (Singh *et al.*, 2008).

H. COMMERCIAL ASPECTS

Demand and usage

Ginkgo has been planted on a large scale in France and USA since the 1980s and plantations are found in the south eastern USA with a density of 10 million Ginkgo trees per 1000 acres. A large amount of Ginkgo sold in the USA, however, comes from plantations in China. The demand of Ginkgo extract for its pharmacological use in different countries is as follows:

Argentina: Standardized extract (Ginkgo NF) is approved for use for peripheral and cerebro-vascular disorders.

Austria: A standardized extract (EGb 761) is approved for cerebral and nutritional insufficiencies, dementia, intermittent claudication, and supportive treatment of hearing deficits.

Brazil: Standardized extracts (Ginkgo NF) are approved for cerebrovascular deficits; peripheral vascular disorders; neurosensorial disorders of vascular origin in the ears, eyes, and nose, and migraine headaches.

Canada: Extract form is unacceptable as food ingredient. It is permitted as a homeopathic drug requiring premarket authorization and assignment of a Drug Identification Number (DIN). Thirty-two Ginkgo homeopathic preparations are listed in the Drug Product Database (DPD).

Denmark: Standardized Ginkgo extracts (most of which comply with Ginkgo NF) are approved for memory and concentration deficits, tiredness, continuing dizziness, tinnitus in the elderly, tendency to cold extremities, and intermittent claudication.

France: Marketed under the brand name Tanakan® (EGb 761) is a prescription medication. The standardized extract (EGb 761) is approved for treating symptoms of cerebral insufficiencies, intermittent claudication, Raynaud's disease, certain dizziness and/or tinnitus syndromes, and retinal conditions due to probable ischemia.

Germany: Only semi-purified normalized (standardized) dry extracts (35–67:1) with less than 5 ppm ginkgolic acids are approved drugs of the German Commission E; Active Ingredient Classification ASK No. 05939. Dry extract, (35–67:1) is official in the *German Pharmacopoeia,* standardized to contain no less than 22.0% and no more than 27.0% flavone glycosides, as well as not less than 5.0% and no more than 7.0% terpene lactones, of which 2.8–3.4% are ginkgolides A, B, and C and 2.5–3.2% are bilobalide. Extract is licensed for the treatment of cerebral dysfunction with attendant memory loss, dementia, poor concentration, vertigo, tinnitus of vascular origin, and for intermittent claudication. It is marketed both as a prescription and a non-prescription drug.

Mexico: Standardized extract (Ginkgo NF) is approved for treating diminished mental capacities, dementia syndromes, dizziness, and tinnitus.

Spain: A standardized extract (Ginkgo NF) is approved for cerebral insufficiencies (such as dizziness, headache, and memory deficits), and peripheral vascular disorders.

Sweden: Classified as natural remedy, requires premarket authorization. Since January 2001, six Ginkgo products are listed in the Medical Products Agency (MPA) "Authorised Natural Remedies," and a monograph is published with the approved indication: "Herbal remedy for the treatment of long-standing symptoms in elderly people such as difficulties of memory and concentration, vertigo, tinnitus and feeling of tiredness. Prior to treatment other serious conditions should be ruled out by doctor".

Switzerland: Herbal medicine with positive classification (List D) by the *Interkantonale Konstrollstellefar Heilmittel* (IKS) and corresponding sales Category D with sale limited to pharmacies and drugstores, without prescription. There are nine Ginkgo mono-preparation phytomedicines, nine poly-preparations and three Ginkgo homeopathic preparations listed in the Swiss Codex2000/01. Standardized Ginkgo extracts, most of which comply with Ginkgo NF, are approved for concentration difficulties, forgetfulness, and dizziness (due to arteriosclerotic complaints).

U.K.: Not entered in the *General Sale List* (GSL). Standardized extracts, are available.

U.S.: Dietary supplement (USC, 1994). Dried leaf containing no less than 0.8% flavonol glycosides is official in the *National Formulary* 19th edition (USP 25-NF 20, 2002).

Standardizedextracts are commonly sold (Gopichand and Meena, 2015).

Market trends

According to USA National Health and Nutrition Survey (NHANES) data, there has been a steady decline in the prevalence of use in men and women: from 1999–2002 (3.9%), 2003–2006 (3.0%), to 2007–2010 (1.6%). Researcher also reported from a survey of users of complementary and alternative medicine that 11.3% of supplement-users had used ginkgo.

Trade data

Western medicinal interest in *G. biloba* has grown dramatically since the 1980s owing to its potent action on cardiovascular system of human beings, particularly on the cerebral vascular activity. Over 7 billion dollars are spent annually on botanical medicines and Ginkgo ranks first among herbal medications. Fifty million *G. biloba* trees are grown, especially in China, France, and South Carolina, USA, producing 8000 tons of dried leaves each year to meet the commercial demand for *G. biloba* products. Ginkgo is among the most sold medicinal plants and the annual consumption in 2001 was between 4.5 million pounds and 5.1 million pounds of dried leaves. The use of *G. biloba* has been growing at a very rapid rate worldwide at 25% per year in the open world commercial market. Germany, Switzerland and France have respectively 31%, 8% and 5% of the world commercial market. Presently there are around 140 *G. biloba* products in the global market and it is estimated that in the coming five years its utilization is expected to grow threefold. *G. biloba* is sold in the form of leaf powder extract, and as a tincture to the pharmaceutical and herbal companies.

According to National Bureau of Statistics of China(2005), the total volume of Ginkgo nuts supplied by China has been estimated as 3,848 and 4,199 ton in the years 2003 and 2004, respectively. China alone is believed to account for more than 90% of the world production of Ginkgo nuts.

Major users

Major users of *G. biloba* in the form extracts or as crude drug are China, Japan among Asian countries.Other users are Argentina, Austria, Brazil, Canada, Denmark, France, Germany, Mexico, Spain, Sweden, Switzerland, U.K and U.S.(Chaudhary *et al.*, 2010).

I. FUTURE DIRECTIONS FOR RESEARCH

GBE 761 is widely used and is available as a standardized extract. It yielded consistent results and reliable dosing for the treatment of a wide range of conditions, particularly in cognitive, cardio-, and cerebro-vascular diseases. Morerigorous clinical trials are required to translate the promising preclinical findings to definitive and reliable clinical data. The common dose of GBE 761 or its individual constituents were found to be consistently higher in preclinical studies as compared to that of clinical trials (100 mg/kg compared to <2 mg/kg, respectively). Some clinical trials have used doses exceeding 300 mg daily, while other trials have used substantially lower doses. While the individual constituents of GBE 761 are standardized, it appears that the dose is not. This makes interpreting its clinical effects difficult. The individual constituents of GBE 761 possess therapeutic efficacy but requires further studies for standardizing dosage to optimize attainment of desired activity (Nash and Shah, 2015).

Patents

Ayroles G, Rossard RM, Cadiou M, Montmiral C.Method for obtaining an extract or *Ginkgo biloba* leaves. United States Patent No.US4,981,688, Paris, France, United States Patent, 1991.

Bombardelli E, Mustich G, Bertani M. Extracts of *Ginkgo biloba* and their methods of preparations United States Patent No. US005637302A, Milan, Italy, United States Patent, 1997.

Bottero, Carlo. The use of *Ginkgo biloba* extract in preparation of a composition for lowering cholesterol.European Patent Application No. Ep1,952,815,A1, Shanghai, China, European Patent Office, 2006.

Christen Y. Use of *Ginkgo biloba* extracts in order to promote muscle mass to the detriment of fatty mass. United States Patent No.US20050202107A1, Paris, France, United States Patent Application, Publication, 2005.

Christen Y. Use of extracts of *Ginkgo biloba* for preparing a medicament intended to treat sarcopenia United States Patent No. US00713814B2, Paris, France, United States Patent 2006.

Marshall JL, Papadopoulos V. *Ginkgo biloba* extract as a treatment for therapeutic induced neurotoxcity.United States Patent No.US20070269539A1, Boston, US, United States Patent Application Publication, 2007.

Xie DL, Wang N, Gao O, Zhang GA, Shao BP, Jin XW, Huang XS. *Ginkgo biloba* composition method to prepare the same and uses thereof United States Patent No.US6187314B1, Shanghai, China, United States Patent, 2001.

Zou Y, Zou Q. *Ginkgo biloba* L. leaves cigarette. United States Patent No.US6776169B1, Shandong, United States Patent, 2004.

References

1. Adler S.R., Fosket J.R. 1999. Disclosing complementary and alternative medicine use in the medical encounter: a qualitative study in women with breast cancer. *J Fam Pract*. 48: 453–458.
2. Akdere H., Tastekin E., Mericliler M., Burgazli K.M.2014. The protective effects of *Ginkgo biloba* EGb761 extract against renal ischemia-reperfusion injury in rats. *Europ. Rev. Med. Pharmacol. Sci.* 18: 2936-2941.
3. Alessandro L., Baldan B., Pavanello A., Casadoro G.2015. BMC Evol. Biol. 15 (139): 1-13.
4. Bai Y. 2015. *Ginkgo biloba* extract induce cell apoptosis and G_0/G_1 Cycle arrest in gastric cancer cell. *Int J Clin Exp Med.* 8(11): 20977-20982.
5. Banachewski T., Brandeis D., Heinrich H., Albrecht B., Brunner E., Rothenberger A. 2011. Association of ADHD and conduct disorder: Brain electrical evidence for the existence of a distinct subtype. *J. Child Psychol. Psychiat. Allied Disciplines.* 44: 356–376.
6. Bao B, 2008. Regulation of autoimmune inflammation by pro-inflammatory cytokines. *Immunol Lett.* 120: 1-5.
7. Bauer B. 2013. Fossil ginkgophyte seedlings from the Triassic of France resemble modern Ginkgo biloba, BMC Evolutionary Biology. 13(177): 1-8.
8. Beek T.A., Montoro P. 2009. Chemical analysis and quality control of *Ginkgo biloba* leaves, extracts and phyto pharmaceuticals. *J Chromatogr A*. 1216: 2002-32.
9. Bernatoniene, 2010. Does *Ginkgo biloba* Reduce the Risk of Cardiovascular Events? *Circ Cardiovasc Qual Outcomes*. 42-47.
10. Ch'ng P.E. 2013. Some physical properties of ginkgo nuts and kernels, *Internat. Agrophysics*. 27: 485-489.
11. Chan P.C, 2007. *Ginkgo Biloba* Leave Extract: Biological, Medicinal, and Toxicological Effects. *J. Env. Sci. & Health Part C*. 25: 211–244.
12. Chan P.C, Xia Q., Fu P.P. 2007. *Ginkgo biloba* Leave Extract: Biological, Medicinal, and Toxicological Effects, *J. Environ.Sci.Health* Part C. 25: 211–244.
13. Chase J. 2007. The Ginkgo: A True Living Fossil (*Ginkgo biloba*), Summer- Forests for Oregon.
(http://www.oregon.gov/odf/urbanforests/docs/featuredtreeginko.pdf.)
14. Chaudhary S.A., Gadhvi, K.V., Chaudhary, A.B., 2010. Comprehensive review onworld herb trade and most utilized medicinal plant. Int. *J. Appl. Biol. Pharm.Technol.* I (2), 510–517, ISSN 0976-4550.

15. Chen Y.J., Tsai K.S., Chiu C.Y., Yang T.H., Lin T.H. Fu W.M., Chen C.F., Yang R.S., Liu S.H. 2013. EGb761 Inhibits Inflammatory Responses in Human Chondrocytes and Shows Chondro protection in Osteoarthritic Rat Knee, *J. Orthoped. Res.* 1032-1038.
16. Chi C, 2015. Effect of ginkgolide B on brain metabolism and tissue oxygenation in severe haemorrhagic stroke. *Int J ClinExp Med.* 8(3): 3522-3529.
17. Christoph S., Hopfield J., Lau H., Klein J. 2015. Effects of *Ginkgo biloba* extract EGb 761, Donepezil and their combination on central cholinergic function in aged rats, *J Pharm Pharm Sci*. 18(4): 634-646.
18. Cisowska J.K., Flaczyk E., Jeszka M. 2010. Antioxidant Activities of *Ginkgo biloba* extracts: Application in freeze stored meat dumplings. *Acta Sci Pol Technol Aliment.* 9(2): 161-170.
19. Demirezer, 2014. Adulteration determining of pharmaceutical forms of *Ginkgo biloba* extracts from different international manufacturers. *Rec Nat Prod.* 8(4): 394-400.
20. Diaset M.C., Furtado K.S, Rodrigues M.A.M., Barbisan L.F., 2013. Effects of *Ginkgo biloba* on chemically-induced mammary tumors in rats receiving tamoxifen, *BMC Complement. Alternat. Med.* 13: 93.
21. Ding C., Chen E., Zhou W., Lindsay R.C. 2004. A method for extraction and quantification of ginkgo terpene trilactones, *Anal Chem*. 76: 4332 6, DOI: 10.1021/ac049809a.
22. Drieu K., Jaggy H., Beek T.A. 2000. *Ginkgo biloba*. In: Medicinal and Aromatic Plants: Industrial Profiles, Amsterdam: CRC Press. Ginkgo biloba; p. 35.
23. Echenard V, Lefort F, Calmin G, Perroulaz R, Belhahri L., 2008. A new and improved automated technology for early sex determination of *Ginkgo biloba*, *Arboricul Urban*. 34: 300-307.
24. Mesallamy HO, Metwally NS, Soliman MS, Ahmed KA, Abdel Moary MM, 2011. The chemopreventive effect of *Ginkgo biloba* and *Silybum marianum* extracts on hepatocarcinogenesis in rats, *Cancer Cell Int*, 11; 38-49.
25. Elatrash A.M., Haleim S.Z.A. 2015. Protective Role of *Ginkgo biloba* on Monosodium Glutamate: Induced Liver and Kidney Toxicity in Rats, *RJPBCS,* 6(1): 1437-1440.
26. Feng X., Zhang L., Zhu H. 2009. Comparative anticancer and antioxidant activities of different ingredients of *Ginkgo biloba* extract (EGb extract), *Planta Med.* 75(8): 792-796.
27. Feucht C., Patel D.R. 2011. Herbal medicines in pediatric neuropsychiatry, *Paediat. Clinics of North America*. 58: 33–54.
28. Fiehe K., Arenz A., Drewke C., Hemscheidt T., Williamson R.T., Leistner E. 2000. Biosynthesis of 4'-*O*-Methylpyridoxine (Ginkgotoxin) from Primary Precursor. *J. Nat. Prod.* 63(2): 185–189.
29. Fu L.G., Jin J.M. 1992. Chinese Plant Red Data Book: Rare and Endangered Plants, Vol. 1; Beijng, China: *Science Press*. 474-5.
30. Galia J.A, Rosato M., Rossello J.A. 2012. Early evolutionary colocalization of the nuclear ribosomal 5S and 45S gene families in seed plants: evidence from the living fossil gymnosperm *Ginkgo biloba Heredity*. 108: 640–646.
31. Gauthier S., Schlaefke S. 2015. Efficacy and tolerability of Ginkgo biloba extract GBE 761® in dementia: systematic review and meta-analysis of randomized placebo-controlled trials, *Clin Intervent Aging*. 9: 2065-2077.
32. Gavrilova I.S.2014. Efficacy and safety of *Ginkgo biloba* extract GBE 761® in mild cognitive impairment with neuropsychiatric symptoms: a randomized, placebo-controlled, double-blind, multicenter trial. *Int J Geriatr Psychiatry*. 29: 1087–1095.
33. Gong W., ZengZ., Chen YY., Chen C. 2008. Glacialrefugia of *Ginkgo biloba* and human impact on its genetic diversity: evidence from chloroplast DNA, *J. Integr. Plant Biol*. 50: 368-374.

34. Gong W., Chen C, Dobes C, Fu C.X., Koch M.A. 2008. Phylogeography of a living fossil: Pleistocene glaciations forced *Ginkgo biloba* L. (Ginkgoaceae) into two refuge areas in China with limited subsequent postglacial expansion, *Mol. Phylogenetics Evol.* 48:1094–1105.
35. Gopichand, Meena, R.L. 2015, Standardization of propagation and agro techniques in *Ginkgo biloba* L. - A medicinally important plant, *J. Med. Plants Studies.* 3(4): 6-15.
36. Harnly, 2012. Detection of Adulterated *Ginkgo biloba* Supplements Using Chromatographic and Spectral Fingerprints, *JAOAC Int.* 95(6): 1579–1587.
37. He B., Gu Y., Xu M., Wang J., Cao F and Xu L. 2015, Transcriptome analysis of *Ginkgo biloba* kernels, Front. Plant Sci. 6 (819) :1-10.
38. Herrschaft H., Nacu A., Likhachev S., Sholomov I., Hoerr R., Schlaefke S. 2012. *Ginkgo biloba* extract GBE 761(R) in dementia with neuropsychiatric features: A randomised, placebo-controlled trial to confirm the efficacy and safety of a daily dose of 240 mg *J. Psychiatric Res.* 46: 716–723.
39. Ho L.J., Hung L.F., Liu F.C., Hou T.Y., Lin L.C., Huang C.Y., Lai J.H. 2013. *Ginkgo biloba* Extract Individually Inhibits JNK Activation and Induces c-Jun Degradation in Human Chondrocytes: Potential Therapeutics for Osteoarthritis. *Plos One.* 8(12): 2033-2040.
40. Hoenerhoff M.J., Pandiri A.R., Snyder S.A., Hong H.H.L., Ton T.V., Peddada S., Shockley K., Witt K., Chan P., Rider C., Kooistra L., Nyska A., Sills R.C. 2013. Hepatocellular Carcinomas in B6C3F1 Mice Treated with *Ginkgo biloba* Extract for Two Years Differ from Spontaneous Liver Tumors in Cancer Gene Mutations and Genomic Pathways. *Toxicol. Pathol.* 41: 826-841.
41. Hsieh L., Duhai Z.2004. Analysis for the origin of *Ginkgo* population in Tianmu Mountains. *Sci. Silvae Sin.* 40: 28-31.
42. Ihl R., Tribanek M., Bachinskaya N. 2012. Efficacy and tolerability of a once daily formulation of *Ginkgo biloba* extract GBE 761(R) in Alzheimer's disease and vascular dementia: Results from a randomised controlled trial. *Pharmacopsychiatry.* 45: 41-46.
43. Isah T, 2015. Rethinking *Ginkgo biloba* L.: Medicinal uses and conservation. *Phcog Rev.* 9:140-8.
44. Jang H.S., Roh S.Y., Jeong E.H., Kim B.S., Sunwoo M.K. 2015.Ginkgotoxin Induced Seizure Caused by Vitamin B6 Deficiency. *J. Epilepsy Res.* 5:2.
45. Jeon Y.J., Jung S.N, Yun J., Lee CW., Choi J., Lee Y.J., Kwon B.M. 2015. Ginkgetin inhibits the growth of DU-145 prostate cancer cells through inhibition of signal transducer and activator of transcription 3 activity. *Cancer Sci.* 106: 413–420.
46. Jin B, 2012a. Structure and function of the tentpole in the reproductive process of *Ginkgo biloba* L. *Plant SignalingBehavior.* 7(10): 1330-1336.
47. Jin B, 2012b. The behavior of pollination drop secretion in *Ginkgo biloba* L., *Plant Signaling Behavior*; 7:(9) 1168-1176.
48. Jing L, 2012. *Ginkgo biloba* Extract EGB761 Protects against Aging-Associated Diastolic Dysfunction in Cardiomyocytes of D-Galactose-Induced Aging Rat, *Oxidat. Med. Cellular Longevity.* 2-7.
49. Kim J.S., Lee H. 2013. Vertigo due to posterior circulation stroke, *Semin Neurol.* 33: 179-84.
50. Klin K.B., Piechal A., Joniec I., Pyrzanowska J., Tyszkiewicz E.W. 2009. Pharmacological and biochemical effects of *Ginkgo biloba* extract on learning, memory consolidation and motor activity in old rats, *Acta Neurobiol Exp.* 69: 217–231.
51. Koczka N., Moczar Z., Stefanovits BÉ., Ombodi A., 2015. Differences in antioxidant properties of ginkgo leaves collected from male and female trees, *Acta Pharm.* 65(1): 99-104.

52. Koh P.O. 2012. *Gingko biloba* extract (EGb 761) attenuates ischemic brain injury-induced reduction in Ca (2+) sensor protein hippocalcin. *Lab Anim Res*. 28 :199–204.
53. Kumar H., Song S.Y., More S.V., Kang S.M., Kim B.W., Kim I.S., Choi D.K. 2013. Traditional Korean East Asian Medicines and Herbal Formulations for Cognitive Impairment, *Molecule*.18: 14670-14693.
54. Kumar, A., Singh, S., Pandey, A. 2009. General microflora, arbuscular mycorrhizal colonization and occurrence of endophytes in the rhizosphere of two age groups of *Ginkgo biloba* L.of Indian Central Himalaya. *Indian J microbiol*. 49. 134-41. 10.1007/s12088-009-0017-x.
55. Kyriakis J.M., Avruch J. 2012. Mammalian M.A.P.K signal transduction pathways activated by stress and inflammation: a 10-year update, *Physiol Rev*. 92: 689–737.
56. Leslie A.B. 2010. Flotation preferentially selects saccate pollen during conifer pollination, *New Phytol*. 188: 273-9.
57. Li W., Luo Z., Liu X., Fu L., Xu Y., Wu L., Shen X. 2015. Effect of *Ginkgo biloba* extract on experimental cardiac remodeling, *BMC Complementary Alter. Med,* 15: 277-285.
58. Lin C.P., Wu C.S., Huang Y.Y., Chaw S.M. 2012. The complete chloroplast genome of *Ginkgo biloba* reveals the mechanism of inverted repeat contraction. *Genome Biol. Evol.* 4(3): 374–381.
59. Liou C.J., Lai X.Y., Chen Y.L., Wang C.L, Wei C.H., Huang W.C. 2015. Ginkgolide C suppresses adipogenesis in 3T3-L1 adipocytes via the AMPK signaling pathway. *Evid. Based Complement Alternat. Med.* pp.1-10.
60. Little SA, 2013.Branch architecture in *Ginkgo biloba*: Wood anatomy and long shoot–short shoot interactions,*American J. Bot.*; 100(10); 1923–1935.
61. Liu P.L., Chen Y.H., Chou S.H., Yang M.C., Cheng Y.J. 2014..*Ginkgo biloba* extract decreases non-small cell lung cancer cell migration by down regulating metastasis- associated factor heat-shock protein 27. *PLOS One*. 9:1331-1338.
62. Lu J.M, Yan S., Jamaluddin S., Weakley S.M., Liang Z., Siwak E.B., Yao Q., Chen C. 2012. Ginkgolic acid inhibits HIV protease activity and HIV infection in vitro. *Med Sci Monit*.18(8): 293-298.
63. Lu Z.W., Popova E.V., Wu C.H., Lee E.J., Hahn E.J., Paek K.Y. 2009. Cryopreservation of Ginkgo biloba cell culture: Effect of pre-treatment with sucrose and ABA. *Cryo Letters*. 30: 232-43.
64. Ma J., Duan W., Han S., Lei J., Xu Q., Chen X., Jiang Z., Nan L., Li J., Chen K., Wang L.H.Z., Li X., Wu E., Huo X. 2010. Ginkgolic acid suppresses the development of pancreatic cancer by inhibiting pathways driving lipogenesis, *Oncotarget*. 6 (25) 890-919.
65. Mahadevan S, 2008. Multifaceted Therapeutic Benefits of *Ginkgo biloba* L.: Chemistry, Efficacy, Safety, and Uses. *J. Food Sci.*, 7, R14-19.
66. McCarney R., Fisher P., Iliffe S., van Haselen R., Griffin M., van der Meulen MJ. 2008.*Ginkgo biloba* for mild to moderate dementia in a communitysetting: a pragmatic, randomised, parallel-group, double-blind. Placebocontrolledtrial. *Int J Geriatr Psychiatry*. 23: 1222-1230.
67. Mdzinarishvili A., Sumbria R., Lang D., Kleina J, 2012.*Ginkgo* Extract EGb761 Confers Neuroprotection by Reduction of Glutamate Release in Ischemic Brain, *J Pharm Pharmaceutics*. 15:(1) 94 – 102.
68. Mohanta T.K., 2012. Advances in *Ginkgo biloba* research: Genomics and metabolomics perspectives.*Afri. J Biotechnol*, 11(93): 15936-15944.
69. Mullaicharam A.R, 2013. Areview on evidence based practice of*Ginkgo biloba*in brain health. *Int. J Chem. & Pharma Anal.*1(1): 24-30.
70. Napolitano, 2012.Complete ^{1}H NMR spectral analysis of ten chemical markers of *Ginkgo biloba. Magn Reson. Chem*. 50(8): 569–575.

71. Nash K.M., Shah Z.A. 2015. Current Perspectives on the Beneficial Role of *Ginkgo biloba* in Neurological and Cerebrovascular Disorders, *Integr Med Insights*. 10: 1–9.
72. Nicolai S.P, Kruidenier L.M, Bendermacher B.L, Prins M.H, Teijink J.A, 2013. *Ginkgo biloba* for intermittent clau-dication. *Cochrane Database Syst Rev.*6;(6):CD006888. doi: 10.1002/14651858.CD006888.pub3.
73. Nuhu A. A, 2014.*Ginkgo biloba*: A 'living fossil' with modern day phytomedicinal applications. *J. Applied Pharmaceut. Sci*. 4:(3) 96-103.
74. Malischewsky, P.G. 2014. A very special Fractal: Gingko of Jena, *Geofisica International*. 53-1: 95-100.
75. Popova E.V., Lee E., Wu C., Hahn E., Paek K. 2009. A simple method for cryopreservation of *Ginkgo biloba* callus, *Plant Cell Tissue Organ Cult.*97: 337-43.
76. Purohit V.K., Phondani P.C., Rawat L.S., Maikhuri R.K., Dhyani D., Nautiyal A.R.2009. Propagation through rooting of stem cuttings of *Ginkgo biloba* Linn. A living fossil under threat. *J Am Sci.* 5: 139-44.
77. Puttalingamma V. 2015.*Ginkgo biloba*"living fossil", wonderful medicinal plant- A Review. *International J. Advan. Res.* 3: (3) 506-511.
78. Ran K, 2014. *Ginkgo biloba* extract post-conditioning reduces myocardial ischemia reperfusion injury. *Genetics Mol. Res.*, 13(2): 2703-2708.
79. Rider C.V., Nyska A., Cora M.C., Kissling G.E., Smith C., Travlos G.S., Hejtmancik M.R., Fomby L.M., Colleton C.A., Ryan M.J., Kooistra L., Morrison J.P., Chan P.C. 2014.Toxicity and Carcinogenicity Studies of *Ginkgo biloba* extract in Rat and Mouse: Liver, Thyroid, and Nose are Targets. *ToxicolPathol*. 42 (5): 830–843.
80. Royer D.L., Hickey L.J., Wing S.L. 2003. Ecological conservatism in the "living fossil" *Ginkgo*, *Palaeobiol*. 29: 84-104.
81. Sadaf H.M., 2016. Status of Maiden Hair Tree-*Ginkgo biloba*, Living Fossils Becoming Endangered. *Nova J. Cancer* Res. 1(1): 1-4.
82. Saker S.A., Mahran H.A., Abdel-Maksoud A.M. 2011. Suppressive effect of *Ginkgo biloba* extract (EGb761) on tops in induced ovarian toxicity and oxidative stress in albino rats. *J. Appl. Pharmaceut. Sci.* 1(4): 46-54.
83. Schmid W., Balz J. 2003. Cultivation of *Ginkgo biloba* L. on three continents. *Acta Hort (ISHS)*. 676: 177-80.
84. Schwarzkopf T.M., Koch K.A., Klein J. 2013. Neurodegeneration after transient brain ischemia in aged mice: beneficial effects of bilobalide, *Brain Res*. 1529: 178–187.
85. Shan-Tan M., YuJ.T., Tan C.C., Wang H.F., Meng X.F., Wang C., Jiang T., Zhuc X.C., Tan L. 2015. Efficacy and Adverse Effects of *Ginkgo biloba* for Cognitive Impairment and Dementia: A Systematic Review and Meta-Analysis. *J. Alzheimer Dis.* 43: 589–603.
86. Sierpina V.S. 2003. *Ginkgo biloba*, American Family Physician. 68: 5: 923-926.
87. Singh B., Kaur P., Gopichand, Singh R.D., Ahuja P.S. 2008, Biology and chemistry of *Ginkgo biloba*. *Fitoterapia*. 79: 401-18.
88. Snitz B.E., O'Meara E.S., Carlson M.C., Arnold A.M., Ives D.G. 2009. *Ginkgo biloba* for Preventing Cognitive Decline in Older Adults. *JAMA*. 302: (24) 2663–2670.
89. Sohier C, 2002. Plant biotechnology: An avant-garde research for an ancestral tree the *Ginkgo biloba*. *Ann Pharm Fr.* 60: 22 7.
90. Szulc M., Radwan-Oczko M., 2012. The role of *Porphyromonas gingivalis* in atherosclerosis, *Kardiochir Torakochi*. 9: 100-5.
91. Tan C.C., Yu J.T., Wang H.F., Tan M.S., Meng X.F, Wang C, 2014. Efficacy and safety of donepezil, galantamine, rivastigmine, and memantine for the treatment of Alzheimer's disease: A systematic review and meta-analysis. *J Alzheimers Dis*. 41: 615-31.
92. Tang C.Q., AngY.Y., HsawaM.O., Omohara A. M., Hara M., Heng S.C, Fan S., 2011. Population structure of relict *Meta sequoia glyptostroboides* and its habitat fragmentation and degradation in south-central China,*Biol. Cons.,* 144: 279–289 .

93. Tommasi F., Scaramuzzi F. 2004. *In vitro* propagation of *Ginkgo biloba* by using various bud cultures. *Biol Plant*. 48: 297-300.
94. Tredici P.D. 2007.The Phenology of Sexual Reproduction in *Ginkgo biloba*: Ecological and Evolutionary Implications. *Bot. Rev.* 73: 267-278.
95. Tsai J.R., Liu P.L., Chen Y.H., Chou S.H., Yang M.C., Cheng Y.J., Hwang J.J, Yin W.H., Chong I.W. 2014. *Ginkgo biloba* Extract Decreases Non-Small Cell Lung Cancer Cell Migration by Down regulating Metastasis- Associated Factor Heat-Shock Protein 27. *Plos One*. 9(3): 1331-1339.
96. Ude C, Schubert-Zsilavecz M, Wurglics M. 2013. Ginkgo biloba extracts: a review of the pharmacokinetics of the active ingredients. *Clin Pharmacokinet.*;52(9):727-749.23703577
97. Vardy J., Dhillon H.M., Clarke S.J., Olesen I., Lesile F., Warby A., Beith J., Sullivan A., Hamilton A., Beale P., Rittau A., McLachlan A. 2013. Investigation of herb-drug interactions with *Ginkgo biloba* in women receiving hormonal treatment for early breast cancer. *Springer Plus*. 2(126): 1-5.
98. Vupputuri G.A., Nichols H.L., Joski P.M., Corp M.L., 2011. Risk of progression of nephropathy in a population-based sample with type 2 diabetes. *Diabetes Res. Clinical Practice*. 91(2): 246–252.
99. Wang X., Ning Y., Xu F., Ye J., Cheng S., Li X., 2014a. Microscopic observation of different tissues from *Ginkgo biloba,. J Pharm Chem. Biol. Sci*. 2(3): 172-175.
100. Wang J, Yu J.T, Wang H.F, Meng X.F, Wang C, Tan C.C, Tan L, 2014b. Pharmacological treatment of neuropsychiatric symptoms in Alzheimer's disease: A systematic review and meta-analysis. *J NeurolNeurosurg Psychiatry.* 86(1):101-9. doi: 10.1136/jnnp-2014-308112.
101. Wang Y., Wang R., Wang Y., Peng R., Wu Y., Yuan Y. 2015. *Ginkgo biloba* extract mitigates liver fibrosis and apoptosis by regulating p38 MAPK, NF-êB/IêBá, and Bcl-2/Bax signaling. *Drug Design, Develo. Therapy*. 9: 6304-6317.
102. Wang C.Z., Li W.J., Tao R., Ye J.Z., Zhang H.Y. 2015. Antiviral Activity of a Nanoemulsion of Polyprenols from Ginkgo Leaves against Influenza A H3N2 and Hepatitis B Virus *in vitro. Molecules*. 20: 5138-5145.
103. Wei J.M., Wang X., Gong H., Shi Y.J., Zou Y. 2013. Ginkgo suppresses atherosclerosis through down-regulating the expression of connexin 43 in rabbits. *Arch Med Sci*. 9(2): 340-346.
104. Weinmann S., Roll S., Schwarzbach C., Vauth C., Willich SN. 2010. Effects of *Ginkgo biloba* in dementia: Systematic review and meta-analysis. *BMC Geriatr*; pp.10:14. doi.org/10.1186/1471-2318-10-14.
105. Wu X., Mao G., Zhao T., Zhao J., Li F., Yang L. 2011. Isolation, purification and in vitro anti-tumor activity of polysaccharide from *Ginkgo biloba* sarcotesta. *CarbohybdPolym*. 86(2): 1073-1076.
106. Wu C.S., Chaw S.M., Huang Y.Y. 2013. Chloroplast phylogenomics indicates that *Ginkgo biloba* is sister to cycads. *Genome Biol. Evol.* 5(1): 243–254.
107. Xiang Y., Xiang Z. 1999. Ancient *Ginkgo biloba* report 3: Investigation on ancient *Ginkgo biloba* remnant population in Guiyang, *Guizhou*. 17: 221-30.
108. Xie C.X., Zhao M.S., Fu C.X., Zhao Y.P. 2013. Development of the first chloroplast microsatelliteloci in *Ginkgo biloba* (Ginkgoaceae). *Applications Plant Sci.* 1(8): 1300019, 1-3 doi: 10.3732/apps.1300019.
109. Yao X, 2013. Simultaneous Quantification of Flavonol Glycosides, Terpene Lactones, biflavones, Proanthocyanidins, and Ginkgolic Acids in *Ginkgo biloba* leaves from Fruit Cultivars by Ultrahigh-Performance Liquid Chromatography Coupled with Triple Quadrupole Mass Spectrometry. *BioMed Res. International* Volume. 1-12.

110. Yoshitake T., Yoshitake S., Kehr J. 2010. The *Ginkgo biloba* extract GBE 761(R) and its main constituent flavonoids and *Ginkgolides* increase extracellular dopamine levels in the rat prefrontal cortex, *Br J Pharmacol.* 159: 659-68.

111. Zhang S.J., 2012. Effect of Western medicine therapy assisted by *Ginkgo biloba* tablet on vascular cognitive impairment of none dementia. Asian Pacific *J. Trop. Med.* 661-664.

112. Zhao J., Geng T., Wang Q., Si H., Sun X., Guo Q., Li Y., Huang W., Ding G., Xiao W. 2015. Pharmacokinetics of Ginkgolide B after Oral Administration of Three Different Ginkgolide Formulations in Beagle Dogs. *Molecules.* 20: 1-11.

113. Zhao X., Yao H., Yin H.L., Zhu Q.L., Sun J.L., Ma W., Shi YQ., Liang Z.G., Li B.X. 2013. *Ginkgo biloba* Extract and Ginkgolide Antiarrhythmic Potential by Targeting hERG and ICa-L Channel. *J Pharmacol Sci.* 123: 318 – 327.

114. Zhao Y., Paule J, Fu C, & Koch MA, 2010. Out of China: Distribution history of*Ginkgo biloba* L.,*Taxon*, 59 (2): 495–504.

115. Zhou LX, Zhu Y. 2012. Influence of *Ginkgo biloba* extract on the proliferation, apoptosis of ACC-2 cell and Surviving gene expression in adenoid cystic carcinoma of lacrimal gland. *Asian Pacific J. Trop. Med.* pp. 897-900.

116. Zhou X.L., Yang M., Xue B.G., He H.T., Zhang C.M., Liu M.M., Zhang L.J., Fei R., 2014. Anti-inflammatory action of *Ginkgo biloba* leaf polysaccharide via TLR4/NF-êâ signaling suppression. *Biomed. Res.* 25:(4) 449-454.

Index

D

E

F

G

H

I

Q

R

S

T

U

V

W

X